Out of the Shadows: Contributions of Twentieth-Century Women to Physics

Why are there not more prominent female physicists? Throughout history women have been denied access to higher learning and scientific laboratories. Today, traditional barriers faced by women in higher education have been breached. However, the female pioneers who overcame discrimination and became major players in their fields remain largely in the shadows. The importance of their work, achievements and contributions to science deserve recognition; the names of these pioneers deserve to be known.

Out of the Shadows will help to bring a more gender-balanced perception of physicists. Here are detailed descriptions of important, original contributions made by women in the century from 1876 to 1976. Many female physicists and mathematicians, historically excluded from participation in science, emerged in this time. It was a period in which there was great progress in science and in the liberation of women from centuries of repression. This book documents both aspects of recent history. Many of the authors here are themselves distinguished scientists who have been actively engaged in the fields of physics about which they write. They write about modern developments in their fields with remarkable clarity and readability. The fields are as diverse as astrophysics, bio-molecular structure, chaos theory, geophysics, nuclear physics, particle physics, and surface physics. The writers include detailed accounts of important discoveries, place them in their historical context, give references to original papers, suggest books for further reading, and provide scientific biographies. The discoveries are well-established and fundamental to modern physics. It is not well known, however, that many were made by women.

NINA BYERS received her Ph.D. in Physics from the University of Chicago in 1956, where she was taught by many eminent physicists

T0291449

including Enrico Fermi, Maria Goeppert Mayer and Murray Gell-Mann. Since then, she has undertaken research and teaching at many institutions including the University of Birmingham, the University of Oxford, UK, Stanford University and the University of California at Los Angeles. She has held numerous posts in scientific societies including Councillor-at-Large and Member of the Panel on Public Affairs of the American Physical Society (APS), President of the APS Forum on Physics and Society and President of the APS Forum on History of Physics as well as Member-at-Large of Section B (physics) of the American Association for the Advancement of Science. She has published many papers in scientific journals, reporting research results in theoretical physics in the fields of elementary particle physics and superconductivity. More recently, she has published papers and given invited talks on the history of physics on such subjects as Emmy Noether, Enrico Fermi and Leo Szilard, and Women in Physics in Fermi's Time. She is the initiator and organizer of an electronic archive making available on the internet the contributions of twentieth-century women to physics http://cwp.library.ucla.edu/.

GARY A. WILLIAMS received his Ph.D. in Physics from the University of California, Berkeley, in 1974. In 1975 he joined the faculty of the University of California at Los Angeles, where he has been ever since. He is the author of over 100 research articles in the field of low temperature physics, specializing in experimental and theoretical studies of the superfluid phase transition of liquid helium.

Out of the Shadows

Contributions of Twentieth-Century Women to Physics

Edited by

NINA BYERS AND GARY WILLIAMS
University of California, Los Angeles

CAMBRIDGE
UNIVERSITY PRESS

CAMBRIDGE UNIVERSITY PRESS
Cambridge, New York, Melbourne, Madrid, Cape Town, Singapore,
São Paulo, Delhi, Dubai, Tokyo, Mexico City

Cambridge University Press
The Edinburgh Building, Cambridge CB2 8RU, UK

Published in the United States of America by Cambridge University Press, New York

www.cambridge.org
Information on this title: www.cambridge.org/9780521169622

© Cambridge University Press 2006

This publication is in copyright. Subject to statutory exception
and to the provisions of relevant collective licensing agreements,
no reproduction of any part may take place without the written
permission of Cambridge University Press.

First published 2006
First paperback edition 2010

A catalogue record for this publication is available from the British Library

ISBN 978-0-521-82197-1 Hardback
ISBN 978-0-521-16962-2 Paperback

Cambridge University Press has no responsibility for the persistence or
accuracy of URLs for external or third-party internet websites referred to in
this publication, and does not guarantee that any content on such websites is,
or will remain, accurate or appropriate.

This book is dedicated to all the women whose contributions should be in it and are not. There are very many. Among them are the women who remain in the shadows of the history of physics, the women who would have become physicists if the way had been open to them, and younger women whose contributions to physics have been made after 1976 and who continue to advance science.

More particularly we dedicate this book to the more than two hundred twentieth-century women whose contributions to physics are documented in the CWP website <http://cwp.library.ucla.edu>, forty of whom are subjects of the chapters of this book. All of these were inspirations for the creation of this book.

Nina Byers and Gary Williams

"This book fills a vacuum in the history of physics. For the first time we have in one place clear accounts of careers and contributions to physics of forty distinguished women from a variety of fields. In particular the authors are informed insiders with intimate knowledge of their fields who often provide fresh information about their subjects. Let us hope that this book will inspire physicists to include these women in their lectures and textbooks so that no one will ever again badger women students with taunts like 'What's a nice girl like you doing in Physics 55?'"

Margaret W. Rossiter
MacArthur Prize Fellow 1989–1994
Marie Underhill Knoll Professor of the History of Science
Cornell University

"As this inspiring gallery of heroines makes plain, there's no such thing as female science – just female scientists, including some very great ones. Their achievements span a vast range of mathematics, physics, and astronomy. In Out of the Shadows, experts lucidly explain what they did, and the lives they led. I was mesmerized, and edified."

Frank Wilczek
Nobel Prize in Physics 2004
Herman Feshbach Professor of Physics
Massachusetts Institute of Technology

"Out of the Shadows gives us fascinating accounts of some of the ground-breaking achivements of women physicists and astronomers, many of whom have never received the recognition they truly deserve. It is a much needed book. In it a reader can learn, for example, about how Henrietta Swan Leavitt provided the first method of measuring inter-galactic distances, and how Cecilia Payne-Gaposchkin, in studies of spectra from stars, discovered that most of the luminous matter in the universe consists of hydrogen and helium. Both of these were advances crucial to the development of astrophysics and modern cosmology. This wonderful book beautifully illustrates that scientific talent has absolutely nothing to do with gender."

Jerome I. Friedman
Nobel Prize in Physics 1990
Institute Professor Emeritus
Massachusetts Institute of Technology

Contents

Contributors

Jean-Pierre Adloff
Laboratoire de Chimie Nucléaire
Centre des Rescherches Nucléaires
France

Noémie Benczer-Koller
Department of Physics and Astronomy
Rutgers University
Piscataway, NJ 08854
USA

Bruce A. Bolt
Department of Earth and Planetary Sciences
University of California, Berkeley
Berkeley, CA 94720
USA

Christiane Bonnelle
Université Pierre et Marie Curie (Paris VI)
75005 Paris
France

Andrzej Buras
Physik Department
Technische Universtät München
D-85747 Garching
Germany

Nina Byers
Department of Physics and Astronomy
University of California at Los Angeles
Los Angeles, CA 90095
USA

David G. Cassel
Physics Department
Cornell University
114 Newman Lab
Ithaca, NY 14853
USA

David B. Cline
Department of Physics and Astronomy
University of California at Los Angeles
Los Angeles, CA 90095
USA

Ferdinand V. Coroniti
Department of Physics and Astronomy
University of California at Los Angeles
Los Angeles, CA 90095
USA

Bryce DeWitt
Department of Physics
University of Texas at Austin
Austin, TX 78712
USA

Mildred S. Dresselhaus
Department of Physics
Massachusetts Institute of Technology
Cambridge, MA 02139
USA

Gene Dresselhaus
The Francis Bitter National Magnet Laboratory
Massachusetts Institute of Technology
Cambridge, MA 02139
USA

Freeman J. Dyson
Institute for Advanced Study
Einstein Drive
Princeton, NJ 08540
USA

Paolo Franzini
Dipartimento di Fisica, Università di Roma
Piazzale Aldo Moro 2
I-00185 Rome
Italy

Jonathan R. Friedman
Department of Physics
Amherst College
Amherst, MA 01002
USA

A. Michael Glazer
Clarendon Laboratory
University of Oxford
Parks Road
Oxford OX1 3PU
UK

Jenny P. Glusker
Institute for Cancer Research
Fox Chase Cancer Center
Philadelphia, PA 19111
USA

Alfred Scharff Goldhaber
C. N. Yang Institute for Theoretical Physics
State University of New York at Stony Brook
Stony Brook, NY 11794–3840
USA

Michael Gutperle
Department of Physics and Astronomy
University of California at Los Angeles
Los Angeles, CA 90095
USA

Leopold Halpern
Department of Physics
Florida State University
Tallahassee, FL 32306
USA

George B. Kauffman
Department of Chemistry
California State University, Fresno
Fresno, CA 93740
USA

Christine Kelsey
Department of Earth Sciences
University of Cambridge
Downing Street
Cambridge CB2 3EQ
UK

Peggy Aldrich Kidwell – Curator of Mathematics
Division of Information Technology and Society
National Museum of American History
Smithsonian Institution
Washington, DC 20560
USA

Hélène Langevin – Joliot
Institut de Physique Nucléaire d'Orsay
CNRS-Université Paris 11
91405 Orsay Cedex
France

Andrei Linde
Department of Physics
Stanford University
Stanford, CA 94305
USA

Joan Mason
Department of History and Philosophy of Science
University of Cambridge
Free School Lane
Cambridge CB2 3RH
UK

Judith Milledge
Department of Earth Sciences
University College London
Gower Street
London WC1E 6BT
UK

Steven A. Moszkowski
Department of Physics and Astronomy
University of California at Los Angeles
Los Angeles, CA 90095
USA

Abraham Pais
Rockefeller University
New York, NY 10021
USA

John Peoples, Jr.
Fermi National Accelerator Laboratory (Fermilab)
Batavia, IL 60510
USA

Pierre Radvanyi
Institut de Physique Nucléaire d'Orsay
CNRS-Université Paris 11
91405 Orsay Cedex
France

Helmut Rechenberg
Max-Planck-Institut für Physik (Werner-Heisenberg-Institut)
Föhringer Ring 6
80805 München
Germany

Lewis Rothberg
Department of Chemistry
University of Rochester
Rochester, NY 14627-0216
USA

Vera C. Rubin
Department of Terrestrial Magnetism
Carnegie Institution of Washington
Washington, DC 20015
USA

Robert J. Rubin
Mathematical Research Branch
National Institutes of Health (NIDDK)
Bethesda, MD 20892
USA

Maurice M. Shapiro
University of Maryland
Naval Research Station
Washington, DC 20375
USA

Ruth Lewin Sime
Sacramento City College
Sacramento, CA 95822
USA

Frieda A. Stahl
Department of Physics and Astronomy
California State University, Los Angeles
Los Angeles, CA 90032
USA

Virginia Trimble
Department of Physics and Astronomy
University of California, Irvine
Irvine, CA 92697
USA

Jean L. Turner
Department of Physics and Astronomy
University of California at Los Angeles
Los Angeles, CA 90095
USA

Cecil J. Waddington
School of Physics and Astronomy
University of Minnesota
Minneapolis, MN 55455
USA

Gary Williams
Department of Physics and Astronomy
University of California at Los Angeles
Los Angeles, CA 90095
USA

Ruth M. Williams
Girton College and Department of Applied Mathematics and
Theoretical Physics
Wilberforce Road
Cambridge CB3 0WA
UK

Chun Wa Wong
Department of Physics and Astronomy
University of California at Los Angeles
Los Angeles, CA 90095
USA

James W. York, Jr.
Physics Department
Cornell University
Ithaca, New York 14853
USA

Foreword

Freeman J. Dyson

Institute for Advanced Study, Princeton, New Jersey

This book is a portrait gallery of women who have made outstanding contributions to physics. There are forty of them. It was a good idea to collect their lives and achievements in one volume. Any aspiring woman physicist can see here that the choice of role models is wide. She may choose an experimenter like Chien-Shiung Wu, an observer like Jocelyn Bell Burnell, or a theorist like Cécile DeWitt-Morette. I hope this book will be widely read by teachers as well as students. My only regret is that many outstanding younger physicists whose careers are still on the rise are not included, such as Claire Max, Ellen Williams, Penny Sackett and Sara Seager. They would be more appropriate role models for an average modern teenager than the grand and remote figures of the past such as Marie Curie (see Chapter 4) and Emmy Noether (see Chapter 8). Curie and Noether are wonderful role models if you only dream of dedicating your heart and soul to science. Max and Williams and Sackett and Seager are better if you are also worried about making a living and raising a family under modern conditions.

I am lucky to have had six of the forty as colleagues and friends, besides many other outstanding women who did not make the list. When I was a young student in England in 1942, I heard Mary Cartwright (see Chapter 15) lecture about the pathological behavior of nonlinear amplifiers. The radars in World War II were driven by amplifiers, which behaved badly when pushed to high power levels. The Royal Air Force blamed the manufacturers and sent the radars back for repair. Cartwright showed that the manufacturers were not to blame. The Van der Pol equation was to blame. The Van der Pol equation is the standard equation describing a nonlinear amplifier. Cartwright studied the solutions of the Van der Pol equation and discovered the unexpected phenomenon that is now called chaos. As the power is increased, the periodic solutions go through an infinite sequence of period doublings and finally become aperiodic. The aperiodic solutions have disastrous effects on the radar, but have a beautifully intricate topological structure. These discoveries were published at the end of the war, but nobody paid much attention to her papers and she went on to other things. She became famous as a pure mathematician. Twenty years later, chaos was rediscovered by the meteorologist Ed Lorenz and became one of the most fashionable parts of physics. In recent years I have been calling attention to Cartwright's work. In 1993, I received an indignant letter from Cartwright, scolding me because I gave her more credit than she thought she deserved. I still claim that she is one of the original discoverers of chaos. She died, full of years and honors, in 1998.

When I started my career as a physicist in the 1940s, we were all survivors of World War II. It did not make much difference whether you were male or female. Rich and poor, women and men, we all got equal rations of food, and sharing of burdens was taken for granted. I came to Cornell University as a student in 1947, when most of the experimental physicists were busy building a new synchrotron and had no time for doing experiments. Hans Bethe used to say that it was only the shortage of steel that kept experimental physics alive in the USA at that time. One of the physicists who were keeping experimental physics alive was Vanna Cocconi. She did not wait for the synchrotron to be finished. She did experiments in particle physics using cosmic rays, driving a truck full of cosmic-ray detectors to the top of Mount Evans in Colorado. She discovered the phenomenon of nuclear spallation, observing the large bursts of neutrons emitted when a high-energy cosmic ray collides with a heavy nucleus. She and her husband, Giuseppe, shared the burdens of driving the truck, building cloud chambers and electronic circuits, raising a baby and running a household. Once a week, I watched with admiration the spectacle of Giuseppe scrubbing the kitchen floor while Vanna fed the baby. For any young woman who chooses Vanna as a role model, the first priority should be to find a husband like Giuseppe. Both of them later became leaders in the European particle-physics laboratory (CERN) at Geneva, Vanna as an experimenter, Giuseppe as a theorist. In the 1960s, Vanna helped set up the large bubble chamber at CERN. Later, she ran a beautiful experiment studying multiple production of pions in proton–proton collisions. Just as a highly excited heavy nucleus emits a shower of neutrons, a highly excited proton emits a shower of pions. She demonstrated the autocorrelation of pions due to quantum statistics. This verified that pions obey Bose–Einstein statistics, and also gave a direct measurement of the size of the fireball out of which the pions emerge. Vanna died in 1998. If this book ever goes into a second edition, she should be included.

After a year at Cornell, I came to the Institute for Advanced Study in Princeton, one of ten young physicists invited by our director Robert Oppenheimer. When we arrived, our office building was still under construction and Oppenheimer was away in Europe, so we all sat around a big table in Oppenheimer's office. That gave us a good chance to get to know one another. Two of us were women, Sheila Power from Ireland and Cécile Morette (see Chapter 29) from France. It was immediately obvious that Cécile was the brightest and most enterprising of the bunch. Before anyone else, she seized on the new idea of Richard Feynman, the path-integral description of quantum processes, and made it her life's work to develop path-integrals into a rigorous mathematical discipline. At a time when Feynman's idea was still regarded as crazy or irrelevant by the majority of physicists, she forced us to take it seriously.

Theoretical physics in France was at that time at a low ebb. The grand old men, Louis de Broglie and his contemporaries, were no longer active. The lives of the younger generation had been disrupted by the war. Cécile decided that the time had come to rejuvenate French physics, and she saw a way to do it. She raised enough money to buy a delapidated farm near the village of Les Houches in the French Alps, and founded the Les Houches Summer School.

Cowsheds were converted into dormitories, and students came from all over Europe to learn modern physics. She ran the school with enormous competence and enthusiasm. I had the pleasure of teaching there for the summer of 1954, and taught the most gifted group of students that I ever encountered. Many of the students in my class became professors and leaders of research in various countries. The most gifted of all was Georges Charpak, who won a well-earned Nobel Prize for physics in 1992. He and Cécile were kindred spirits, organizing hikes in the mountains when the weather was good, organizing games and charades in the dormitory when it was bad and arguing incessantly about physics.

Besides the Les Houches Summer School that she ran herself, Cécile also played a part in the founding of another institution, the Institut des Hautes Etudes Scientifiques (IHES) at Bures-sur-Yvette on the southern edge of Paris. IHES is a French equivalent of the Institute for Advanced Study at Princeton. I remember vividly the day in 1948 when Cécile, then newly arrived at Princeton, brought a distinguished looking French visitor to lunch at the Institute. She told the visitor that France needed an Institute for Advanced Study too, privately funded and independent of government. She made arrangements for him to meet with Robert Oppenheimer. The visitor was impressed by Cécile's vision. He said he would try to do something about it. His name was Léon Motchane. A few years later he became the founder and first director of IHES. The Les Houches Summer School and IHES did the job for which Cécile designed them. They opened the door for a new generation of young people to become the leaders of French physics. Both institutions are still flourishing today. All it took to accomplish this revolution was one young woman with courage and imagination. When the revolution was still a dream in Cécile's head, we used to tease her, saying that she was a reincarnation of Joan of Arc. She said, no, she was not Joan of Arc, because she had no intention of being burned at the stake. To avoid being burned at the stake, she settled in America and started a new career at the University of North Carolina, so that she had an independent base from which to organize the revolution in France.

I have described these three women, Mary Cartwright, Vanna Cocconi and Cécile DeWitt-Morette, because they are the most memorable of my female mentors. I knew them and admired them when I was young and impressionable. Unfortunately, none of them was American-born. Fifty years ago when I was a student, physics in America was far more male-dominated than physics in Europe. Today, America still lags behind Europe in the advancement of women. I recently collected statistics of the numbers of women and men who have been members of the School of Natural Sciences at the Institute for Advanced Study in Princeton. The fraction of women has not increased since I first came to the Institute in 1948. It was then, and has remained ever since, about ten percent. In American history there was no Joan of Arc.

A few years ago, the Russian mathematician Mikhail Monastyrsky wrote a little book with the title, *Modern Mathematics in the Light of the Fields Medals*. It is a historical survey, describing the people who won Fields Medals for outstanding contributions to twentieth-century mathematics. A Fields Medal for a mathematician is the equivalent of a Nobel Prize for a physicist. Monastyrsky wrote his book first in Russian and then translated it himself into English. He told me that he had great difficulty getting it published, both in Russia and in America. The Russian publishers said, "Too many Jews." The American publishers said, "Not enough women." Political correctness makes different demands in different countries. His book has now been successfully published in both languages, but the problem of the under-representation of women remains. Monastyrsky's history does not provide inspiring role models for the next generation of women mathematicians. This book does better. It provides plenty of role models for the next generation of women physicists. It says to any young woman aspiring to a career in physics, "Welcome to the club."

Introduction

Nina Byers

Physics Department, University of California, Los Angeles

People tend to think that physicists are men. This book will help to bring a more gender-balanced perception of physicists. You will find here detailed descriptions of important, original contributions made by women in the century from 1876 to 1976. Many female physicists and mathematicians, historically excluded from participation in science, emerged in this time. It was a period in which there was great progress in science and in the liberation of women from centuries of repression. This book documents both aspects of recent history. Many of the contributors are distinguished scientists who have been actively engaged in the areas of physics about which they write. They describe with remarkable clarity and readability modern developments in fields as diverse as astrophysics, biomolecular structure, chaos theory, geophysics, nuclear physics, particle physics and surface physics. Each chapter provides detailed accounts of important discoveries made by a particular woman, places them in their historical context, gives references to the original papers for further reading, and suggestions, and provides a scientific biography. The discoveries are well established and fundamental to modern physics. It is not well known, however, that they were made by women.

There are only a few women in the history of physics in past centuries. They are upper-class women such as Emilie du Châtelet (1706–49), Laura Bassi (1711–78) and Mary Somerville (1780–1872), women who had a passion for physics and could study it. But, generally, most women in those centuries past had no opportunity to do this. With the proliferation of printing presses in the mid nineteenth century, and the opening of universities to women, this changed, and women took up the study of physics and mathematics.[1] Some were endowed with extraordinary talent and ability. This book is about forty such women. Some people have said that the examples of these women are daunting to the young who may be thinking of going into science. On the contrary, I have found them inspiring. For a number of years I taught a course at UCLA entitled "Women in Physics and Mathematics," which drew young men and women interested in science. The course was about the women who are the subjects of essays in this book. It was clear to us all that their intellects were far beyond those of ordinary people, but we didn't compare ourselves with them, rather their lives and accomplishments inspired us.

The essays here are generally presented in three sections; the first, and usually the longest, is a detailed exposition of major discoveries and their significance; the second is a brief biography focusing mainly on the subject's life in science; the third gives biographical details and references to original papers and suggested further reading. The authors were commissioned to write short essays of only about 3000 words. The chapters, therefore, are brief accounts of the work and lives of these women. They give references to longer appreciations, and biographies, where they exist. For example, several of the women, now deceased, were Fellows of the Royal Society (London) and Biographical Memoirs have been published in the RS yearly collection.

The women lived very full and diverse lives. Some married and had children and others never did. In the early days, it was far more difficult than it is now to reconcile a normal family life with a scientific career. For example, in 1906 the Dean of Barnard College of Columbia University obliged Harriet Brooks (Chapter 6) to resign her job teaching physics when she revealed that she planned to marry. The Dean said, "The College cannot afford to have women on the staff to whom the college work is secondary; the College is not willing to stamp with approval a woman to whom self-elected home duties can be secondary." And Harriet Brooks said, "I think it is a duty I owe to my profession and to my sex to show that a woman has a right to the practice of her profession and cannot be condemned to abandon it merely because she marries."[2] Some of the earliest physicists such as Marie Curie (Chapter 4) and Hertha Ayrton (Chapter 1) had both marriage and a scientific career, though Ayrton was refused election to The Royal Society (London) on grounds that she was a married woman.

These women emerged as scientists in a historical period in which opportunities for women to study and work were opening up in Europe and America. All were encumbered with serious gender discrimination in one form or another, which generally hindered their ability to get on with their work. This aspect of their lives is only briefly mentioned in essays here or left out altogether. This is mainly due to limitations of space. The authors' focus was on scientific achievement, rather than on barriers the women had to overcome to do their work. To compensate for this, at least in part, in this introduction I would like to augment chapters of this book with some examples of the egregious gender discrimination that are, or should be, part of the historical record.

Many of the women in this book worked unpaid or in poorly paid positions. An outstanding example of an unpaid worker who made important discoveries is Marietta Blau (Chapter 10). She worked from 1923 to 1938 in the Institut für Radiumforschung in the University of Vienna. She had a place to work in the lab and some support for equipment, but no salary. As a Jew and a woman, her applications for paid positions were rejected. In his classic *Atomic Physics* text Max Born wrote, "A great advance was made by two Viennese ladies, Misses Blau and Wambacher, who discovered a photographic method of recording tracks of particles."[3] He explained that "the grains of a photographic emulsion are sensitive not only to light but also to fast particles; if a plate exposed to a beam of particles is developed and fixed the tracks are seen under the microscope as chains of black spots. Their quality depends very much on the size of the grains, and special emulsions with very small and dense grains have been developed." Much more about this and its significance for nuclear, cosmic ray and particle physics is given in Leopold Halpern and Maurice Shapiro's essay. Blau fled Nazi persecution when Germany occupied Austria in 1938. Shortly thereafter her methods were adopted by C. F. Powell in Bristol. (See endnote 2.) He subsequently improved them and won the

1950 Nobel Prize in Physics "for his development of the photographic method of studying nuclear processes and his discoveries regarding mesons made with this method." He is justly famous for his contributions, but adequate credit to Blau for her pioneering work is lacking. She was known and highly respected in Europe before World War II but, as Leopold Halpern and Maurice Shapiro recount in their chapter, did not receive appropriate recognition and employment after the war.

Another remarkable experimentalist is Agnes Pockels. Her early work was unpaid and, as Gary Williams explains in Chapter 3, laid the foundation for modern quantitative studies of liquid surfaces and films. She learned physics from her brother's texts, but was herself denied a university education. Apparently, her family felt it would be inappropriate for her to attend the university with her brother. For some twenty years, she pursued an experimental program in her kitchen with apparatus she devised and built herself. When she read that Lord Rayleigh was doing some of the same experiments, she wrote him a letter describing her work and her results. He published her letter in the March 12, 1891 issue of *Nature*, introducing it with the following comments: "I shall be obliged if you can find space for the accompanying translation of an interesting letter which I have received from a German lady, who with very homely appliances has arrived at valuable results respecting the behaviour of contaminated water surfaces. The earlier part of Miss Pockels' letter covers nearly the same ground as some of my own recent work, and in the main harmonizes with it. The later sections seem to me very suggestive, raising, if they do not fully answer, many important questions. I hope soon to find opportunity for repeating some of Miss Pockels' experiments." As Gary Williams writes, her work laid foundations for the work of Irving Langmuir, who was awarded a Nobel Prize in 1932 "for his discoveries and investigations in surface chemistry." It also underlies the important results on surface films achieved by his co-worker Katharine Blodgett (Chapter 13).

Although the discoveries described in these chapters underlie many well-established and fundamental elements of modern physics, most of the women are not well known. One reason for this is that references to their publications in scientific journals are not cited as frequently as one might expect. This is due to operation of the Matthew Principle, named by the sociologist R. K. Merton.[4] He found that highly acclaimed scientists' papers are by far the most frequently cited in scientific journals. Historically, females were not highly acclaimed and often remain unmentioned. One outstanding example of the Matthew Principle are references in the literature to the discovery of the period-luminosity relation of Cepheid variable stars. As is explained by Jean Turner in Chapter 5, this discovery enabled the first measurements of intergalactic distances. It was made by Henrietta Leavitt. Her discovery is reported in the *Harvard College Observatory Circular*, March 3, 1912. However, the report was published under the name of Edward C. Pickering, Director of the Harvard College Observatory.[5] The first sentence of the circular reads, "The following statement regarding the periods of 25 variable stars in the Small Magellanic Cloud has been prepared by Miss Leavitt." The remaining content is Leavitt's extensive, detailed presentation of her data and analysis. For decades, and probably even now, references to the discovery cite Pickering.

Several other very important discoveries are described here: for example nuclear fission (Lise Meitner, Chapter 7), the solid inner core of the earth (Inge Lehmann, Chapter 9) and the spatial structure of biomolecules (Dorothy Hodgkin, Chapter 22). Historically, the discovery of nuclear fission may be of prime importance. Often only Nobel laureate Otto Hahn is credited with this discovery. The relatively infrequent reference to Lise Meitner in this connection is another example of the Matthew Principle. As Ruth Sime writes in Chapter 7, this discovery was made by Hahn, Fritz Strassmann and Meitner. The work was begun and carried out in Meitner's laboratory in Berlin. Meitner, a Jewish woman, had had to flee Nazi persecution in 1938.[6] The discovery was announced by Hahn and Strassmann in 1939.[7] It is now generally acknowledged that physicist Meitner played a crucial role, and many contend that it was a mistake for the Nobel Committee to award the Prize to Hahn rather than to Hahn, Meitner and Strass-mann.

Inge Lehmann's discovery of the hard inner core of the earth was certainly important. Before Lehmann announced her discovery, people believed there was simply the molten iron/nickel core. In 1935, well before high-speed computers were available, Lehmann analyzed observational data on seismic waves generated by deep focus earthquakes, and found that there was a hard inner core. Bruce Bolt gives a very clear account of this in Chapter 9. Lehmann published this discovery in a paper that may have the shortest title of any in scientific literature.[8] She was Denmark's only seismologist for 20 years. In modern times most discoveries in physics are made collaboratively. However, historically, an unusually large fraction of female physicists worked alone. Many had been obliged to do so as a result of various forms of gender discrimination. Furthermore, funding for assistants, students and/or young colleagues was difficult, if not impossible, to obtain. Lehmann's nephew Niels Groes remembers, "It was not easy for a woman to make her way into the mathematical and scientific establishment in the first half of the twentieth century. As she said, 'You should know how many incompetent men I had to compete with – in vain.'" (See endnote 2.)

One sees from the essays in this book that, as soon as the restrictions barring women from studying and working in institutions of higher learning began to be breached, female scientists of the highest caliber appear. An outstanding example is one of the greatest mathematicians of the twentieth century, Emmy Noether (Chapter 8). She is universally credited with the development of modern abstract algebra. She was one of the first two women admitted as students to the University of Erlangen. They enrolled in 1904. Margaret Maltby (Chapter 2) was the first woman to receive a doctorate in physics from the University of Göttingen, awarded in 1895. No doubt she was a good physicist, but she had to abandon research for teaching. In those days, paid positions for a female physicist to do research were, as far as I know, non-existent. Maltby took a job teaching physics in Barnard

College, Columbia University's adjunct college for women. She had a long and distinguished career as a teacher and mentor to many young women. The fact that so many gifted and talented female scientists emerged in the early days indicates that many more were there also studying and wishing to pursue scientific careers. Opportunities to do so, however, were few and far between.[9]

Marie Curie (Chapter 4) is perhaps the most famous woman in the book. Her fame is in part due to the widespread publicity she encouraged in the interest of obtaining funding for her laboratory.[10] The success of this effort, particularly in the United States, enabled her to bring into her lab and support many male and female physicists. However, it seems to me that her fame today tends to be misplaced, because it rests more on the fact that she was a female physicist rather than on her brilliant insights and amazing experimental achievements. She was a very gifted physicist, as is apparent from her early studies of radioactivity which she initiated for her doctoral research. From these, she obtained and published important confirmation of the then still controversial idea that atoms are fundamental constituents of matter. Shortly thereafter her husband, Pierre, left his lab to join her in further studies. Together they laid the foundations of nuclear physics, as Abraham Pais beautifully describes in Chapter 4. People, in my view inaccurately, often describe her role in this collaboration as secondary. Her husband insisted this was not the case, refusing to accept the Nobel Prize the committee wished to bestow on him and Henri Becquerel if she were not also a recipient. Pierre died suddenly in 1906, only eight years after their collaboration began. Grief stricken, she spoke for years afterward of the importance of Pierre in her work. For example, on the occasion of her second Nobel Prize,[11] she says many of the discoveries for which she was given the award were made "by Pierre Curie in collaboration with me . . . The chemical work aimed at isolating radium in the state of the pure salt . . . was carried out by me, but it is intimately connected with our common work . . . I thus feel that . . . the award of this high distinction to me is motivated by this common work and thus pays homage to the memory of Pierre Curie." Many who knew her had the highest regard for her scientific acumen and ability. Indeed Rudolf E. Peierls[12] tells of meeting her at the 1933 Solvay Conference: "The subject was the atomic nucleus . . . I was very impressed with her poise and command of physics. She was then 66 years old yet she was absolutely up to date on all technical details, and often corrected her daughter Irène and her son-in-law Frédéric Joliot on points of fact." Irène and Frédéric Joliot-Curie were also Nobel laureates.[13] Aside from Marie and Irène Curie, three more of the women in this book are Nobel laureates – Maria Goeppert Mayer (Chapter 18), Dorothy Crowfoot Hodgkin (Chapter 22) and Rosalyn Yalow (Chapter 27). And there are several more women whose scientific achievements, it is thought, should also have been so honored. Some of them are Lise Meitner (Chapter 7), Marietta Blau (Chapter 10) and Chien-Shiung Wu (Chapter 24).

It has been my good fortune to have been able to work closely with some of the women in this book. When I was a graduate student in the University of Chicago, Maria Goeppert Mayer was on my Ph.D. committee. Though she was a world-renowned, highly respected theoretical physicist, author of many important papers in the literature, she was unpaid. She had a special appointment as "Volunteer Professor." Her husband, Joe, was Professor of Chemistry, and because of a nepotism rule she was denied a regular faculty appointment. She was finally offered an ordinary, paid professorship only a few years before she was awarded a Nobel Prize for Physics. Later, I met her in San Diego, where she was Professor in the University of California. In a brief conversation about her having worked for so long unpaid, she revealed, "I almost became bitter." Maria Mayer's 1948 discoveries of the magic numbers and nuclear shell model, are regarded by those who know them well as truly astonishing. Knowledgeable people were not surprised that a theoretical physicist with her acumen and insight might have made these discoveries. S. A. Moszkowski, author of Chapter 18, was a Ph.D. student of Mayer. When asked if he thought she was a genius, he said yes. When asked what kind of genius – a distinction the mathematician Mark Kac made between the ordinary kind whose results we might also achieve given enough energy, time and hard work, and the magician whose creative gifts go beyond that – Moszkowski replied she was a magician. She was legendary as one of the stars in the University of Chicago physics faculty.

One of the stars in the constellation of great scientists of our time, and all time, is Dorothy Crowfoot Hodgkin (Chapter 22). Among the great achievements for which she will always be remembered are her discoveries of the structure of penicillin, of the antipernicious anaemia factor, vitamin B_{12}, and of the diabetic hormone insulin. Her work was foundational for modern structural chemistry, molecular biology and genetics. I had the good fortune to work alongside her when I was elected to the Governing Body of Somerville College, Oxford, in 1967 as Official Fellow and Tutor in Physics. Dorothy, as she was universally called by her friends, associates, students and co-workers, was Professorial Fellow and Wolfson Research Professor. She regularly attended Governing Body meetings when she was in Oxford though she was no longer teaching undergraduates. Part of the work of the Governing Body was to elect fellows and tutors, and this necessitated rejections of some applicants. Dorothy never had a negative word to say about anyone. She shared her wisdom, remarkably, without ever denigrating anyone. She had an inner beauty of spirit that seemed to make all in her presence better than they might otherwise be. Anne Sayre described Dorothy very well for those of us who knew her when she wrote that:

> Few were her equal in generosity of spirit, breadth of mind, cultivated humaneness, or gift for giving. She should be remembered not only for a lifetime's succession of brilliantly achieved structures. While those who knew her, experienced her quiet and modest and extremely powerful influence, learned from her more than the positioning of atoms in the three-dimensional molecule, she will be remembered not only with respect, and reverence, and gratitude, but more than anything else, with love.

In Chapter 22, Jenny Glusker describes with clarity the progression of Dorothy's scientific work: her determinations of the three-dimensional structures of biochemically important molecules for which the chemical formula was uncertain or unknown, namely cholesterol, penicillin and vitamin B_{12}. It is a wonderful chapter, close study of which rewards the reader with an understanding of the scientific advances Dorothy achieved. Great respect is also due Dorothy for her courage and selflessness when one realizes that during most of her life she was afflicted with crippling rheumatoid arthritis. Since its onset at age twenty-eight, she must have suffered severe pain. However, those of us who knew her were unaware of this. Though the crippling effect of the arthritis was apparent, one's awareness of it was quickly overcome by the beauty of her spirit and intellect. The great British sculptor, Henry Moore, was asked to draw a portrait of Dorothy, and he chose to make the drawing of her hands, shown here.

FIGURE I A portrait of Dorothy Crowfoot Hodgkin's hands, by Henry Moore. From the Art Gallery of Ontario.

Among the many awards Dorothy Crowfoot Hodgkin received was a Nobel Prize "for her determinations . . . of the structures of important biological substances." She was the third woman to receive a Nobel Prize after Marie Curie who received two in 1903 and 1911, and her daughter Irène Joliot-Curie (1935). She was given the highest civilian honor awarded in Great Britain, the Order of Merit, in 1965, and was the second woman to receive this honor, the first being Florence Nightingale in 1907.

This book grows out of a website (see note 2), which was begun in 1995 to mark the occasion of the centenary of the American Physical Society (1899–1999). Asked to compile a list of distinguished female physicists to be included with distinguished male physicists in a poster to mark the occasion, my colleague Steve Moszkowski and I expected to make up a list of 15 or 20. For completeness I posted on the World Wide Web a request to colleagues around the world for nominations for the list. To our astonishment, more than 200 nominations came in.[14] Most of the women were unknown to us. The nominations contained documentary evidence of notable contributions. We wished to restrict our attention at first to those that could be regarded as original and important contributions. Bibliographic and historical research was needed to sort these out of the hundreds of nominations. Help was needed, and a project funded by the Alfred P. Sloan Foundation, the University of California, the American Physical Society, Joan Palevsky and other donors at UCLA, was initiated to research and document accomplishments of women who have made original and important contributions to physics. Students, graduates and undergraduates, and colleagues joined the effort. We were able to create web pages for 83 outstanding women. In doing the historical research we came into contact with colleagues who had known and worked with many of the women. Margaret O'Gorman, then Editor of the Book Division of the Institute of Physics Press (IOPP), commissioned these scientists and others intimately familiar with the achievements of 40 of the women to write about them for a companion volume for the second edition of IOPP's *Twentieth Century Physics*, and I agreed to work with her as an editor. Because of the authors' deep and intimate knowledge of the women's scientific achievements, the chapters are unique contributions to scientific literature and the history of physics. It is remarkable that the authors took time from their busy research schedules to write these essays. IOPP's plan to publish the second edition of *Twentieth Century Physics* and this companion to it did not materialize but, fortunately, the nearly complete collection of essays was taken in for publication by Cambridge University Press. The book will fill a gap in scientific literature by virtue of the completeness and scientific accuracy of its descriptions of important discoveries.

ACKNOWLEDGEMENTS
First and foremost, co-editor Gary Williams and I wish to express gratitude to the authors for the time and attention they gave these chapters. Nearly all of them are research physicists for whom the writing of historical texts is a departure from their normal work, and they were willing to make this sacrifice in tribute to the women about whom they have written realizing women's scientific achievements generally have not been given the attention they deserve. Gary Williams collected and edited nearly half the chapters and, in addition, wrote or co-wrote three himself. Gillian Lindsey was the original commissioning editor whose ideas shaped the book and who, with suggestions from me and my colleague Chun Wa Wong at UCLA, invited and guided the authors' contributions. Publication of this book is due in large measure to the encouragement and support of Ted Greenwood, Program Director at the Alfred P. Sloan Foundation. I wonder how books like this, with the attendant bibliographic research we were able to provide the authors, were compiled before the advent of computers. This book owes much to the IT support services provided by Lingyu Xu and Craig Murayama of the UCLA Physics Department. Mary Jo Roberts and Karin Nachtigal at UCLA and Barbara Drauschke in the Harvard University Physics Department provided invaluble help and support in preparation and editing of the book. We would like to express our appreciation with deep gratitude to Simon Mitton, Senior Editor at Cambridge University Press. He recognized that the essays fill a gap in the literature of science describing authoritatively, with precision and clarity, some important contributions

to physics that have been made by women. Our colleague, Steven Moszkowski, along with his wife, Esther Kleitman, gave us support and assistance in our editorial work. The essays in this book span a wide range of fields of physics and, as an eminent nuclear physicist, Moszkowski's help and advice was very valuable. Finally, we wish to thank Juliet Popkin of the Popkin Literary Agency for her patient and good-natured guidance that saw us through the territory from writing to publication. One of us (N.B.) wishes to thank the physics departments of UCLA and Harvard where so much of her work was done.

NOTES

1. See T. Woody, *A History of Women's Education in the United States* (New York: Octagon Books, 1980). There is included in this classic work of two volumes also an extensive history of women's education worldwide. See also, A. H. Koblitz, *A Convergence of Lives. Sofia Kovalevskaia: Scientist, Writer, Revolutionary* (New Brunswick N.J., 1993)., who writes about the many women who entered Russian universities in the 1860s to study physics and mathematics.

2. See http://cwp.library.ucla.edu.

3. M. Born, *Atomic Physics* (New York: Hafner Publishing Company, 1935); pp. 36–7 in 7th edition, revised by author and R. J. Blin-Stoyle, 1961.

4. R. K Merton, *Isis* **79** (1988) 606. He named this effect after a biblical parable in the Book of Matthew, which illustrates the phenomenon that "them that has gets."

5. To view this circular see http://cwp.library.ucla.edu/articles/leavitt/leavitt.note.html.

6. R. L. Sime, *Lise Meitner: A Life in Physics* (Los Angeles: University of California Press, 1996).

7. O. Hahn and F. Strassmann, Uber den Nachweis und das Verhalten der bei der Bestrahlung des Urans mittel Neutronen enstehenden Erdalkalimetalle. *Die Naturwissenschaften* **27**: 11 (1939). Meitner does not appear as an author of this paper because she fled Berlin July 13, 1938 to escape Nazi persecution of Jews and professional women.

8. I. Lehmann, P', *Union Geodesique at Geophysique Internationale, Serie A, Travaux Scientifiques* **14** (1936) 87.

9. See, e.g., M. W. Rossiter, *Women Scientists in America: Struggles and Strategies to 1940* (Johns Hopkins University Press, 1982).

10. See, e.g., S. Quinn, *Marie Curie: A life* (New York: Simon & Schuster, 1995).

11. The Nobel Prize in Chemistry was awarded to her alone in 1911. Her first Nobel Prize was in Physics, awarded in 1903 jointly to Marie, Pierre and J. Becquerel.

12. R. Peierls, *Bird of Passage: Recollections of a Physicist* (Princeton, New Jersey: Princeton University Press, 1985) pp. 100–1.

13. Awarded the Nobel Prize in Chemistry in 1935 for their discovery of artificial radioactivity.

14. See http://cwp.library.ucla.edu/lists/origCit.html. Materials received are in the CWP archive in the Department of Special Collections of the UCLA Library.

NOTE ADDED IN PROOF

Margaret Rossiter (The Matthew Matilda Effect in Science, *Social Studies of Science*, Vol 23, No. 2 (May, 1993), 325–341) extends the observations of Merton[4] noting the gender-based systematic bias in scientific information and recognition.

I   Hertha Ayrton (1854–1923)

Joan Mason (deceased)
Department of History and Philosophy of Science, Cambridge

FIGURE I.I Photograph of Hertha Ayrton.

IMPORTANT CONTRIBUTIONS

Hertha Ayrton came to science through invention: her biographer Evelyn Sharp traced her mechanical ability to her father, a watchmaker. As a student she constructed a simple sphygmomanometer, for measuring the pulse in the wrist. Later, she designed a line divider, for marking off equal parts on a line or for scaling drawings. She lectured on this to the Physical Society, patented it, and it was marketed by an instrument maker.

Hertha Ayrton's taming of the electric arc

Arc lights had been used in lighthouses from the mid 1800s, and by the 1890s were taking over from gas for street lighting. To strike the arc, a

Out of the Shadows: Contributions of Twentieth-Century Women to Physics,
eds. Nina Byers and Gary Williams. Published by Cambridge University Press.
© Cambridge University Press 2006.

high voltage was established between two carbon rods a short distance apart in air, so that carbon evaporated from the positive electrode, forming a crater at the tip. A high current was carried by the ionized carbon atoms and counter-streaming electrons as they moved between the two rods, forming what would now be called a plasma in the electrode gap region.

Arcs were plagued with problems of humming, hissing, sputtering and instability, and needed constant adjustment of the electrode separation as the carbon evaporated. Hertha Ayrton showed that the major problems were caused by oxidation of the carbon anode at its tip. The hissing was due to air rushing in, enlarging the crater and promoting unstable conditions. She achieved the reproducible conditions that had eluded other experimenters by shaping the ends to prevent oxidation, and by careful control. The Admiralty and War Office were able to standardize searchlight carbons, following her instructions.

With a stable arc, observing a magnified image on a screen, she monitored the changes in the tips of the electrodes, and the resulting changes in the voltage for a given current and arc length. She could then demonstrate a linear relationship between the voltage drop (V) across the arc, its length (L) and the current $(I$ amps$)$

$$V = a + bL + (c + dL)/I$$

with the coefficients a–d depending on the electrodes. This is known as the Ayrton equation.

She published her results in twelve papers, in *The Electrician*, during 1895–6. The Institution of Electrical Engineers (IEE) invited her to read her paper on "The Hissing Arc" in 1899, probably the second read by a woman to the IEE.[1] Unusually, they elected her directly to full membership (MIEE – equivalent to the present-day Fellowship), and no other woman was elected MIEE until 1958. They awarded her a prize of £10. Guglielmo Marconi was a fellow recipient.

When she demonstrated the electric arc to the IEE in 1899, the *Daily Telegraph* remarked that much of the attractiveness of her lecture was in:

> the feminine running commentary . . . as . . . when a peculiar modification of the negative carbon that appears to be due to hissing, the formation of a mushroom, . . . failed to make its prompt appearance. 'Mushrooms', observed the lady dryly, 'won't grow to order'; and again 'It's coming now, but I'm afraid it's on the side you can't see.'

The Royal Society, also, took notice, inviting her to demonstrate her experiments with the electric arc at their annual *conversazione* in 1899. The *Daily News* remarked:

> What astonished the lady visitors . . . was to find one of their own sex in charge of the most dangerous-looking of all the exhibits – a fierce arc light enclosed in glass. Mrs. Ayrton was not a bit afraid of it. She calmly showed how, by excluding air from the light, its unpleasant hissing could be prevented; and throwing a picture of the burning carbon on the screen, she made the reason evident to the dullest.

She gave a paper at the International Electrical Congress in Paris in 1900, in French; and her paper on "The Mechanism of the Electric Arc" was read before the Royal Society by John Perry in 1901. Her book *The Electric Arc* (1902) explained her results in detail, and described the historical background, back to Volta's pile in 1800, which Humphry Davy used to produce white sparks between electrodes of charcoal. *The Electric Arc* became a standard work until the 1920s, when it was overtaken by newer electrical technology.

She took out many patents for searchlight carbons, and also for cinema projectors. Movies were called "the flicks" because of the flickering of the arc, and the electrodes she patented in 1913 gave a steadier light. She later patented anti-aircraft searchlights, developed for the Admiralty, and arc lamp technology.

Hertha Ayrton's studies of sand ripples, and vortices in water and air

In 1901, while temporarily living at the seaside, Ayrton began to study the formation of sand ripples and bars by the motion of water above them (not by breaking waves, nor by the retreating tide).[2] For her experiments, she used the landlady's zinc bath, soap dishes and baking tins. Back home in London she continued her research at the top of the house, in the attic, producing stationary waves of different wavelengths by rocking glass vessels between 4 and 44 inches wide. She used permanganate, paint or metal powder to visualize the eddies and vortices. From 1904, she was able to read her own papers to the Royal Society, and she gave papers also to the British Association and the Physical Society. Her attempts at a mathematical description aroused controversy, and caused her to delay publication. It is now known that complicated viscous effects and (chaotic) turbulence are involved, and these were not understood at the time.

The Ayrton fan

When the Germans began to use chlorine as a poison gas in the trenches in World War I, early in 1915, she realized that this could be dispersed by a fan or flapper. Her studies of vortices in water gave her insight into vortices in air, and as a good physicist she knew that scale was unimportant. She experimented in her drawing room laboratory, pouring smoke from burning brown paper so as to roll along the floor; cooling the smoke, through pipes and chambers, so that it did not rise too high.

She found that she could readily clear model dugouts and pill-boxes of smoke with a few flaps of a card stuck to a matchbox representing the parapet of a trench. The fan (or paddle) she developed was a square foot of flexible blade on a T-shaped handle. When she smacked this down in face of the advancing gas the crossbar hit the ground first and the blade flattened out, sending a puff of air forward. This became a cylindrical vortex through friction against the ground, just as smoke rings formed at the end of a gun or funnel (or at the lips of a smoker) become a ring that revolves as the central part of the puff advances faster than the edges. A smoke ring enlarges as it travels, inducing a blast of air to pass through it, and this drove the gas back. When poison gas was first used in war, chlorine, a heavy gas, was blown across by the wind. The Ayrton fan was quite capable of rolling it back into the open, even of clearing dugouts of gas.

Hertha Ayrton could stand up to academic mathematicians, but was now up against officialdom to whom her home modeling seemed a long way from the battlefield. Intelligent soldiers were impressed, as was Alexander Trotter. Several reports commended her fans, but other officials shirked the responsibility of a decision about something so unusual, and simple, advocated by a woman. Safest was to label the report "Confidential" and pigeonhole it.

Many inventors suffered such disheartening opposition. The word "fan" may have put people off. She knew how to handle them – they were not supposed to be waved about – others needed instruction. But one of her assistants from the Central College went out to demonstrate their use, persuaded some officers of their value, and supplies of gas-repelling fans were sent to the front.

After battling with the War Office for two years, she asked Trotter if he could help her with the Admiralty. Although it was 30 years since he had dealings with them, he managed to persuade the Electrical Engineer in-Chief, C. H. Wordingham, to see a demonstration late in 1917, and he decided the fans would be valuable for emergency ventilation of ships. Overall, she managed to win acceptance of the fan, and organized its production, so that over 100 000 were supplied to the Western Front.

BIOGRAPHY

Hertha Ayrton (1854–1923), mathematician, physicist, engineer and suffragist, was born Sarah Phoebe Marks on April 28, 1854, at 6 Queen

Street, Portsea, near Portsmouth, on the English south coast. Her family was poor, having fled the pogroms against the Jews in Poland. Her father, an itinerant watchmaker and jeweller, died when she was seven. Her mother, with six young boys and a baby daughter, very generously accepted her sister's offer to look after Sarah in London, where they kept a school. From age nine to sixteen, Sarah was educated with and by her talented Hartog cousins, who included Numa (later the first Jewish Senior Wrangler).

At sixteen, Sarah became a governess in London, taking in needlework to earn extra money to support her mother, looking after her invalid sister so that she too could be educated. With her friend Ottilie Blind she studied in the evenings for the Cambridge local examinations, now open to women. Ottilie called her Hertha, earth-goddess, and Sarah adopted this name. They joined the suffrage movement, and her ambition for higher education reached the ears of the suffragist Barbara Bodichon, one of the founders of Girton College (in 1869). Hertha became a protégée of Bodichon, and her friend George Eliot, who called her "Marquis" (Marky). As George Eliot described Mirah, heroine of *Daniel Deronda*, she thought of Hertha's singing of Hebrew hymns.

Mathematics was Hertha's first love: she studied mathematics at Girton during 1877–81, coached by Richard Glazebrook, and for many years supplied problems and solutions to *Mathematical Questions* and the *Educational Times*. She was a lively Girtonian. She led the Choral Society, founded the Fire Brigade, and also a Mathematical Club with Charlotte Scott, Girton's first Wrangler.

Back in London she taught mathematics, ran a working girls' club and held a singing class for laundry girls. She met Will Ayrton in 1882, and perhaps met his wife Matilda who died of consumption in 1883 leaving a young daughter, Edith. Matilda was one of the "seven against Edinburgh" in 1869, an early battle for women's medical education, encouraged by Will. She qualified as an M.D. in Paris, following Elizabeth Garrett Anderson, whose daughter Louisa was one of Hertha's pupils.

Professor Ayrton was a pioneer of electrical science and technology. He ran courses in practical electricity at Finsbury, the first Technical College, with John Perry; then they founded an Electrical Engineering department at the City and Guilds Central Institution (later merged into the Imperial College at South Kensington). Their inventions ranged from voltmeters and ammeters to contact systems for electric railways. Ayrton became President of the Physical Society (1890–2) and of the Institution of Electrical Engineers (1892). He was elected to the Royal Society in 1881, and received the Society's Royal Medal in 1901. Ayrton thus bridged the worlds of physics and engineering, as Hertha was to do.

Hertha Marks enrolled in Ayrton's evening classes in 1884, they were married in 1885, and their daughter, Barbara, was born in 1886. Hertha lived domestically for a while, giving some popular lectures on electricity, which was to lighten the drudgery of women's lives.

Her work on the electric arc began through an accident, as Alexander P. Trotter wrote in *Nature*:

> During the Chicago Electrical Conference of 1893 a paper written by Ayrton, describing an unfinished and inconclusive research, was accidentally destroyed. No copy existed and Mrs. Ayrton took up the research and found that her husband had been at work on wrong lines. Her results were published in *The Electrician* and afterwards in book form in 1902.

She solved major problems of instability of carbon arcs, widely used for lighting, and patented her improved electrodes. The Institution of Electrical Engineers elected her to Membership in 1899 (the only woman Member until 1958). In 1902, she published her definitive work *The Electric Arc*, which she dedicated to Mme. Bodichon.

John Perry and others took the audacious step of proposing her as candidate for Fellowship of the Royal Society in 1902. Her achievement was worthy of this, but no woman had ever been proposed, not even Mary Somerville. The Royal Society's lawyers pronounced that a married woman was not a "person" eligible for fellowship under the

Charter.[3] As Bodichon had protested half a century earlier,[4] a married woman's 'person', under the law, was covered by that of her husband, or if unmarried by that of her father, until she reached the age of majority.[5] Such arguments were quashed by the Sex Disqualification Removal Act (1919), but still no woman was proposed again until 1944.

While Will Ayrton was convalescing at the seaside in Margate, Kent, in 1901, Hertha finished off her book *The Electric Arc*, and then began to study the formation of sand ripples by wave motions of the water. After her husband's death in 1908, she moved her laboratory down to the drawing room, as can be seen in the photograph "Mrs. Ayrton in her Laboratory." (See Fig. 1.2.)

Remarkably, for a medal is worth more than a fellowship, the Royal Society awarded her its Hughes Medal, in 1906, for her work on the electric arc, and on sand ripples. Although this medal is awarded annually "for original discovery in the Physical Sciences," no woman has received it since 1906; only eight more women have won a Royal Society medal, of which about a dozen are awarded each year.[6]

She was a champion of women's rights. In chairing the Physical Science Section of the International Congress of Women in London in 1899, she encouraged women to take up applied science. She supported the militant suffragists, marching in all the suffrage processions. In 1912–13, Mrs. Pankhurst and others, recovering from hunger strike in prison, were nursed in Hertha's home. When they recovered, the police took them back to prison, under the "Cat and Mouse Act." Hertha was "very proud" when her daughter, Barbara, went to prison in 1912. Her friend Marie Curie (see Chapter 4), also a militant suffragist, brought her daughters for summer holidays with her in 1912 and 1913. Her daughter Barbara Bodichon Ayrton Gould (1886–1950) became Labour member of Parliament for Hendon North in 1945. Her stepdaughter, Edith, who married the novelist Israel Zangwill, was an accomplished author: the heroine of her novel *The Call* (Allen and Unwin, London, 1924) was a suffragist, and an aspiring scientist.

Mrs Ayrton in her Laboratory.

LONDON EDWARD ARNOLD & Cº

FIGURE 1.2 Mrs Ayrton in her laboratory.

Hertha Ayrton was a founder of the International Federation of University Women in 1919, and of the National Union of Scientific Workers in 1920. She died on August 26, 1923.

She was clearly a woman of courage, energy and scientific ability. She was not religious, but through her Jewish family and fellow suffragists, she had access to intellectual life in London; and with Will Ayrton's help, was able to ignore significant social barriers, of class, religion and gender in Victorian and Edwardian society.

NOTES

1. Marion Greenwood read her paper (on food vacuoles in infusoria) in person to the Royal Society in 1895. It was published in *Phil. Trans. Roy. Soc. (B)* 185 (1895), 355–83.
2. Her paper "The origin and growth of the ripple mark" can be read on the internet at http://cwp.library.ucla.edu/articles/ayrton/ Ayrton_ripple/ripple.html.
3. *Roy. Soc. Minutes of Council 1898–1903*, **8**, 270, minute 12.
4. Barbara Bodichon (1854), "A Brief Summary, in Plain Language, of the Most Important Laws of England concerning Women," republished in Candida Ann Lacey (1987), *Barbara Leigh Smith Bodichon and the Langham Place Group* (NY and London: Routledge and Kegan Paul).
5. Albie Sachs and Joan Hoff Wilson (1978), *Sexism and the Law: a study of legal bias and male belief in Britain and the United States* (Oxford: Martin Robertson), pp. 14ff.
6. Year Book of the Royal Society, 1999.

IMPORTANT PUBLICATIONS

Marks, S., The uses of a line-divider. *Philos. Mag.*, *Ser. 5*, **19** (1885) 280–5.

Ayrton, H., The electric arc. *Proc. R. Soc. Lond.*, *Ser. A*, **118** (1901) 410.

The electric arc, *Philos. Trans. R. Soc. Lond.*, *Ser. A*, **199** (1902), 299–336.

The Electric Arc (London: The Electrician Printing and Publishing Co. Ltd, 1902).

On the non-periodic or residual motion of water moving in stationary waves. *Proc. R. Soc. Lond.*, *Ser. A*, **80** (1908) 252–260.

The origin and growth of ripple-mark. *Proc. R. Soc. Lond.*, *Ser. A*, **84** (1910), 285–310.

Local differences of pressure near an obstacle in oscillating water. *Proc. R. Soc. Lond.*, *Ser. A*, **91** (1915), 405–510.

On a new method of driving off poisonous gases. *Proc. R. Soc. Lond.*, *Ser. A*, **96** (1919–20), 249–56.

FURTHER READING

Sharp, E., Hertha Ayrton 1854–1923, a memoir (1926). *Girton College Register 1869–1946* (1946) 8.

Mrs. Ayrton, *The Girton Review*, no. 67 (1923) 2–3.

Mason, J., Hertha Ayrton and the admission of women to the Royal Society of London. *Notes. Rec. R. Soc. Lond.*, **45**:2 (1991) 201–20.

Malley, M., Hertha Ayrton. *Dictionary of Scientific Biography* (New York: Scribner, 17 II, 1990) 40–2.

Appleyard, R., *The History of the Institution of Electrical Engineers, 1871–1931.* (1939) pp. 167–8.

Trotter, A. P., *Early Days of the Electrical Industry, and other Reminiscences of Alexander P. Trotter* (The Institution of Electrical Engineers, 1948) pp. 125–9.

Mrs. Ayrton's work on the electric arc. *Nature*, **113** (1924) 48–9.

Armstrong, H., obituary in *Nature*, **112** (1923) 800–1.

Mather, T. *Nature*, **112** (1923) 939.

2 Margaret Eliza Maltby (1860–1944)

Peggy Aldrich Kidwell

National Museum of American History, Smithsonian Institution, Washington, DC.

FIGURE 2.1 Photograph of Margaret Eliza Maltby. Probably taken in 1891 when she was a student at Massachusetts Institute of Technology (MIT). Photo courtesy of the MIT Museum in the contribution of Katherine R. Sopka to the volume *Making Contributions: An Historical Overview of Women's Role in Physics* published by the American Association of Physics Teachers (ISBN 0-917853-09-1).

IMPORTANT CONTRIBUTIONS

Margaret Maltby contributed to physics in three very different ways. First, she carried out research both in acoustics and then in the relatively new discipline of physical chemistry. Second, as a graduate student and then a research assistant in Germany she showed by personal example that women could contribute to physics and to

Out of the Shadows: Contributions of Twentieth-Century Women to Physics, eds. Nina Byers and Gary Williams. Published by Cambridge University Press.
© Cambridge University Press 2006.

scholarship more generally. Third, as the head of the physics department at Barnard College, she attempted to shape the physics curriculum to suit the needs of a wide range of students, and became a valued teacher and role model for generations of women in her classes.

As an undergraduate, Maltby relished her exposure to classical music. Appropriately enough, her first published research, carried out when she was a special student of Charles R. Cross at MIT, concerned acoustics. Specifically, she was interested in the minimum number of vibrations required to distinguish two different frequencies of sound. The physicists Felix Savart and Friedrich Kohlrausch had found that at least two cycles of a sound wave were required to distinguish its pitch from that of another sound wave. In an 1884 article, Cross disagreed, suggesting that Savart and Kohlrausch's procedure used very impure tones. Cross proposed an alternate method, which gave waves that were sinusoidal in form and were heard as a pure tone. In 1892, Maltby and Cross published two papers describing their own investigations. They reported that it was possible to distinguish as little as 42/100 of a cycle of a C3 tuning fork (256 Hz) from 84/100 of a C4 tuning fork (512 Hz). Cross and Maltby presented their results to Boston's American Academy of Arts and Sciences, and reported them in both its *Proceedings* and in *Technology Quarterly*.

In 1893, Maltby left MIT for Göttingen University, where she studied in the newly established laboratory of the Prussian physical chemist Walther Nernst. Nernst was an enthusiastic advocate of the theory of ionic dissociation, that is to say the view that salts in solution separated into charged ions, which then moved under the influence of electric fields. The Swede Svante Arrhenius, building on ideas of J. H. van't Hoff, had advanced the dissociation hypothesis in 1887 and used it to explain a wide range of phenomena.[1] Early research on dissociation focused on solutions that conducted current relatively well. In her dissertation, Maltby studied solvents that were poor conductors such as alcohol, ether and pure water. She took pains to avoid the effects of polarization and controlled her experiments carefully for temperature. In 1895, this research won Maltby a Ph.D. in physics from Göttingen, the first doctorate in physics granted to a woman by a German university. She also received an award from the Association for Collegiate Alumnae that allowed her to stay on for an additional year. This was the first award of the ACA for study in physics. During this time, Maltby once again used oscillating currents and a Wheatstone bridge, this time to balance the currents in a resistor and a capacitor and thence to compute a new value for the velocity of electromagnetic waves.

After completing her postdoctoral year, Maltby returned to teach for a year at Wellesley College in Massachusetts and then spent another year teaching physics and mathematics at Lake Eire College for Women in Painesville, Ohio. In 1898, she returned to Germany as assistant to Friedrich Kohlrausch, who by then was President of the newly established Physikalisch-Technischen Reichsanstalt, an agency of the Prussian government in Charlottenburg, near Berlin. Maltby was the first female staff member at the Physikalisch-Technischen Reichsanstalt. She and Kohlrausch carried out detailed studies of the electrical conductivity of aqueous solutions of alkali chlorides and nitrates. They considered changes in conductivity with temperature and with concentration, and found that conductivities were somewhat lower than those previously accepted as the most accurate, especially in the most dilute solutions. Kohlrausch and Maltby argued that up to a concentration of 0.002 moles per liter, the assumption that ions move independently made in the dissociation hypothesis was precisely true; in more concentrated solutions the conductivity was lower than predicted by theory. Physical chemists would cite their data for decades.

In 1899, Maltby returned to the United States, where she studied theoretical physics for a year with Arthur G. Webster of Clark University in Worcester, Massachusetts. Then, in the fall of 1900, she finally found what would prove to be a permanent position. She wrote to Svante Arrhenius in January, 1901:

Since the beginning of the year (October 1st) I have been teaching
Chemistry at Barnard College in New York City. Barnard is a
young college for women and has recently been adopted by
Columbia University. All women students who have not the
degree of A.B. or B.S. must do their work at Barnard, but graduate
students can take their courses in the university. Thus you see
Barnard is a true college for undergraduate study. The courses are
given almost entirely by University professors and instructors, but
a few have been engaged who give no courses in the university. I
am one of those few and am in fact the only woman instructor.
There are perhaps half a dozen tutors and assistants who are
women, but no instructors except myself. I have practically the
entire charge of the department, for the Head Professor of
Chemistry at Columbia has given it over to me. I do not expect to
be in the chemistry department long, for it is just a temporary
arrangement. I am to be transferred to the physics department as
soon as possible (perhaps not for a year or so) and because of the
fine prospects in physics I consented to take the work in
chemistry for a time.[2]

During her first year at Barnard, Maltby taught general chemistry,
organic chemistry and physical chemistry. As she had never taught
any of these subjects, she found it exhausting and had no time for
research.

In 1903, Marie Reimer was hired to head the Barnard College chemistry department, and Maltby finally had the opportunity
to teach physics. When Barnard was founded in 1889, it used the
same entrance examinations, as well as some of the same faculty, as
Columbia University. However, under Maltby's supervision, Barnard
developed a distinct curriculum in physics. The two-semester introductory course included not only light, heat, sound and electricity,
as at Columbia, but mechanics and magnetism. Intermediate courses
offered further work on all of these topics. Finally, Maltby taught a

course on physical chemistry. Columbia had a much larger number of upper level courses in the physics department, but left physical chemistry to the chemists.

Four aspects of the curriculum Maltby developed deserve further mention. First, she firmly believed that introductory science courses should include laboratory work, arguing that sciences like physics and chemistry were particularly suited for general study not only because they were of fundamental importance, but because they allowed students to observe phenomena as they took place. Deducing processes that may have gone on historically, as in geology or evolutionary biology, was far less satisfying as a means of instruction. Hence Maltby devoted considerable time to assuring that the physics department received a $500 annual appropriation for laboratory equipment. Discussing apparatus for courses on sound and light, she emphasized that it could be shared with other departments such as psychology.[3] When salary expenditures had been less than anticipated, she urged that the funds be spent on apparatus.[4] When her pleadings brought special grants, as with funds for the purchase of an electromagnet in 1906 and money for apparatus relating to light in 1907, she was enthusiastically grateful.[5] As one might expect, the space required by the physics department grew apace. Fiske Hall, intended as the Barnard science building when it was built in 1898, was initially used as a dormitory. However, the physics laboratories soon were established on the second floor, and the department expanded to include a basement room in 1907.[6]

Second, Maltby reportedly sought to tie instructions on the physics of sound to the interests of musicians. Although the description of this course in the *Barnard College Announcement* makes no mention of music students, her obituary in the *New York Times* states, "Before her retirement, Dr. Maltby introduced a course in the physics department designed for music students and dealing with the physical basis of music, believed to have been the first such course."[7]

Maltby's attempt to broaden the Barnard curriculum to include recent experimental discoveries in physics is documented. In the spring of 1909 she offered a new upper-level course called "Electricity" that discussed "special subjects in electricity, as alternating currents, discharge of electricity through gases, electrical phenomena of the aether, radio-activity."[8] Maltby might have felt it useful to review these subjects before spending the academic year 1909–10 on sabbatical, a tour that included six months at the Cavendish Laboratory of Cambridge University. In any event, the course became a fixture in the Barnard curriculum, though offered at a lower level and taught by the lecturers and assistants whom Maltby recruited from Columbia's graduate school. Beginning in the spring of 1912, and continuing until her retirement, Maltby also taught advanced topics in a course with the ambiguous title "Theory of Electricity."

Finally, Maltby was one of the first American scientists to include a course in the history of physics in the offerings of a physics department. Several scholars associated with Columbia University took an active interest in the history of science in the early twentieth century. In the history department, James Harvey Robinson guided the dissertation research of a Barnard graduate, Martha Ornstein; future Columbia history professor, Lynn Thorndike; and a future Goucher College professor, Dorothy Stimson. At Teacher's College, David Eugene Smith wrote a two-volume history of mathematics. This activity may have encouraged Maltby to offer a course entitled "History of the Development of Some Fundamental Theories in Physics." According to the 1918–19 *Barnard College Announcement*: "This course is designed to give the students a more comprehensive view of the development of certain fundamental theories, of the experiments which have been crucial, and the bearing of experimental evidence from various fields of physics upon these theories."[9] Maltby offered the course every fall from 1918 through 1930.

BIOGRAPHY

Born on the family farm in Bristolville, Trumbell County, Ohio, and christened "Minnie Eliza," Maltby would legally change her first name to Margaret in 1889. She attended the preparatory school for

Oberlin College in her native state for a year before matriculating at Oberlin in 1878 and graduating in 1882. Unsure of her vocation, she spent a year studying at the Art Students' League in New York City, taught high school for a time in Ohio, and then enrolled as a special student at MIT, graduating with a B.S. in physics in 1891. That same year, Oberlin granted her an M.A. Maltby also taught physics at Wellesley from 1889 to 1893. Awarded a graduate fellowship by MIT, she went to study physical chemistry at Göttingen University in Germany, obtaining her Ph.D. in 1895. After a year's postdoctoral study in Germany, she returned briefly to teach at Wellesley and then at Lake Erie College in Ohio. She spent the academic year 1898–9 in Germany as an assistant at the Physikalisch-Technische Reichsanstalt, and then returned to study theoretical physics for a year at Clark University. In the fall of 1900, Maltby was appointed instructor in chemistry at Barnard College; in 1903 she moved to the physics department. Maltby remained on the faculty at Barnard for the rest of her career, rising gradually through the ranks of adjunct professor (1903), assistant professor (1910) and finally, at the ripe age of 52, associate professor (1913). She retired in the spring of 1931.

Maltby believed firmly in encouraging women to pursue academic careers. In 1896, when Svante Arrhenius's wife deserted him, Maltby wrote offering sympathy but at the same time urged "Don't please surrender the struggle for the intellectual, social and moral advancement of women, because one woman did not prove to be a devoted wife."[10] In the summer of 1906, Harriet Brooks, a physics graduate at Columbia who had tutored at Barnard the previous two years, announced that she planned to marry (see Chapter 6). Laura Gill, the Dean of Barnard College, insisted that Brooks resign. Maltby protested, noting in a letter to Gill that Brooks was a skillful teacher and commenting that "Neither you nor I would like to give up our active professional lives suddenly for domestic life or even for research alone."[11] Gill was unrelenting. In fact, Brooks broke off her engagement, but also decided to leave New York rather than return to

Barnard. In 1921, Frances Orr, another Columbia graduate student who taught physics at Barnard, announced that she planned to marry. By this time, Dean Gill had retired. Frances Orr, now Mrs. Severinghaus, was allowed to teach at Barnard another six years.[12] Maltby also encouraged women scholars as a member of the Committee on Fellowships of the Association for Collegiate Alumnae (later the American Association of University Women [AAUW]) from 1912 to 1929 and as chair of this committee from 1913 to 1924. In 1926, the AAUW created a fellowship in her honor. In 1929, Maltby prepared a small volume in which she listed the names of women granted fellowships over the first forty years of the program, gave a brief account of their careers and included a list of publications.

Maltby joined the American Association for the Advancement of Science in 1898. In 1900, she was elected a regular member of the newly established American Physical Society. In 1906, James M. Cattell published the first edition of his listing of American scientists, *American Men of Science*. Cattell starred the names of the 1000 scientists he thought most eminent. Maltby was one of 150 physicists starred – the only woman physicist so honored.

In addition to pursuing her career in physics, Maltby was a dedicated traveler and remained devoted to art and music. In 1901, she adopted Philip Randolph Meyer, the orphaned son of a close friend. This, as she would write to an Oberlin classmate many years later, added the "human feminine touch" to her life.[13] Maltby referred to Philip as her nephew, and they would remain close even after he married and fathered three children. Maltby remained a devoted traveler. She and Philip not only went to England on her first sabbatical, but traveled to Belgium, France, Switzerland and Italy. She spent a sabbatical semester in 1917 touring the American west and in the spring of 1924 visited Morocco, Algeria, Tunisia and Europe. The collapse of the stock market in 1929 curtailed her travel considerably, but she continued to attend operas and symphonies and visit friends, despite severe arthritis. She died at the age of 83 at the Columbia-Presbyterian Medical Center in New York City.

NOTES

1. Crawford, E., *Arrhenius: From Ionic Theory to the Greenhouse Effect* (Canton, MA: Science History Publications, 1996).
2. M. E. Maltby to Svante Arrhenius, January 1, 1901, Svante Arrhenius Collection, Center for History of Science, Royal Academy of Sciences, Stockholm.
3. M. E. Maltby to Laura Gill, September 28, 1906 and February 27, 1907, Dean's Office Departmental Correspondence, 1906–8, Box 1, Barnard College Archives. I thank Jane Lowenthal for her generous assistance in locating materials at Barnard College.
4. M. E. Maltby to Laura Gill, September 28, 1906, Dean's Office Departmental Correspondence, 1906–8, Box 1, Barnard College Archives.
5. M. E. Maltby to Laura Gill, December 21, 1906 and February 8, 1907, Dean's Office Departmental Correspondence, 1906–8, Box 1, Barnard College Archives.
6. Miller, A. D. and Myers, S., *Barnard College: The First Fifty Years* (New York: Columbia University Press, 1939); Laura Gill to M. E. Maltby, June 6, 1907, Dean's Office Departmental Correspondence, 1906–8, Box 1, Barnard College Archives.
7. "Dr. M. E. Maltby Long at Barnard," *New York Times*, May 6, 1944.
8. M. E. Maltby to Dean Brewster, January 17, 1908, Dean's Office Departmental Correspondence, 1906–8, Box 2, Barnard College Archives.
9. *Barnard College Announcement*, 1918–19, p. 100.
10. M. E. Maltby to S. Arrhenius, December 24, 1896, Svante Arrhenius Collection.
11. M. E. Maltby to Laura Gill, July 24, 1906, Dean's Office Departmental Correspondence, 1904–6, Box 3, Barnard College Archives.
12. Rossiter, M., *Women Scientists in America: Struggles and Strategies to 1940* (Baltimore: Johns Hopkins University Press, 1982).
13. M. E. Maltby to Newton W. Bates, May 23, 1922, Oberlin College Archives. I thank Roland M. Baumann and Tammy L. Martin for locating materials in the Oberlin College Archives.

IMPORTANT PUBLICATIONS

Cross, C. R. and Maltby, M. E., On the least number of vibrations necessary to determine pitch. *Proc. Am. Acad. Art. Sci.*, **27** (1892) 222–35.
On the least number of vibrations necessary to determine pitch. *Tech. Quart.*, **5** (1892) 213–28.

Maltby, M. E., Methode zur Bestimmung grosser elektrolytischer Widerstande. Doctoral dissertation, Georg-Augustas-Universität zu Göttingen (1895).

Methode zur Bestimmung grosser elektrolytischer Widerstande. *Z. Phys. Chem.*, **22** (1895) 133–58.

Methode zur Bestimmung der Periode electrischer Schwingungen. *Ann. Phys. Chem.*, **61** (1897) 553–77.

Kohlrausch, F. and Maltby, M. E., Das elektrische Leitvermögen wässriger Lösungen von Alkali-Chloriden und Nitraten. *Wissenschaftliche Abhandlungen der Physikalisch-Technischen Reichanstalt*, **3** (1900) 157–227.

Maltby, M. E., The relation of physics and chemistry to the college science courses. *Columbia Univ. Quat.*, **18**:1 (1915) 56–62.

History of Fellowships Awarded by the American Association of University Women (New York: Columbia University Press, 1929).

FURTHER READING

Barr, E. S., Anniversaries in 1960 of interest to physicists. *Am. J. Phys.*, **28** (1960) 474–5.

Ferris, H. and Moore, V., *Girls Who Did: Stories of Real Girls and Their Careers* (New York: E. P. Dutton & Company, 1927) pp. 213–26.

Harrison, S. W., Margaret Eliza Maltby (1860–1944). In *Women in Chemistry and Physics: A Bibliographic Sourcebook*, eds. L. S. Grinstein, R. K. Rose and M. H. Rafailovich (Westport, CT: Greenwood Press, 1993) pp. 354–60.

Wiebusch, A. T., Margaret Eliza Maltby. In *Notable American Women 1607–1950*, eds. E. T. James, J. W. James and P. S. Boyer, **2** (1971) pp. 487–8.

3   Agnes Pockels (1862–1935)

Gary A. Williams

*Department of Physics and Astronomy, University of California,
Los Angeles*

FIGURE 3.1 Portrait of Agnes
Pockels.

SIGNIFICANT CONTRIBUTIONS

The work of Agnes Pockels formed the basis of what today is known as
surface science, the branch of physics and chemistry that studies the
properties of liquid and solid surfaces. In the late 1800s very little was
known about surfaces. There were qualitative visual observations of
such phenomena as the presence of oily films on the surface of water,

Out of the Shadows: Contributions of Twentieth-Century Women to Physics,
eds. Nina Byers and Gary Williams. Published by Cambridge University Press.
© Cambridge University Press 2006.

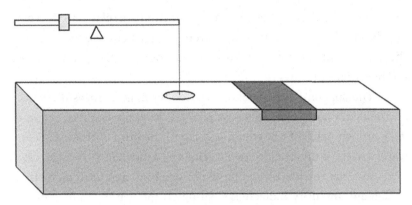

FIGURE 3.2 The Pockels trough, filled to the brim with water.

but no quantitative measurements had been carried out. By devising a simple apparatus that was crude but effective (see Figure 3.2), Pockels was able to measure the surface tension of water, and showed that it was greatly reduced by the presence of organic impurities such as soap, stearic acid or olive oil floating on the surface. She studied the variation of the surface tension as the surface concentration of impurities changed. The impurity films she studied were less than one full molecular layer thick. These were landmark observations, and were all the more remarkable for the fact that Pockels had no formal training or advanced degree in science, and worked alone in her kitchen with an apparatus fashioned out of household items.

Much of the later work in the study of molecular surface layers and films was based directly on Pockels' discoveries and experimental methods. In 1917, the chemist Irving Langmuir of the General Electric laboratories undertook measurements with a trough apparatus quite similar to Pockels's, but with several key improvements that allowed very detailed investigations into the properties of the surface molecules (in later measurements he was assisted by Katharine Blodgett, who is profiled in Chapter 13). For this work Langmuir was awarded the Nobel prize in chemistry in 1932.

Pockels's work has also proven to have important implications in biology. Cell walls are now known to be double layers of polymeric

molecules, and their structure and organization are related to the surface layers that Pockels first characterized. This connection reflects
the fact that life probably first originated from organic scum floating
on the surface of water.

The apparatus that Pockels devised for making measurements
on the water surface is shown in Fig. 3.2. It employs a long rectangular trough made of tin, which is filled to the brim with water. The
water surface is divided into two sections by a movable tin bridge rest-
ing across the trough. By sliding the bridge back and forth along the
trough any impurity molecules floating on the water surface can be
manipulated. At the start of an experiment the bridge is moved from
one end of the trough to the other end, sweeping the impurities with
it, ensuring that the water surface initially is very clean. A quantity of
impurity molecules can then be introduced on the left-hand surface
of the trough at such low concentrations that the separation between
molecules can be quite large, forming a low-density two-dimensional
gas. By slowly moving the bridge to the left to reduce the surface area,
the molecules are compressed and their density increases, until they
are close enough together (within a few Ångstroms) that they begin to
repel each other. At that point their density becomes constant, forming a single monolayer of molecules floating on the surface.

To monitor the surface impurities, Pockels measured the surface tension of the water. The surface tension of a liquid is a measure
of the molecular forces at the surface, since the creation of a surface
from a bulk fluid requires breaking the molecular bonds that hold the
liquid together. If the bonds are made weaker by impurity molecules
less force will be required to create the surface, lowering the surface
tension. Pockels measured the surface tension by placing a small disk
on the water surface, and then observing how much lifting force on
the disk was needed to just break the disk away from the surface.
Dividing this force by the perimeter of the disk gives the surface tension. (It is interesting to note that since she had no access to scientific
equipment, the disk she used was initially a common ceramic button,
6 mm in diameter, held by a fine thread.) The force on the disk was

applied by a small weight sliding on a pivoted lever arm, which could be calibrated from the position of the weight on the lever arm.

Pockels's measurements showed that the surface impurities act to lower the surface tension. As the impurity molecules are compressed together by the bridge, the surface tension decreases further, until the point where repulsive forces between the molecules cause their density to become constant. At that point the surface tension also becomes constant, defining what is now commonly known as the "Pockels point." The molecules cannot be compressed any further, and each molecule occupies a minimum area on the surface, known to be about 20 square Ångstroms for many common impurity molecules.

Pockels also used the apparatus for several other important measurements. By oscillating the surface at one end with a stick, she generated a water wave traveling down the trough. The attenuation of the wave was found by measuring how far it could propagate before dying out; she discovered that the presence of the surface molecules greatly increased the attenuation from that of clean water. She also discovered that the surface films could be visualized by sprinkling the surface with a fine powder such as lycopodium. The surface molecules prevent the powder from entering the water, which makes the extent of the surface film visible.

BIOGRAPHY

Agnes Pockels was born in 1862 in Venice, Italy, where her father was stationed with the Royal Austrian Army. Malaria was then prevalent in northern Italy; illness forced her father to retire in 1871 and the family moved to the town of Brunswick in Germany. Pockels attended the Municipal High School for Girls, where she displayed a strong interest in the natural sciences. She hoped to continue her studies, but at the time German universities did not admit women. In any event, she was obliged to care for her parents, who continued to be in poor health. Her brother Friedrich, however, was able to attend the university in Göttingen, and eventually earned a Ph.D. in physics in

1888. He supplied her with physics textbooks, enabling her to study the subject on her own.

As the unmarried caregiver of her parents and their household, Pockels spent much time in the kitchen cooking and cleaning. Being observant and curious, she closely studied the nature of water surfaces and how they are affected by thin contaminant films. Soap and oil films are, of course, a common aspect of kitchen cleaning, but little was known then about how they affected the water surface. In 1878, at age 18, she began to devise simple measuring techniques to try to quantify the surface properties. Two years later she had constructed the trough and balance device shown in Fig. 3.2, and began making observations with the device over a period of several years. Pockels tried to interest physicists in Germany about her findings, but got no response. In 1888, she learned that Lord Rayleigh in England had begun investigating some of the same questions, as he had published several related abstracts and articles. Although she must have felt trepidation about writing to one of the most eminent scientists of the day, she nonetheless overcame her fear and sent him a long letter detailing her work. Raleigh was quite impressed, and sent the letter on to the journal *Nature*, adding a preface to the letter recommending its publication:

> I shall be obliged if you can find space for the accompanying translation which I have received from a German lady, who with very homely appliances has arrived at valuable results respecting the behavior of contaminated water surfaces. The earlier part of Miss Pockels' letter covers nearly the same ground as some of my own recent work, and in the main harmonizes with it. The later sections seem to me very suggestive, raising, if they do not fully answer, many important questions. I hope soon to find opportunity for repeating some of Miss Pockels' experiments.

After the publication of her letter in *Nature*, Pockels continued to refine her apparatus and make measurements. She exchanged a number of further letters with Rayleigh, and published, on her own, several more papers in *Nature* and in German journals. Her work began to be recognized in Germany: she notes in her journal in 1893 that "Professor Voigt offers me facilities in the Physical Institute." She was able to continue her surface investigations for the next decade, interspersed with her duties nursing her family and managing the home. After 1902, however, she could no longer keep up her experiments, due to the demands on her time. Her father died in 1906 and her mother in 1914. The death of her brother, in 1913, and World War I cut her off from the literature and from ongoing developments in the field. Her own health and eyesight were also beginning to deteriorate. Nonetheless, Pockels did publish several later papers based on the earlier measurements, with the last one appearing in 1933. In her later years she lived a quiet life in Brunswick, keeping busy with friends and social activities. Postwar Germany was quite impoverished, but she had made fortunate investments with relatives in California that allowed her to maintain an adequate, if modest, lifestyle.

Near the end of her life, Pockels began to receive formal recognition of her highly original work. In 1931, she received the Laura Leonard Prize jointly with H. Devaux (who also had carried out surface tension experiments following those of Rayleigh). And in 1932, when she was nearly 70, she was awarded an honorary doctorate from the Carolina-Wilhelmina University of Brunswick. To honor her 70th birthday a special review of her work was published in the journal *Kolloid-Zeitschrift* by W. Ostwald, who noted that ". . . every colleague who is now engaged on surface layer or film research will recognize that the foundations for the quantitative method in this field . . . had been laid by [her] observations fifty years ago."

IMPORTANT PUBLICATIONS

Pockels, A., Surface tension. *Nature*, **43** (1891) 437.

The dependence of wetting of solids on the time of contact. *Koll. Zeit.*, **62** (1933) 1.

FURTHER READING

Ostwald, W., The works of Agnes Pockels on boundary layers and films. *Koll. Zeit.*, **62** (1933) 1.

Giles, C. H., and Forrester, S. D., The origins of the surface film balance. *Chem. and Ind.*, **2** (1971) 43.

Elder, E. S., Agnes Pockels – indeed a lady. *Chemistry*, **47** (1974) 10.

Derrick, M. E., Agnes Pockels, 1862–1935. *J. Chem. Ed.*, **59** (1982) 1030.

4 Marie Curie (1867–1934)

Abraham Pais (deceased)
Rockefeller University, New York

FIGURE 4.1 Photograph of Marie Curie. From ACJC-Curie and Joliot-Curie fund, with permission.

Marie Curie (November 7, 1867–July 4, 1934), baptized Maria Sklodowska, was born in Warsaw, the fifth child of Wladyslaw Sklodowska, a physics teacher, and Bronislava née Boguska. In her high-school years she read French, German and Polish poets, all in the original language, gave lessons to earn money, and in 1886 became a governess for three years. In 1891, she moved to Paris to study physics at the Sorbonne, graduating in 1893 at the top of her class. In April 1894, she met the physicist Pierre Curie (1859–1906), whom she married in July 1895. By then his career was already well underway. He

Out of the Shadows: Contributions of Twentieth-Century Women to Physics,
eds. Nina Byers and Gary Williams. Published by Cambridge University Press.
© Cambridge University Press 2006.

had done important work on piezo-electricity, on symmetries of crystals and on magnetism. They had two daughters, Irène (1896–1956) and Eve (1894–). Marie was a devoted mother, bathing her two babies every day herself.

Soon after Marie had completed her first paper, dealing with magnetism of tempered steels, she heard of Becquerel's discovery (1896) of "uranic rays," which she discussed with her husband. "The study of this phenomenon seemed to us very attractive . . . I decided to undertake the study of it . . . In order to go beyond the results reached by Becquerel, it was necessary to employ a precise quantitative method."[1] For this purpose she developed a new tool, an early form of the parallel plate ionization chamber: two condenser plates, 3 cm apart, each 8 cm in diameter. A modest 100 volt potential difference between the plates and a sensitive electrometer was all she needed for her work. Her first paper on this new subject contains two major new points.[2]

(1) She not only reconfirmed Becquerel's earlier findings for uranium but also discovered a new active substance: thorium. "Thorium oxide is even more active than metallic uranium." While she discovered the activity of thorium independently, she was not the first to do so. Unbeknownst to her, the German physicist Gerhard Carl Schmidt, from Erlangen, had reported the same result three weeks earlier.[3] The appellation "uranic" for the new rays was certainly too narrow. In the title of her next paper the term "radioactive substance" makes its first appearance in the world literature.[4] I note right away that in a sequel to this paper,[5] one finds the following remark: "One of us [M.C.] has shown that radioactivity is an atomic property." It is the first time in history that radioactivity is explicitly linked to *individual* atoms. Twenty years later we find a remark in a textbook about "Madame Curie's theory that radioactivity is an intrinsic property of the atom."[6]

(2) Marie's first paper clearly shows that her insight into what constituted a "normal" increase of radioactivity with increasing uranium content was sufficiently quantitative to note that two minerals,

pitchblende (rich in uranium oxide) and chalcite (rich in uranyl phosphate), behaved anomalously: "[They are] much more active than uranium itself. This fact is very remarkable and leads one to believe that these minerals contain an element which is much more active than uranium." With this conjecture she introduced another novelty into physics: *radioactive properties are a diagnostic for the discovery of new substances*, the second of the major points of her first paper on radioactivity.

Her next assignment was obvious: to verify whether her idea of a new element was indeed true. "I had a passionate desire to verify this hypothesis as rapidly as possible. And Pierre Curie, keenly interested in the question, abandoned his work on crystals . . . to join me in the search." He was never again to return to his crystals. Jointly, they investigated pitchblende by ordinary chemical methods. In July, 1898, they announced that by precipitating with bismuth they had been able to isolate a product about 400 times as active as uranium. "We believe that [this product] . . . contains a not yet observed metal . . . which we propose to call polonium, after the country of origin of one of us."

On October 17, Marie Curie wrote in a private notebook: "Irène can walk very well and no longer goes on all fours."[7]

The analysis of pitchblende continued, Gustave Bémont, a laboratory chief at Pierre's school, aiding the couple in their labors. Meanwhile, the Austrian government had presented them with a gift of 100 kg of residues of pitchblende, material (believed to be without commercial value) obtained from mining operations in Joachimsthal. Separation by chemical methods remained their main procedure. They worked "in an old, by no means weatherproof, barrack in the yard of [Pierre's] school."[8] These were no longer table-top operations. As Marie wrote later: "It was exhausting work to move the containers about . . . to transfer the liquids, and to stir for hours at a time, with an iron bar, the boiling material in the cast iron basin." On December 26, 1898, they announced that, by precipitating with barium, they had found yet another radioactive substance in pitchblende: radium, a discovery that led Rutherford to remark: "The spontaneous emission of radiation from this element was so marked that not only was it difficult at first to explain but also, what was more important, still more difficult to explain away."[9]

On January 5, 1899, Marie wrote in her notebook: "Irène has fifteen teeth!"

Ever since Becquerel's initial discovery in 1896, those in the know had been surprised at the persistence with which "uranic rays" kept pouring out energy. In 1910, Marie Curie reminisced as follows about those early days: "The constancy of the uranic radiation caused profound astonishment to those physicists who were the first to be interested in the discovery of H. Becquerel. This constancy appears in fact to be surprising: the radiation does not seem to vary spontaneously with time . . ." In order to appreciate this statement fully, three facts should be borne in mind:

(1) The radiation emitted by uranium when unseparated from its daughter products does indeed represent, to a very high degree, a steady state of affairs.
(2) It took two years from Becquerel's initial discovery until the first parent–daughter separation was affected.
(3) It took another two years until it was firmly established that radioactivity does diminish with time.

Speculation on the origin of radioactive energy started with Marie Curie's very first paper on radioactivity (the one in which she announced her discovery of the activity of thorium, in 1898). There, cautiously, she suggests the possibility that the energy might be due to an outside source: "One might imagine that all of space is constantly traversed by rays similar to Roentgen rays, only much more penetrating and being able to be absorbed only by certain elements with large atomic weight, such as uranium and thorium."

In that same year, 1898, Marie Curie discovered polonium, for which the liberated energy per unit weight of separated material was even larger than for uranium and thorium. Thus, the question of the origin of this energy became an even more burning one and she returned to it, listing a number of possible answers.[10] Here we find the first mention that one might have to face a contradiction with the law of conservation of energy. Furthermore, she emphasized that the assumption of an external source would be nothing but an evasion of energy nonconservation – unless the nature of the external source were determined: "Any exception to Carnot's principle [i.e., the law of conservation of energy] can be evaded by the intervention of an unknown energy which comes to us from space. To adopt such an explanation or to put in doubt the generality of the Carnot principle are in fact two points of view which to us amount to one and the same as long as the nature of the energy here invoked stays entirely 'dans le domaine de l'arbitraire.'" She also pointed out that the interior of the atom could be the energy source. "The radiation [may be] an emission of matter accompanied by a loss of weight of the radioactive substances."

That last article was written before, but published after, the discovery of radium by her, Pierre Curie and Bémont. This last development once again brought the issue to the fore. The radium radiation was even more intense than the polonium, about 100,000 times more active than uranium! The question of nonconservation of energy came up once again: "On réalise ainsi une source de lumiére a vrai dire trés faible, mais qui fonctionne sans source d'énergie. Il y a lá une contradiction tout au moins apparente avec le principe de Carnot."[11]

Nonconservation of energy was never a widely held explanation of these effects. In 1902, the Curies again gave a list of possible interpretations, on which this possibility no longer appears.[12] It should also be stressed that such options as nonconservation of energy or external sources were not proposed lightheartedly. The idea that the atom itself is the source was not easily swallowed at that time, since it meant giving up the concept of the atom as an immutable entity. By 1900, the debate over the reality of atoms was well past its peak; but at that time the question was not universally regarded as settled. The Curies were proponents of the existence of real atoms, as their writings make abundantly clear. But to accept the atom itself as the source of the energy could only mean one thing to them: transmutation. And they could not simply accept this, since to them at that time it seemed in conflict with the principles of chemistry as then known – which indeed it was. In 1900, Marie Curie summed up the dilemma in the following way:

> Uranium exhibits no appreciable change of state, no visible chemical transformation, it remains, or so it seems, identical with itself, the source of energy which it emits undetectable – and therein lies the profound interest of the phenomenon. There is perhaps a disagreement with the fundamental laws of science which until now have been considered as general . . . The materialistic theory of radioactivity is very attractive. It does explain the phenomena of radioactivity. However, if we adopt this theory, we have to decide to admit that radioactive matter is not in an ordinary chemical state; according to it, the atoms do not constitute a stable state, since particles smaller than the atom are emitted. The atoms, *indivisible from the chemical point of view* [author's italics], are here divisible, and the sub-atoms are in motion . . . The materialist theory of radioactivity leads us . . . quite far. If we refuse to admit its consequences, our embarrassment will not lessen. If radioactive matter does not modify itself, then we find ourselves again in the presence of the question: from where comes the radioactive energy? And if the source of energy cannot be found we are in conflict with Carnot's principle, a principle fundamental to thermodynamics . . . We are then forced to admit that Carnot's principle is not absolutely general [and] . . . that the radioactive substances are able to transform heat from the ambient environment into work. This hypothesis undermines the accepted ideas in physics as seriously as the hypothesis of the transformation of the elements does in chemistry, and one sees that the question cannot easily be resolved. (Cette hypothèse porte une atteinte aussi grave aux idées admises en physique que l'hypothése de la transformation des

éléments aux principes de la chimie, et on voit que la question n'est pas facile a résoudre.) M. Curie.[13]

It is important to stress at this point that these fascinating puzzles were never any hindrance to progress in those days. If anything, the contrary is true. The field of radioactivity was young when these questions arose, the tasks were enormous. While these problems were given much thought by the Curies, that never inhibited them from continuing their superb research.

In her doctoral thesis, Marie Curie raised yet another puzzlement. This document has a section entitled "Is atomic radioactivity a general phenomenon?" in which she reported on the analysis of substances other than uranium and thorium compounds: "I undertook this research with the idea that it was scarcely probable that radioactivity, considered as an atomic property, should belong to a certain kind of matter to the exclusion of all other."[14] It is to her credit that of all the materials she examined (including a dozen rare earths) she found no evidence for radioactivity.

I conclude this survey of Marie Curie's early contributions with her remark made in the course of discussions at the first Solvay Conference, held in Brussels in October, 1911. There she observed that thermal, optical, elastic, magnetic and other phenomena all appear to depend on the peripheral structure of the atom, then she continued:

> Radioactive phenomena form a world apart, without any
> connection with the preceding [phenomena]. It seems therefore
> that radioactive phenomena originate from a deeper region of the
> atom, a region inaccessible to our means of influence and probably
> also to our means of observation, except at the moment of atomic
> explosions.[15]

Rutherford's discovery of the nucleus had been announced the preceding May. I do not know whether Marie Curie did not follow up on her wise remark with a comment on the nucleus because she was unaware of this work, or did not believe the results, or failed to see the connection. Nor do I know why Rutherford, who was in the audience, refrained from drawing Curie's attention to the nucleus. However, what may be the long and the short of the matter is that radioactivity prior to 1911 has turned out to be prenatal nuclear physics.

Radium was to make radioactivity known to the public at large. The attendant fame, which was to disturb the quiet and concentrated existence to which the Curies were so deeply attached, would "deafen with little bells the spirit that would think."

Pierre Curie's discoveries of the experimental laws of piezo-electricity and of what became known as the Curie temperature in ferromagnetism are samples of his outstanding contributions to physics. He also labored mightily alongside his wife on the problems of radioactivity. But, insofar as one can rely on the record of published papers, it is Marie Curie, a driven and probably obsessive personality, who should be remembered as the principal initiator of radiochemistry. That is abundantly clear, it seems to me, from her April, 1898 paper, discussed at length earlier.

The year 1898 was heroic in the Curies' careers. Further important work was to come, yet what followed was to a considerable extent painstaking elaboration of their early discoveries. The finest appreciation of their early work was given by Ernest Rutherford, arguably the greatest experimental physicist of the twentieth century: "I have to keep going, as there are always people on my track. The best sprinters in this road of investigation are Becquerel and the Curies in Paris, who have done a great deal of very important work on the subject of radioactive bodies during the last few years."[16]

As to their later life: in 1900, Pierre was appointed assistant professor at the Sorbonne; Marie, teacher at a high school for girls. Continued intense research and teaching duties were so exacting that she had no time, until 1903, to complete her Ph.D. thesis, a masterful summary of her work to date. She received her degree on June 25 with the distinction "trés honorable," an elegant understatement. That day was memorable to her for another reason as well: in the evening she met Rutherford for the first time.

In November of that year, the Curies were in London to share the Humphrey Davy Medal of the Royal Society and were informed that they would share the 1903 Nobel Prize with Becquerel, "in recognition of the extraordinary services they have rendered by their joint researches on the radiation phenomena discovered by Professor Henri Becquerel." The Curies' prize was the subject of the following comment by the *New York Times* of December 11, 1903: "The discoverers of radium have, it is understood, not profited financially from the work as greatly as might have been expected, and their admirers throughout the world will be delighted to hear of this windfall for them." They never took out patents for any of their discoveries.

In December, 1904, Nobel laureate Marie Curie was named assistant to Pierre at the Faculté des Sciences of the Sorbonne. Until then she had been working without pay.

The Curies were too unwell and overworked to attend the ceremony in person. When they finally went to Stockholm, in June, 1905, only Pierre delivered a Nobel Lecture, as his wife sat and listened. Meanwhile, in 1904 a special chair had been created for him at the Sorbonne.

It appears that Pierre suffered from radiation sickness during the last years of his life, which came to a cruel end on April 19, 1906. He was not yet 47 years old. What happened on that day has been described by his wife.

> [As he] was crossing the rue Dauphine, he was struck by a truck coming from the Pont Neuf and fell under its wheels. A concussion of the brain brought instantaneous death. So perished the hope founded on the wonderful being he thus ceased to be. In the study room to which he was never to return, the water buttercups he had brought from the country were still fresh.

In the introduction to his collected works published in 1908,[17] Marie wrote of the shattered hopes for further research, in the French she mastered so exquisitely: "Le sort n'a pas voulu qu'il en fut ainsi, et nous sommes contraints de nous incliner devant sa décision incompréhensible."[18]

Marie's younger daughter Eve has written of her mother's intense grief after this event and of her determined efforts to avoid speaking of Pierre in later years. George Jaffe has left us his impressions of Madame Curie in 1911, when he spent a year at her laboratory as a Carnegie Scholar: "The stranger saw and admired the power of the intellect. He could not know what was going on behind the air of self-constraint or almost impassiveness which became characteristic of Madame Curie after the death of her husband."

Shortly after Pierre's death, Marie was named his successor at the Sorbonne. It was the first time in that venerable institution's more than 600-year-long history that a woman was appointed to a professorship. The Paris papers treated this as a major event. On Monday, 5 November, 1906, at 1:30 in the afternoon, Marie began her inaugural lecture, continuing a discourse on radioactivity at precisely the point where Pierre had left off in his last lecture. In 1910, she declined the Legion of Honor.

Marie Curie's life in 1911 was marked by two events, one humiliating, one gratifying. In the autumn of that year, news items appeared in the French press, quoting private letters, and alleging the existence of an affair between Mme. Curie, widow, and the physicist Paul Langévin, married, a disgusting public treatment of private matters, probably true.[19] In November, she received word that she had been awarded the Nobel Prize for chemistry for 1911, "in recognition of the part she has played in the development of chemistry: by the discovery of the chemical elements radium and polonium; by the determination of the properties of radium and by the isolation of radium in its pure metallic state; and finally, by her research into the compounds of this remarkable element." In her Nobel Lecture, given on December, 11 1911, she recalled that many of these discoveries were made "by Pierre Curie in collaboration with me . . ." The chemical work aimed at isolating radium in the state of the pure salt ". . . was carried out by me, but it is intimately connected with our common work . . . I thus feel that . . . the award of this high distinction to me is motivated by this common work and thus pays homage to the memory of Pierre Curie." She is the only woman ever to have been awarded two Nobel Prizes.

During World War I, Marie Curie organized and participated in the work of a number of radiology units for diagnostic and therapeutic purposes. She arranged for the equipment of some twenty automobiles with Roentgen apparatus in order that soldiers could be operated on near the battlefield. She also oversaw the installation of some 200 radiological rooms in various hospitals. She trained others as X-ray diagnosticians and often acted as one herself.[20] In 1921, she paid a triumphal visit to the United States, where she was received by President Harding.

In May, 1922, she was named member of the League of Nations' International Committee on Intellectual Cooperation, and took an active part in its work for many years thereafter.

The fifth Solvay Conference, largely devoted to the new quantum mechanics, was held in October, 1927. The printed proceedings of this meeting appeared in 1928. They open with a tribute by Marie Curie to Hendrik Lorentz, who had presided over the October conference and who had died shortly thereafter.

Meanwhile, Marie continued her research activities. In all, she published about 70 papers, the last one in 1933. Soon thereafter, at age 66, she died in a sanatorium in the Haute Savoye after a brief illness. The death report reads as follows: "Mme Pierre Curie died at Sancellemoz on 4 July 1934. The disease was an aplastic pernicious anaemia of rapid feverish development. The bone marrow did not react, probably because it had been injured by a long accumulation of radiations." She was buried at Sceaux, near her husband. She just did not live long enough to hear of the Nobel Prize in chemistry (1935) award to daughter Irène and her husband Frédéric Joliot (1900–58), for the discovery of positron radioactivity.

George Jaffe, who spent the year 1905 at the Curies' laboratory in Paris, has written about them:

There have been and there are, scientific couples who collaborate with great distinction, but there has not been a second union of woman and man who represented, both in their own right, a great scientist. Nor would it be possible to find a more distinguished instance where husband and wife with all their mutual admiration and devotion preserved so completely independence of character, in life as well as in science . . . I was most strongly impressed by their extreme simplicity and modesty together with their extraordinary devotion to their task . . . There was about both of them an unostentatious superiority . . . [Pierre's] disposition made *him* stand aloof when *she* entered upon something like a romantic enterprise: the search for an unknown element.

The complete works of Pierre Curie appeared in 1908, those of Marie Curie in 1954.[21] Of particular interest in the latter book are the elegantly written essays on the status of radioactivity at various times. Biographies by Marie Curie of her husband (in which an auto-biographical sketch is included), by Eve Curie of her mother and by Robert Reid, also of Marie, are especially important.

NOTES

1. Curie, M., *Pierre Curie* (New York: MacMillan, 1929).
2. Curie, M., *Comptes Rendus*, **126** (1898) 1101.
3. Schmidt, G. C., *Annalen der Phys.*, **65** (1898) 141.
4. Curie, M. and Curie, P., *Comptes Rendus*, **127** (1898) 175.
5. Curie, P., Curie, M. and Bémont, G., *Comptes Rendus*, **127** (1898) 1215.
6. Soddy, F., *The Interpretation of Radium*, 4th edn. (New York: Putnam, 1920) p. 83.
7. Curie, E., *Madame Curie* (New York: Doubleday, 1938).
8. Jaffe, G., *J. Chem. Educ.*, **29** (1952) 230.
9. Rutherford, E., *Nature*, **134** (1934) 90.
10. Curie, M., *Rev. Gén. Sci. Pures et Appl.*, **10** (1899) 41.
11. Thus one realizes a source of light [*sic*], quite weak to be sure, but which functions without a source of energy. There is here a contradiction, or so it seems, with the principle of Carnot [the first law!].
12. Curie, P. and Curie, M., *Comptes Rendus*, **134** (1902) 85.

13. Curie, M., *Rev. Scientifique,* **14** (1900) 65.
14. Curie, M., Recherches sur les substances radioactives, Ph.D. thesis, 1903; Engl. transl.: Radioactive substances, A. del Vecchio (New York: Philos. Library, 1961).
15. Curie, M., in *Theorie du Rayonnement et les Quanta,* eds. P. Langevin and M. de Broglie (Paris: Gauthier-Villars, 1911) p. 385.
16. Eve, A., *Rutherford* (Cambridge: Cambridge University Press, 1939) p. 80.
17. Curie, P., *Oeuvres* (Paris: Gauthier-Villars, 1908).
18. Fate has not willed it to be so, and we are forced to bow before its incomprehensible decision.
19. Reid, R., *Marie Curie* (New York: Dutton, 1974).
20. Curie, M., *La Radiologie et al Guerre* (Paris: F. Alcan, 1921).
21. Joliot-Curie, I., *Prace Marii Sklodowskiej-Curie* (Warsaw: Panstwowe Wydawnictwo Naukowe, 1954); text in French.

5   Henrietta Swan Leavitt (1868–1921)

Jean L. Turner

Department of Physics and Astronomy, University of California, Los Angeles

FIGURE 5.1 Photograph of Henrietta Swan Leavitt. Courtesy Harvard College Observatory.

IMPORTANT CONTRIBUTIONS

Henrietta Swan Leavitt is best known for her discovery of the period-luminosity relation for Cepheid variable stars. The relation between the periods of variation of these stars and their intrinsic luminosities allows the determination of their distances. Cepheids have become the preferred "standard candle" for the measurement of galactic and extragalactic distances in the local universe. The impact of this

Out of the Shadows: Contributions of Twentieth-Century Women to Physics,
eds. Nina Byers and Gary Williams. Published by Cambridge University Press.
© Cambridge University Press 2006.

discovery has been profound. Harlow Shapley used the Cepheid period-luminosity relation in his determination of the distance to the Galactic Center, the first real measurement of the size of the Milky Way and probably his most important astronomical contribution. It was used to show that many of the nebulae in the sky are located beyond the reaches of the Milky Way, that they are "island universes," or galaxies, in their own right. Edwin Hubble used Cepheid distances for nearby galaxies in his discovery of the Hubble law, that the redshift of a galaxy is proportional to its distance, an indication of the expansion of the universe.

The Cepheid period-luminosity relation grew out of Leavitt's considerable body of work on variable stars as a member of the research staff at the Harvard College Observatory. She discovered over 2400 variable stars in all, including 1777 in the Magellanic Clouds. She discovered four novae. Leavitt noticed that the distribution of variable stars was nonuniform across the sky, being greatest in the directions of certain nebulae known as "Milky Way Clouds" and the Large and Small Magellanic Clouds (1908; *Harvard Circular*, **141**). We now know that the "Milky Way Clouds" comprise the Galactic Center region, and that the Magellanic Clouds are actually small galaxies themselves. One would indeed expect larger numbers of Cepheids towards these "nebulae" than towards the gaseous nebulae associated with relatively small star clusters in our own galaxy. Her observation was a clue as to the unusual nature of these nebulae. Leavitt may have made one of the first astronomical estimates of statistical completeness for her variable star sample, in which she estimated the percentage of stars missed by her survey due to their faintness (1907; *Harvard Circular*, **127**). She also estimated the specific frequency of variable stars among the observed stellar population (1907; *Harvard Circular*, **135**).

In addition to her studies of variable stars, Leavitt worked on the development of a standard photographic magnitude scale for stellar photometry. In her time, this was viewed by most as her major achievement (Bailey, 1922). One of the lasting contributions of the

Harvard College Observatory of the late nineteenth century was the Harvard Revised Magnitude scale, a standard scale of visual brightness used to determine the comparative brightnesses of stars. When astronomical photography became commonplace, there arose the need for a standard scale for photographic magnitudes, to take account of the enhanced sensitivity of photographic emulsions in the blue part of the spectrum. The development of the photographic magnitude scale was assigned to Leavitt, who eventually headed the department of photographic stellar photometry at the Observatory. Her North Polar Sequence (1912; *Harvard Circular*, **170**, 1917; *Harvard Annals*, **71**) consisted of 96 stars covering 4th through 21st magnitudes near the north celestial pole. Leavitt determined photographic magnitudes for these stars from over 300 photographic plates made with 13 different telescopes using various observing techniques, utilizing the diversity of the observations to minimize systematic errors. The North Polar Sequence was particularly important because these stars can be observed any night of the year from observatories in the northern hemisphere; the polar standard stars could be superimposed on plates with stars of unknown brightness to allow precise magnitudes to be determined for the unknown stars. The secondary sequences that Leavitt completed, over 100 by the time of her death (Gingerich, 1981), formed the basis for early photometric studies, which were used until they were supplanted by photoelectric techniques at mid-century.

Henrietta Leavitt's greatest achievement was the Cepheid period-luminosity relation. The Large and Small Magellanic Clouds are a pair of large nebulae in the southern sky. They are actually small satellite galaxies of the Milky Way, but this was not known in 1912. The photographic plates that Leavitt used in the study were taken at Harvard's southern station, in Arequipa, Peru; the observations and the 1777 variables she discovered were published in 1908 (*Harvard Annals*, **60**). At that time she noticed that a subset of the variable stars in the Small Magellanic Cloud with a particular kind of regular light variation appeared to obey a relation in which the brighter stars had the longer periods. At the time she considered the sample of 16 stars too small to draw inferences, but she continued to pursue the idea. The period-luminosity relation was formally reported in *Harvard Circular* **173**, "Periods of 25 Variable Stars in the Small Magellanic Cloud." Leavitt noted that these variables "resemble the variables found in globular clusters, diminishing slowly in brightness, remaining near minimum for the greater part of the time, and increasing very rapidly to a brief maximum." Variable stars with light curves of this description are now known as Cepheid variable stars. Concerning the plot of apparent magnitudes and periods, she writes:

> a remarkable relation between the brightness of these variables and the lengths of their periods will be noticed . . . Since the variables are probably at the same distance from the Earth, their periods are apparently associated with their actual emission of light, as determined by their mass, density, and surface brightness.

Not only did she notice the relation between periods and observed brightness, she correctly deduced that this was a relation between their periods and true brightnesses, which would allow them to be used as standard candles for the measurement of distances.

The Cepheid period-luminosity relation has played a remarkable role in our understanding of the Milky Way and the universe. Before its application, we did not know the size of the Milky Way, nor our location within it. We were unaware of the existence of galaxies other than our own. Since the application of the period-luminosity relation, we now have an appreciation of the vastness of our galaxy and of the universe, the multiplicity of other galaxies, and a notion of the beginning and evolution of the universe. The period-luminosity relation is still in active use today, nearly 90 years after its discovery. It is at present the preeminent method of measuring the distances to nearby galaxies. It also forms the basis for recent refinements of the Hubble law, by astronomers using the Hubble Space Telescope. Neither Henrietta Leavitt, nor any other astronomer of her time, could have envisioned the far-reaching consequences of her study of "25 Variable Stars in the Small Magellanic Cloud."

Box 5.1   The Cepheid variable stars as distance indicators

One of the most difficult problems in astronomy is the determination of distances. One way to derive the distance to a star is to compare its true luminosity (wattage) to its brightness as observed in a telescope (Watts/meter^2). Since the observed brightness per square meter falls off as the square of the distance, one may thereby calculate the distance to the star. This is most easily formulated using the magnitude scale, a logarithmic measure in which 5 magnitudes are defined to correspond to a factor of 100 in brightness, with lower magnitudes denoting brighter stars. The relation between apparent magnitude, m (observed brightness), and absolute magnitude, M ("true" brightness, defined as the brightness an object would have at the standard distance of 10 parsecs = 32.6 light years), can be written as $M - m = 5 - 5 \log d$, where d is measured in parsecs. Unfortunately, it is generally not possible to use this simple formula to determine the distance to a star because one has to obtain some independent determination of the true luminosity of the star. Without further information, one cannot tell if a star is faint because it is intrinsically faint, or is actually a bright but distant star.

Henrietta Swan Leavitt's discovery that the periods of variation of Cepheid variable stars were related to their true luminosities allows astronomers to use the periods of these stars as that extra bit of information needed to use the inverse square law. Her discovery was possible because she limited her sample to the variable stars in the Small Magellanic Cloud. The Small Magellanic Cloud is a tiny galaxy on the outskirts of the Milky Way. Because the Small Magellanic Cloud is a small and relatively distant stellar system, the stars within it are all at roughly the same distance from Earth. That is, the separations between stars within the Small Magellanic Cloud (ΔR) are proportionally so much smaller than the distance (R) to the Small Magellanic Cloud from Earth that they can be considered to be at essentially the same distance from Earth, with only a small percentage of error (ΔR/R $<<$ 1). For stars that are at the same distance, their relative apparent

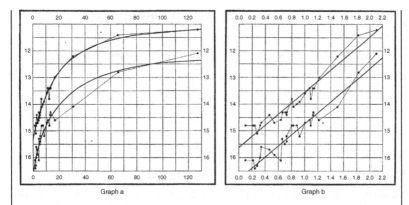

FIGURE 5.2 The Cepheid Period-Luminosity Relation. From *Harvard Circular*, **173** (1912). The abscissa in Graph a is period in days and the ordinate apparent magnitudes, m, for 25 stars in the Small Magellanic Cloud. Graph b is the same, except the abscissa is the logarithm of the periods.

magnitudes are proportional to their absolute magnitudes. Thus, Leavitt's plot of apparent magnitudes in the Small Magellanic Cloud revealed the relation of absolute magnitudes and periods for the Cepheid variable stars.

The plot of Cepheid apparent magnitude versus period from *Harvard Circular*, **173** (1912) is shown in Fig. 5.2. It can be seen from the plots that the brighter the Cepheid variable star, the longer its period of variation. Therefore, if one can determine the period of variation of the Cepheid, which is a very easy quantity to measure, one can calculate the distance to the star, and to the galaxy that contains the star.

To use the Cepheid variable stars as indicators of distance, the relation had to be calibrated. Henrietta Leavitt's relation was between apparent magnitude and period. The distance of at least one Cepheid variable star had to be determined independently in order to be able to assign absolute magnitudes to the ordinates on Fig. 5.2. A number of astronomers, including Ejnar Herzsprung and Harlow Shapley, recognized the significance of Leavitt's discovery and proceeded to work on the calibration. With the absolute calibration of the period-luminosity relation, one could then easily use the variation in the apparent brightness of a Cepheid variable star to determine its distance. This can be done for quite distant Cepheids, since all that is needed to determine their distances is to detect their variation. Fortunately, Cepheid variables are very luminous stars, which can be seen not only across the Milky Way, but also in a number of other galaxies.[1]

BIOGRAPHICAL INFORMATION

Henrietta Leavitt was born on July 4, 1868, in Lancaster, Massachusetts, one of seven children of George Roswell Leavitt, a Congregationalist minister, and Henrietta Swan Kendrick Leavitt. She attended public school in Cambridge, Massachusetts, and later, when her family moved to Cleveland, Ohio, studied in the preparatory department at Oberlin College. She attended Oberlin College, from 1885–6, spending the first year in the Conservatory of Music. She entered the Society for Intercollegiate Instruction (soon to be Radcliffe College) in 1888, earning her A.B. in 1892. Leavitt apparently developed an interest in astronomy during a course in her senior year, since she remained to take another course after graduation. After a period of travel she returned, in 1895, to Harvard College Observatory as volunteer research assistant. Leavitt was hired as a permanent staff member of the Observatory in 1902. At that time the typical pay for computing assistants, generally female, was 25 cents an hour, for a six-day working week of seven-hour days (Mack, 1990). Family and health problems interrupted Leavitt's work from time to time; some of her research on the North Polar Sequence was done from her family home in Wisconsin, as she recovered from a serious illness that befell her in 1908. Like her famous colleague Annie Cannon, she was deaf.

Leavitt's most inspired work was her research on variable stars. It was her study of the variable stars in the Small and Large Magellanic Clouds that led to her discovery of the Cepheid period-luminosity relation. She clearly recognized the key importance of distance in this relation. That she did not pursue this remarkable discovery can probably be understood in the context of the working environment at the Harvard College Observatory in the early twentieth century. While the Director, Edward C. Pickering, was a keen advocate of women's work at the Observatory, he was still essentially a Victorian gentleman. He saw to it that women were assigned work such as the analysis of photographic plates that was not too physically taxing (Jones and Boyd, 1971). Women's opportunities for advancement and pay were commensurate with their constrained research opportunities (Mack, 1990). Another factor influencing the work at Harvard was Pickering's emphasis on data surveys, in the form of sky and spectral catalogs, and in the standards for the measurement of these data, as represented by the Harvard Revised Magnitude System. He viewed the production and publication of data as the primary goal of the Observatory. Toward this end he assigned tasks to his female staff and male junior faculty alike. For much of her career, Leavitt was assigned to work on the system of photographic magnitudes to complement the visual magnitudes of the Harvard Revised Magnitude System. Pickering probably viewed the entrustment of this difficult, but critical and highly visible, task to Leavitt as an honor. A different view was taken by Cecilia Payne-Gaposchkin, who arrived at the Observatory shortly after Leavitt's death, and who has remarked (Mack, 1990):

> It may have been a wise decision to assign the problems of photographic magnitudes to Miss Leavitt, the ablest of the many women who have played their part in the Harvard College Observatory. But it was also a harsh decision, which condemned a brilliant scientist to uncongenial work, and probably set back the study of variable stars several decades.

It is difficult to determine the extent of Leavitt's contributions from publications. Most of her work, as was the common practice at the Harvard College Observatory in the late nineteenth and early twentieth century, was reported formally as an Observatory contribution by the Director, Pickering. However, a careful reading of the Harvard publications suggests Leavitt's great originality and cleverness, qualities not often evident in the communications of other researchers. The sheer abundance of her reports is another measure of how greatly she contributed to the Observatory during her career.

Statements from Leavitt's contemporaries support the notion that she was an exceptionally talented and dedicated scientist. Margaret Harwood, who worked with Leavitt, stated that she had the "best mind at the Observatory." Cecilia Payne-Gaposchkin said, "I think she was the most brilliant of all the women [at Harvard] . . . she was certainly someone who would've considered that duty was foremost. She wouldn't have complained . . ." (Mack, 1990). Dorrit Hoffleit (1971) noted "Miss Leavitt quietly found happiness in doing the necessary groundwork on which others could build their research." Solon Bailey, a colleague, wrote in 1922:

> Miss Leavitt inherited in a somewhat chastened form the stern virtues of her puritan ancestors. She took life seriously. Her sense of duty, justice and loyalty was strong. For light amusements she seemed to care little. She was a devoted member of her intimate family circle, unselfishly considerate in her friendships, steadfastly loyal to her principles and deeply conscientious and sincere in her attachments to her religion and her church. She had the happy faculty of appreciating all that was worthy and lovable in others, and was possessed of a nature so full of sunshine that to her all of life became beautiful and full of meaning.

Henrietta Leavitt was a member of Phi Beta Kappa, The American Association of University Women, The American Astronomical and Astrophysical Society (now the American Astronomical Society), the American Association for the Advancement of Science and an honorary member of the American Association of Variable Star Observers. She died of cancer on December 21, 1921, in Cambridge.

NOTE

1. For further details regarding stellar magnitudes see, e.g., http://www.astro.ucla.edu/~wright/magcolor.htm.

IMPORTANT PUBLICATIONS

Most of Henrietta Swan Leavitt's work appeared in *the Annals of the Astronomical Observatory of Harvard College and Circulars of the Astronomical Observatory of Harvard College*. Until fairly late in her career, these research reports were signed by the Observatory Director, E. C. Pickering, although often explicitly stated as having been prepared by Leavitt. The following are selected publications containing highlights of her work.

New variable stars in Harvard map, Nos. 3 and 6, *Harvard Circular*, **127** (1907). (Statistical completeness of the variable star sample.)

25 new variable stars in Harvard map, Nos. 24, 36 and 42, *Harvard Circular*, **141** (1908). (Estimates of specific frequency of variable stars.)

29 new variable stars near Nova Sagittarii, *Harvard Circular*, **141** (1908).

A catalogue of 1777 variable stars in the two Magellanic Clouds, *Harvard Annals*, **60**, No 4 (1908).

Adopted photographic magnitudes of 96 polar stars, *Harvard Circular*, **170** (1912).

Periods of 25 variable stars in the Small Magellanic Cloud, *Harvard Circular*, **173** (1912).

The North Polar Sequence, *Harvard Annals*, **71** (1917).

Standards of magnitude for the astrographic catalogue, *Harvard Annals*, **85** (1919).

FURTHER READING

Bailey, S. I., In *Popular Astronomy* (1922) p. 196. (Obituary.)

Cannon, A., Stars having peculiar spectra: eight new variable stars. *Harvard Circular*, **231** (1922).

Gingerich, O., Henrietta Swan Leavitt. In *Dictionary of Scientific Biography*, ed. C. C. Gillispie (New York: Scribner, 1981) p. 105.

Haramundamis, K., *Cecilia Payne-Gaposchkin: an Autobiography and other Recollections* (Cambridge, UK: Cambridge University Press, 1984).

Hoffleit, D., Henrietta Swan Leavitt. In *Notable American Women*, eds. E. T. James, J. W. James and P. S. Bowyer, Vol. II (Cambridge, MA: Belknap Press, 1971) p. 382.

Jones, B. Z. and Boyd, L. G., *The Harvard College Observatory: The First Four Directorships, 1839–1919* (1971).

Mack, P., Straying from their orbits: women in astronomy in America. In *Women and Science: Righting the Record*, eds. G. Kass-Simon, and P. Barnes (Bloomington: Indiana University Press, 1990) p. 72. (Interviews with Cecilia Payne-Gaposchkin and Margaret Harwood.)

6   Harriet Brooks (1876–1933)

C. W. Wong

Department of Physics and Astronomy, University of California, Los Angeles

FIGURE 6.1 Miss Harriet Brooks, nuclear physicist, Montreal, QC, 1898. (II-123880.) Notman Photographic Archives, McCord Museum of Canadian History, Montreal.

SOME IMPORTANT CONTRIBUTIONS

Harriet Brooks made original contributions to our understanding of radioactivity during the years 1901–5. She was the first person to observe the recoil of a decaying nucleus. Ernest Rutherford, in whose laboratory Brooks did her best work, considered her the most prominent woman physicist in radioactivity next to Marie Curie (see Chapter 4).

Out of the Shadows: Contributions of Twentieth-Century Women to Physics,
eds. Nina Byers and Gary Williams. Published by Cambridge University Press.
© Cambridge University Press 2006.

Brooks's work on radioactivity began in 1901 with a joint paper with Rutherford on the molecular weight of the radioactive gas or emanation emitted by radium. Brooks determined the molecular weight by measuring the time it took the emanation to disperse from one half of a long cylinder of air to the other half. The reported result of 40–100, though smaller than the actual value of 222, was sufficiently well thought of that when Brooks went to the Curie Institute in 1906–7, she appeared to have repeated the measurement on the emanation from actinium. Her result, mentioned in a published paper from the Curie Institute, was never published.

In 1902, Brooks published a long paper with Rutherford on the properties of radiations and emanations emitted by radioactive substances. They showed that radioactivity was a complicated phenomenon: the radioactivity of certain decay products of radium and thorium decayed at different rates, with the rates being dependent also on the time of exposure to the emanations (i.e., the radon isotopes appearing in these decay series). The study was extended in a later paper published in 1904 by Brooks herself, in which she clarified the differences between the radioactive decays of thorium, radium and actinium after different times of exposure to the emanations. These results helped in the eventual understanding of the complex transformations occurring in series of chains of radioactive decays.

Brooks's most noteworthy result is probably that on nuclear recoil contained in the paper "A volatile product from radium" published in *Nature* in 1904. In this paper, she described the apparent volatility at room temperature of a product of radium decay. A body, made radioactive by exposure to the radium emanation, was found to produce secondary radioactivity on the wall of a vessel in which it was placed. The volatile material was later identified as radium B (lead-214) by Rutherford.

Rutherford mentioned this and the other works of Brooks in his Bakerian Lecture to the Royal Society in England in May, 1904. Ten days after the lecture, J. Larmor, secretary of the Royal Society, wrote to Rutherford to suggest that the phenomenon was associated instead with nuclear recoil on the emission of an energetic alpha particle. Rutherford accepted this interpretation, and gave the correct description of the phenomenon in a footnote in the published version of his Bakerian Lecture, and also on p. 392 of the second edition of his book, *Radio-activity*, published in 1905. Details of Rutherford's recoil interpretation of Brooks's result can be found in a 1975 historical study by Thaddeus J. Trenn in *Historical Studies in Physical Sciences*, **6**, 513.

The phenomenon of nuclear recoil was rediscovered or qualitatively confirmed by Otto Hahn and by Russ and Markower in 1909, when its importance as a general method of radioactive analysis was also recognized. Rutherford continued to call attention to Brooks's contribution on recoil in his 1913 book *Radioactive Substances and Their Radiations*, but dropped the reference in the 1930 book *Radiations from Radioactive Substances*, written with James Chadwick and C. D. Ellis. The quantitative confirmation of nuclear recoil was first obtained in 1932 when C. D. Ellis noted that nuclear recoil energies must be included to reconcile theoretical and measured energies of α-particle groups from the decay of thorium C.

In spite of Rutherford's valiant efforts to give Harriet Brooks credit, knowledge of her contributions fades with time. This is partly because of a certain lack of incisiveness in her papers, and also because her scientific career was so brief. She would have done much worse if she did not have the good fortune to work with the great Rutherford.

BIOGRAPHY

Harriet Brooks was born on July 2, 1876, in Exeter, a small town in western Ontario, Canada. Her father owned a flour mill, but after the mill was destroyed by fire, he was forced to work for the rest of his life as a commercial traveler for a flour company. Harriet Brooks attended the local high school in Seaforth, Ontario, where her interest in mathematics and science might have first started. She entered McGill University in Montreal as an undergraduate in 1894, only ten years after McGill first opened its door to women students.

Harriet Brooks did well in her classes, winning a prize or an award every year. She graduated in 1898 with first-rank honors in mathematics and natural philosophy. She also won a medal in mathematics, and obtained a teaching diploma from the McGill Normal School. A career in teaching was one of the purposes of university education for women in her day.

By a happy coincidence, Rutherford arrived in McGill in September, 1898, at the young age of 27 as the second Macdonald Research Professor of Physics. Brooks was invited to join his well-equipped research group as his first graduate student. Her first research project, on the damping of electrical oscillations, was completed in 1899, and published in the *Royal Society of Canada (Transactions)* that year. The work was then expanded to form her Master's thesis, which was published in the *Philosophical Magazine* in 1901. The M.A. degree granted her in 1901 was the first physics Master's given to a woman student at McGill. It was also the highest professional degree granted, for McGill did not begin to award doctoral degrees in physics until 1909.

If Brooks had been a man, the usual career path for her in the academic world would have been as a demonstrator (graduate teaching assistant) on graduation, perhaps a lecturer on acquiring the Master's degree and eventually, with continued achievement, a professorship somewhere. Given the constraints of being a woman a century ago, Brooks was not unlucky, for while she was completing her studies, the Royal Victoria College, the women's college in McGill, was being founded. In the fall of 1899, a year before the official opening of the college, Brooks became its first tutor in mathematics.

Brooks probably began working with Rutherford on radioactivity some time after he returned from New Zealand with his wife in the fall of 1900. Rutherford himself had started his research on radioactivity in the Cavendish Laboratory in Cambridge in 1897, only a year after Becquerel's discovery of a visible radiation from uranium salts that could darken covered photographic plates. His first paper in the subject, published in 1899 after he had gone to McGill, described and named alpha- and beta-rays from uranium radiation.

In McGill, Rutherford continued his research on radioactivity. Brooks now joined his radioactivity research. She would

continue until she left for graduate studies in Bryn Mawr in Pennsylvania in the fall of 1901. Her research in the 1900–1 academic
year resulted in two published papers, both written jointly with
Rutherford.

In March, 1902, she was awarded a European Fellowship by Bryn
Mawr to spend a year in Cambridge on the strength of her research
papers. With Rutherford's help, Brooks joined J. J. Thomson's group
in the Cavendish Laboratory, and continued to work on radioactive
emanations. She was in frequent contact with Rutherford by letter.
Her work on this and other excited radioactivities would be completed
when she went back to McGill in 1903.

Brooks worked in Rutherford's laboratory again on her return to
McGill. Before completing the work started in Cavendish, she pub-
lished a paper on "A volatile product from radium" in *Nature* in 1904.
This paper is her most important work in physics.

Brooks's interest then appeared to turn away from research, for
she left the Royal Victoria College in March, 1904, for the position of
tutor in physics in Barnard College, Columbia University, in spite of
the absence of any research opportunity there. She stayed there until
1906, when she asked permission from the Barnard dean to stay in her
teaching position after a planned marriage to a physicist in Columbia.
The dean insisted that she should resign on marriage "for the good of
the College and the dignity of the woman's place in the home." Margaret Maltby (see Chapter 2), the head of physics at Barnard, wrote to
support Brooks's request, but to no avail. The dispute became so dis-
tressful to Brooks that even her engagement to marry did not survive
the resulting stress. She resigned from Barnard in September, 1906.
Soon after, she traveled to Europe in the entourage of Maxim Gorky,
the exiled Russian Marxist writer, whom she had met a few months
earlier during his American tour.

Brooks visited the Curie Institute in Paris, probably in early
1907, and did research there. The Curie ledger for 1906–7 mentions
Brooks as an "independent worker" (*travailleur libre*) with A.
Debierne, but gives no dates or details of her appointment. Her

research, which was never published, can be traced from mentions in three published papers of Curie scientists to have been concerned with the properties of the gaseous emanation (radon-219) from actinium and one of its decay products.

Frank Pitcher, whom Brooks married in London in July, 1907, was a demonstrator in physics at McGill when Brooks was a student there. Their relationship at that time did not appear overly friendly. Pitcher was promoted to a lectureship on receiving his M.Sc. degree in 1897, but left McGill in 1899 to work with the Montreal Water and Power Company. He became its general manager in 1903.

Pitcher began a serious pursuit of Brooks after she left Barnard for Europe. When Brooks moved to London in May, 1907, Pitcher came personally in June to press his suit. Brooks was at first ambivalent at the prospect of marriage, perhaps because Rutherford, who had left McGill to take up the physics chair in Manchester in May, had recommended her for a Manchester fellowship since March. This was an opportunity offered to very few persons in science, men or women. Brooks eventually decided on marriage, however, very much to Rutherford's disappointment. Brooks's marriage ended her scientific career.

It is clear from the record that Brooks had obtained noteworthy results at the frontier of research on radioactivity that had impressed Rutherford, J. J. Thomson and others. Within the narrow confines allowed professional women in her times, Brooks was not without prospects: she could go to Manchester, she could return to Paris or find a position elsewhere. A humble and self-effacing person, she eventually came to the conclusion that marriage without profession was to be preferred to profession without marriage.

As Frank Pitcher had become affluent, Harriet Pitcher settled down to a very comfortable life at the upper end of the Montreal society. One brother-in-law, A. S. Eve, professor of physics in McGill, wrote a well-known biography of Rutherford. Another brother-in-law, Sir Charles Gordon, became president and chairman of the Bank of Montreal.

Harriet Pitcher was active in many social organizations, including the McGill Alumnae Society, and the Women's Canadian Club (WCC), a conservative women's group for public affairs in Canada. She was elected to the presidency of the WCC in 1923. She enjoyed gardening and reading. To both family and friends, she was a source of calm, serene and reassuring influences.

Harriet Pitcher died at the age 56 on April 17, 1933, after a lengthy illness that was described as blood poisoning or blood disorder. Her son, Paul, thought that it might have been radiation sickness.

In writing this biographical sketch, I have relied heavily on the book *Harriet Brooks Pioneer Nuclear Scientist* by M. F. and G. W. Rayner-Canham, the only book-length biography on Harriet Brooks.

NOTE

I want to thank Nina Byers for permission to consult the archival file on Harriet Brooks in the Internet project "Contributions of 20th-Century Women to Physics."

IMPORTANT PUBLICATIONS

Rutherford, E. and Brooks, H. T., The new gas from radium. *T. Roy. Soc. Can.*, Ser. 2, **7** (1901) 21.

Rutherford, E. and Brooks, H. T., Comparison of the radiations from radioactive substances. *Philos. Mag.*, Ser. 6, **4** (1902) 1.

Brooks, H. T., A volatile product from radium. *Nature*, **70** (1904) 270.

FURTHER READING
Biography

Eve, A. S., *Rutherford* (Cambridge, UK: Cambridge University Press, 1939).

Millar, D., *The Cambridge Dictionary of Scientists* (Cambridge, UK: Cambridge University Press, 1996).

Rayner-Canham, M. F. and Rayner-Canham, G. W., Harriet Brooks – pioneer nuclear scientist. *Am. J. Phys.*, **57** (1989) 899.

Pioneer women in nuclear science. *Am. J. Phys.*, **58** (1990) 1036.

Harriet Brooks Pioneer Nuclear Scientist (Montreal & Kingston: McGill-Queen's University Press, 1992).

Rossiter, M. W., *Women Scientists in America: Struggles and Strategies to 1940* (Baltimore: Johns Hopkins University Press, 1982).

Women Scientists in America: Before Affirmative Action, 1940–1972 (Baltimore: Johns Hopkins University Press, 1995).

Rutherford, E., Obituary Harriet Brooks (Mrs. Frank Pitcher). *Nature*, **131** (1933) 865.

Trenn, T. J., Rutherford and recoil atoms: the metamorphosis and success of a once stillborn theory. *Hist. Stud. Phys. Sci.*, **6** (1975) 513. (In this historical account of our understanding of nuclear recoil, Trenn describes J. Larmor's contribution to Rutherford's explanation of Brooks's observed volatility in radium decay. He also traces the history of nuclear recoil up to and including C. D. Ellis's quantitative observations on nuclear recoil energies.)

Wilson, D., *Rutherford Simple Genius* (Cambridge, MA: MIT Press, 1983).

Physics

Chadwick, Sir J., *The Collected Papers of Lord Rutherford of Nelson* (New York: Interscience Publishers, 1962).

Ellis, C. D., The association of γ-rays with the α-particle groups of thorium C. *Proc. R. Soc. Lon. Ser-A*, **136** (1932) 396.

Rutherford, E., The succession of changes in radioactive bodies. *Philos. Trans. R. Soc. Lon. Ser- A*, **204** (1904) 169. (This is the published version of Rutherford's Bakerian Lecture of 1904. The footnote referring to Brooks's observation of the volatility of radium B appears on p. 200. This paper has been reprinted in Rutherford's *Collected Papers*.)

Radio-activity, 2nd edn. (Cambridge, UK: Cambridge University Press, 1905).

Radioactive Substances and Their Radiations (Cambridge, UK: Cambridge University Press, 1913).

Rutherford, Sir E., Chadwick, J. and Ellis, C. D., *Radiations from Radioactive Substances* (Cambridge, UK: Cambridge University Press, 1930).

Internet

Byers, N., Contributions of 20th-Century Women to Physics, http://www.physics.ucla.edu/~cwp/.

Weisbard, P. H., Women and Science: Issues and Resources, gopher://silo.adp.wisc.edu:70/11/.uwlibs/.womenstudies/.wom_science/, or http://www.library.wisc.edu/libraries/Womens-Studies/bibliogs/wom_science. (This is one of the series of Wisconsin Bibliographies in Women's Studies.)

7 Lise Meitner (1878–1968)

Ruth Lewin Sime
Department of Chemistry, Sacramento City College, Sacramento, CA

FIGURE 7.1 Photograph of Lise Meitner. Courtesy of AIP Emilio Segrè Visual Archives, Herzfeld Collection.

SOME IMPORTANT CONTRIBUTIONS

Lise Meitner participated in the development of twentieth-century atomic physics from radioactivity to nuclear fission and beyond. In the 1920s Einstein called her "our Madame Curie," and it is true that beyond gender there were many similarities: both struggled for an education and professional acceptance; both left their native countries;

Out of the Shadows: Contributions of Twentieth-Century Women to Physics,
eds. Nina Byers and Gary Williams. Published by Cambridge University Press.
© Cambridge University Press 2006.

both served in World War I; both did important early work with male collaborators; both were heads of internationally recognized research laboratories; both discovered an element (Meitner one; Curie two) and both were experimentalists known for their work in radioactivity. Meitner, however, was closer to theory, often choosing her experiments at the point where experiment and theory advanced together. She achieved prominence in radioactivity before World War I, was a pioneer in nuclear physics in the 1920s and is perhaps best known for her part in the discovery of nuclear fission and its theoretical explanation, which took place in 1938–9.

In Berlin in 1907, Meitner began a collaboration with the chemist Otto Hahn that helped establish the subdiscipline of radiochemistry. By systematically studying the natural radioactive decay series of uranium, thorium and actinium, they identified and characterized several new radioactive species. In 1909, they discovered and developed a physical separation method known as radioactive recoil, in which a daughter nucleus is forcefully ejected from its matrix as it recoils at the moment of decay. As they and other workers in this foundational period helped complete the natural radioactive decay series, patterns in chemical relationships and genetic sequences became apparent. Recognizing that several different radioactive species have identical chemical properties, Frederick Soddy and Kasimir Fajans in 1912–13 independently and nearly simultaneously recognized the existence of isotopes; they also stated the "group displacement laws" that summarized the observed chemical changes associated with alpha and beta decay. These concepts clarified the entire field of radioactivity, integrated the radioactive species into the periodic table, showed that periodicity applies to the heavy elements and, perhaps most fundamental, indicated that charge, rather than mass, determines an element's identity. Using these concepts and their own exceptional radiochemical expertise, Meitner and Hahn embarked on a search for the "missing" element between uranium and thorium. In 1918, they reported the discovery of the long-lived alpha-emitting isotope of element 91, which they named protactinium (Pa).

Meitner and Hahn were especially interested in beta decay. In 1910, they and Otto von Baeyer were the first to photographically record the lines made by beams of beta particles (electrons) that were deflected in a magnetic field. Expecting that the electrons from a pure beta source would be mono-energetic, they observed instead that magnetic beta spectra always display several lines, an indication that beta decay is not energetically homogeneous. In the early 1920s, as radioactivity was evolving into nuclear physics, Meitner and others turned to the magnetic beta spectrum and its associated gamma radiation as a means of probing the nucleus. Convinced that the nucleus, like the atom as a whole, must be quantized and that radioactive decay must resemble atomic radiative processes, Meitner described radioactive decay as the emission of an alpha or beta particle followed by gamma radiation. Accordingly, she ascribed one line in each beta spectrum to the primary (decay) electron and all the other lines to secondary (orbital) electrons that had been ejected from the daughter atom in a gamma-induced photoelectric process. Her interpretation was diametrically opposed by Charles D. Ellis, an associate of Ernest Rutherford, who argued that gamma radiation precedes radioactive decay, that the primary beta spectrum is continuous and that the magnetic line spectrum is entirely due to secondary electrons from the parent atom. In the course of an extended and fruitful scientific controversy, Meitner proved in 1924 that gamma radiation is indeed emitted after radioactive decay. In 1927, however, Ellis proved his contention that the primary beta spectrum is continuous. The result created consternation in the physics community, as it seemed to exempt nuclei from quantization and conservation of energy. In 1930, Wolfgang Pauli proposed as a "desperate remedy" that a neutrino was emitted along with an electron in beta decay. The neutrino was quickly incorporated into theory but not directly observed experimentally for another 20 years.

In the 1920s Meitner was involved in nearly every aspect of experimental nuclear physics, including artificial nuclear reactions, nuclear scattering interactions and cosmic radiation. In 1923, she was the first to describe radiationless transitions among orbital electrons, an effect now named for Pierre Auger, who reported it in more detail two years later. In 1929–30, her measurements of the Compton scattering of high-energy gamma radiation verified the formula of Oskar Klein and Yoshio Nishina, which was based on the relativistic electron theory of P. A. M. Dirac. In these studies, she also found that some of the hard gamma radiation was unaccounted for when scattered by heavy elements. The "missing" radiation had been transformed into electron–positron pairs, and although this effect was predicted by Dirac's theory, it was not understood until positrons were actually discovered in cosmic radiation by Carl D. Anderson in 1932. Shortly thereafter, Meitner and her assistant Kurt Philipp were the first to observe positrons from a non-cosmic source and to observe the formation of electron–positron pairs in a cloud chamber. Following the discovery of the neutron in 1932, Meitner and Philipp also worked on neutron mass determinations, of theoretical importance for deciding whether neutrons were complex or fundamental particles.

In 1934, Meitner embarked on her last major investigation, the "uranium project" that culminated in the surprise discovery of nuclear fission in 1938. The project, based on the experiments of Enrico Fermi and his group in Rome, was an effort to synthesize transuranic elements (artificial elements beyond uranium) by bombarding uranium with neutrons: uranium was expected to capture a neutron and then undergo beta decay to the next higher element and beyond. The project was framed by nuclear physics in methodology, instrumentation and the interpretation of reaction mechanisms, but it required radiochemical analysis for the new radioactive species, and for this Meitner recruited Hahn and the younger chemist Fritz Strassmann. Misled by the physical assumption that all nuclear changes would be small, and the chemical assumption that transuranic elements would have the chemistry of transition metals, the investigators spent four years analyzing a complex mixture of beta emitters, nearly all of which were later found to be smaller nuclei produced by fission. Only in December 1938, when isotopes of barium were discovered among the uranium products, was it understood that the uranium nucleus had split. Although Meitner had been forced to flee Germany and had emigrated to Sweden a few months before the discovery (see later), she continued to collaborate closely with her team through her ongoing correspondence with Hahn. She and her nephew, the physicist Otto Robert Frisch, provided the first theoretical interpretation of the fission process. Based on a liquid-drop model, they recognized that the surface tension of the large uranium nucleus would be small; they pictured the nucleus as an unstable, elongated liquid drop that was ready to divide in two and calculated the energy released as the nucleus split. Their proposed name, "nuclear fission," was immediately adopted by the physics community, and their interpretation became the basis for an extended theory developed by Niels Bohr and John A. Wheeler in 1939. From Meitner's earlier physical measurements of neutron energies and reaction cross-sections, Bohr also deduced that the fissile uranium nuclide must be the rare uranium-235 isotope and not the more common uranium-238, an essential step in the eventual development of nuclear weapons and reactors. In 1943, Meitner refused, as a matter of principle, an invitation to work at Los Alamos on nuclear weapons. She continued her research on nuclear reactions and participated in the construction of Sweden's first nuclear reactor after World War II. In her later years, she said that her "unconditional love for physics" had been diminished by the knowledge that her work had led to nuclear weapons.

BIOGRAPHY

Lise Meitner was born in Vienna on November 7, 1878, into a close-knit, intellectual family with progressive political and social ideals, the third of eight children of Philipp (a lawyer) and Hedwig (née Skovran) Meitner. Although both parents were of Jewish origin, the religion played no part in the children's upbringing and all were baptized as adults, Lise as a Protestant in 1908.

In Austria, public school for girls ended at age 14, and it was poor. Lise prepared to become a teacher of French, but in 1897 when Austrian universities were opened to women, she took two years of intensive private lessons to make up for eight years of missed schooling. She entered the University of Vienna in 1901, where the theoretical physicist Ludwig Boltzmann taught the entire physics sequence; a charismatic teacher and engaging personality, Boltzmann imparted to her "the vision of physics as a battle for ultimate truth, a vision she never lost." Her thesis research under Franz S. Exner was experimental, but she retained a keen interest in theory. In 1906, she was awarded her doctorate. As before, however, the only positions for women were as schoolteachers.

Perhaps as a delaying tactic, Meitner went to Berlin in 1907 to take some courses from Max Planck. Planck was the second great theoretical physicist to have a formative influence on Meitner: he would become her mentor and eventually a close friend. At the same time, she began research with Otto Hahn, a radiochemist just her age. Meitner was reserved, even shy by nature, but all her life she had a pronounced talent for friendship. In Berlin she found a congenial group of young people, mostly physicists, who shared her love for music and the outdoors and became her lifelong friends. She never returned to Vienna.

Between 1907 and 1912 Meitner became well known, having published nearly 20 papers (most with Hahn), but her status remained that of a tolerated outsider with neither position nor pay. The turning point came in 1912, when the Kaiser-Wilhelm-Institut für Chemie (KWI) was established in Dahlem, a Berlin suburb. Hahn was made professor and head of a small section for radioactivity, while Meitner, although still a "guest," was appointed by Planck to be his assistant, a paid position and the first rung on the German academic ladder. In 1913, she was given a position and salary in the KWI, and promoted in 1914. During World War I, Hahn was conscripted and Meitner joined the Austrian army as a volunteer X-ray nurse, a harrowing experience treating severely wounded men near the front. After returning to Berlin in 1916, she was given her own section for physics in the KWI. During the last years of the war Hahn was in Berlin on leave from time to time, but Meitner worked essentially alone on the isolation and identification of protactinium. The discovery was reported in 1918 with Hahn as first author.

In the 1920s Hahn stayed with radiochemistry, but Meitner entered the emerging field of nuclear physics. Her choice was scientific, but professionally it was also beneficial: without a male collaborator, she could establish herself as an independent scientist. In this period, her body of work, in particular her studies of magnetic beta spectra, placed her among the first rank of physicists worldwide. Hahn would later remember that it was her work, as least as much as his, that brought international recognition to their institute.

Meitner's career established a string of firsts for women in German science. In 1920, she was awarded the title of professor in the KWI; in 1922, she underwent *Habilitation* at the University of Berlin and became a *Privatdozentin*, one of the first women to do so in Germany; and in 1926, when she was appointed adjunct professor of physics in the University of Berlin, she was among the very first women university professors in Germany. She was awarded prizes by German, Austrian and American societies and nominated several times for a Nobel Prize.

When the Nazis came to power in 1933, she was dismissed from the University but not from the KWI. She considered emigration, but her success in Germany and the memory of her early struggles made her cling to what she had, much to her later regret. With the Austrian Anschluss in March 1938, however, her dismissal was imminent. She fled Germany in July, 1938, with two small suitcases and ten marks in her purse, and accepted a hastily arranged position in Stockholm at the Nobel Institute for Physics, where she never felt welcome.

In the fall of 1938, she conducted an intense scientific correspondence with Hahn that documents her ongoing participation in the uranium investigation. According to Strassmann, it was at her "urgent" request that they conducted the experiments that led directly to the barium finding and the recognition of nuclear fission. But the political situation in Germany was difficult and Hahn, anxious to protect his Institute and himself, never acknowledged Meitner's continuing collaboration or the importance of her earlier work with the Berlin team. The barium finding was published under the names of Hahn and Strassmann only, and Meitner was not credited for her share in the discovery. Although she and Otto Robert Frisch were widely recognized for their theoretical explanation of the fission process, Hahn alone was awarded the 1944 Nobel Prize in chemistry. The injustice clouded Meitner's reputation, and in later years, especially in Germany, she was often described as subordinate – a stereotype for a woman scientist. Only recently have historical correctives set in, most visibly, perhaps, in the naming of element 109 as Meitnerium (Mt) for Lise Meitner.

Lise Meitner moved from Stockholm to Cambridge, England, in 1960. She died on October 27, 1968, a few days before her 90th birthday.

NOTE

Lise Meitner's list of honors is very long. Most are listed on the CWP website http://cwp.library.ucla.edu.

IMPORTANT PUBLICATIONS

Hahn, O. and Meitner, L., Die Muttersubstanz des Actiniums, ein Neues radioaktives Element von Langer Lebensdauer. Z. Phys., **19** (1918) 208.

Meitner, L., Die γ-Strahlung der Actiniumreihe und der Nachweis, dass die γ-Strahlen erst nach erfolgtem Atomzerfall emittiert werden. Z. Phys., **34** (1925) 807.

Die Kernstruktur. Handb. D. Phys., **22** (1926) 124.

Meitner, L. and Orthmann, W., Über eine absolute Bestimmung der Energie der primären β-Strahlen von Radium E. Z. Phys., **60** (1930) 143.

Meitner, L. and Hupfeld, H. H., Über das Absorptionsgesetz für kurzwellige γ-Strahlung. Z. Phys., **67** (1931) 147.

Uber die Streuung kurzwellige γ-Strahlung an schweren Elementen. Z. Phys., **75** (1932) 705.

Meitner, L. and Philipp, K., Die bei Neutronenanregung auflretenden Elektronenbahnen. Naturwiss, **21** (1933) 286.

Die Anregung positiver Elektronen durch γ-Strahlung von Th C". Naturwiss, **21** (1933) 468.

Meitner, L., Hahn, O. and Strassmann, F., Über die Umwandlungsreihen des Urans, die durch Neutronenbestrahlung erzeugt werden. Z. Phys., **106** (1937) 249.

Meitner, L. and Frisch, O. R., Disintegration of uranium by neutrons: a new type of nuclear reaction. *Nature*, **143** (1939) 471.

Meitner, L., Fission and the nuclear shell model. *Nature*, **165** (1950) 561.

FURTHER READING

Meitner, L., The status of women in the professions. *Physics Today*, **13**:8 (1960).

Looking back. *Bull. Atom. Sci.*, **20**:11 (1964) 2.

Frisch, O. R., Lise Meitner 1878–1968. *Biographical Memoirs of the Fellows of the Royal Society London*, **16** (1970) 405.

Krafft, F., *Im Schatten der Sensation: Leben und Wirken von Fritz Strassmann* (Weinheim: Verlag Chemie, 1981).

Watkins, S. A., Lise Meitner and the beta-ray energy controversy. *Am. J. Phys.*, **51** (1983) 551.

Sime, R. L., The discovery of protactinium. *J. Chem. Educ.*, **63** (1986) 653.

Lise Meitner's escape from Germany. *Am. J. Phys.*, **58** (1990) 262.

Lise Meitner: A Life in Physics (Berkeley: University of California Press, 1996).

Crawford, E., Sime, R. L. and Walker, M., A Nobel tale of wartime injustice. *Nature*, **382** (1996) 393.

A Nobel tale of postwar injustice. *Physics Today*, **50**:9 (1997) 26.

Sime, R. L., Lise Meitner and the discovery of nuclear fission. *Sci. Am.*, **298**:1 (1998) 80.

8 Emmy Noether (1882–1935)

Nina Byers

FIGURE 8.1 Photograph of
Emmy Noether.

SOME IMPORTANT CONTRIBUTIONS

Introduction

In the early twentieth century Emmy Noether was one of the great
mathematicians who, like David Hilbert and Felix Klein, took a
keen interest in physics. The separation of theoretical physics and
mathematics did not exist in that era. In 1915, she was invited by
Hilbert to join his group in Göttingen, and shortly after her arrival
solved a problem in the general theory of relativity that Hilbert and

Out of the Shadows: Contributions of Twentieth-Century Women to Physics,
eds. Nina Byers and Gary Williams. Published by Cambridge University Press.
© Cambridge University Press 2006.

Klein along with many physicists including Albert Einstein had not been able to resolve. The solution is given in a 1918 paper that not only solved the problem but in addition profoundly influenced the development of modern physics. The paper reveals the fundamental connection between conservation laws and symmetries. A theorem familiar to all physics students, and often referred to as Noether's Theorem, is Theorem I proved in this paper. It is one of her many theorems.[1] Most are in the realm of pure mathematics.

Modern abstract algebra, one of the most important developments in twentieth-century mathematics, is credited largely to her innovations. It has been extensively employed by physicists, and we here count it as a contribution she made to physics. Besides being a creative mathematician, she was a highly acclaimed teacher and lecturer. The distinguished mathematician Hermann Weyl, an associate of hers in Göttingen, wrote that "one cannot read the scope of her accomplishments from individual results of her papers alone; she originated above all a new and epoch-making style of thinking in algebra."[2] In a memorial appreciation, Albert Einstein wrote: "In the realm of algebra, in which the most gifted mathematicians have been busy for centuries, she discovered methods which have proved of enormous importance."[3] However, it is Noether's 1918 paper, *Invariante Variationsprobleme*, that applies most directly to physics. This and an earlier paper on invariant differential forms[4] were written when the newly completed general theory of relativity was a subject of particular interest in Göttingen. In her *Collected Papers* this work seems to be something of a detour from the main line of her research, which was the development of abstract algebra. This chapter would be incomplete without a review of her contributions to this subject and to pure mathematics more generally.

Mathematics

In their definitive history of modern mathematics, Bourbaki attributed the creation of modern algebra to the group working around E. Noether and E. Artin in Göttingen in the period 1921–33.[5] Her first publication in this area is dated 1910. Nathan Jacobson, in his introduction to her collected works, writes:

> Abstract algebra can be dated from the publication of two papers by Noether, the first, a joint paper with Schmeidler *Moduln in nichtkommutativen Bereichen* and *Idealtheorie in Ringbereichen*. The truly monumental work *Idealtheorie in Ringbereichen* belongs to one of the mainstreams of abstract algebra, commutative ring theory, and may be regarded as the first paper in this vast subject . . .

After Emmy Noether's sudden, untimely death when she was at the height of her powers, distinguished mathematical colleagues gave memorial addresses describing at length her contributions. Among these is the memorial address by the Russian topologist P. S. Alexandrov in which he wrote:

> It was she who taught us to think in terms of simple and general algebraic concepts – homomorphic mappings, groups and rings with operators, ideals, . . . theorems such as the 'homomorphism and isomorphism theorems', concepts such as the ascending and descending chain conditions for subgroups and ideals. The notion of groups with operators was first introduced by Emmy Noether and has entered into the daily practice of a wide range of mathematical disciplines . . . We need only glance at Pontryagin's work on the theory of continuous groups, the recent work of Kolmogorov on the combinatorial topology of locally compact spaces, the work of Hopf on the theory of continuous mappings, to say nothing of van der Waerden's work on algebraic geometry, in order to sense the influence of Emmy Noether's ideas. This influence is also keenly felt in H. Weyl's book, *Gruppentheorie und Quantenmechanik*.[6]

van der Waerden's book *Modern Algebra* and Weyl's *Gruppentheorie und Quantenmechanik* were read and prized by generations of physics students.

Physics

Noether's Theorem: history

The connection between symmetries and conservation laws is generally referred to by physicists as Noether's Theorem. This is proved in her 1918 paper *Invariante Variationsprobleme*. The paper has an interesting history. It is published in the *Proceedings of the Royal Society of Göttingen* with Felix Klein, not Noether, as presenter. Noether was most likely not at the meeting in which her paper was presented.[7] Women were not accepted into exclusive scientific societies such as that one at that time. Nevertheless, her results were immediately and widely accepted. She had solved in an elegant and general way the problem of energy conservation in the general theory of relativity, a theory that lacks a law of local energy conservation. She showed this is as it should be due to the fact that the theory is invariant under general coordinate transformations. For theories to which only special relativity applies and one has time translational invariance, local as well as total energy conservation obtains. Her work deepened understanding of energy conservation. The physicist Feza Gursey wrote "Before Noether's Theorem the principle of conservation of energy was shrouded in mystery, leading to the obscure physical systems of Mach and Ostwald. Noether's simple and profound mathematical formulation did much to demystify physics." The paper, however, went much farther. It exposed more generally dynamical consequences of symmetries. In doing this, it played a key role in the discovery of the fundamental laws governing the observed interactions of the elementary particles.[8] Symmetries were found among the so-called elementary particles discovered in the post-World War II period (~1946–76). The understanding that these led to dynamical consequences helped to formulate the theory now known as the Standard Model. The underlying idea that the law of conservation of energy was connected with symmetry under time translations had been discussed by a number of authors. Noether brilliantly generalized this with far-reaching consequences.

Noether's paper was written shortly after Hilbert's elegant formulation of Einstein's general theory of relativity as a Lagrangian field theory. The widespread use of Lagrangian field theories in theoretical physics derives largely from this work of Hilbert and Noether. The Lagrangian description of the dynamics of physical systems is based on Hamilton's principle of least action. (See Box 8.1.) Hamilton's principle states that a physical system evolves such that the action is a maximum or a minimum with respect to path variations. The action is a path integral whose integrand is the Lagrangian for the system. Symmetries of the system are transformations that leave the action unchanged. One says the action, A, is invariant with respect to those transformations. For systems with a group of symmetries, solving for the dynamics using Hamilton's principle leads to invariant variational problems. Noether's paper analyzes such problems and proves two theorems, I and II, about their solutions. The converses are also proved.

Box 8.1   Symmetries, the Lagrangian and the action

If L(q, dq/dt) is the Lagrangian of a system, the action A is a path integral of it; i.e.,

$$A = \int dt \ L(q, dq/dt).$$

Hamilton's principle states that the path that the system follows is one that makes A an extremum (maximum or minimum).

Symmetry transformations leave A unchanged (invariant). For example, if L describes an isolated system in empty space, rotations of the system leave A unchanged. The group of such spatial rotations is a finite continuous (Lie) group.

Noether's Theorem I and II
Noether's Theorem I and II apply when the action A is invariant (doesn't change) under continuous groups of symmetry

transformations (Lie groups). The theorems show that there are dynamical consequences that follow. There are two theorems because she distinguishes Lie groups of two different types; *finite* continuous groups and *infinite* continuous groups. These two types are used extensively in modern physics. They are called gauge groups; *global* gauge groups are of the *finite* type and *local* gauge groups are of the *infinite* type. Theorem I is Noether's result for the finite case. It states the connection between symmetries and conservation laws. A well-known example is that of rotational symmetry. If the action is rotationally invariant, the dynamics are constrained by conservation of angular momentum. Theorem II, on the other hand, applies when the action is invariant with respect to an infinite continuous group. An example is the general theory of relativity whose basic symmetry is invariance under general coordinate transformations (diffeomorphisms). Mathematically the difference between finite and infinite continuous groups is as follows. Finite continuous group elements are specified by N parameters where N is a finite (or countably infinite) number whereas infinite continuous group elements are specified by functions and, possibly, their derivatives. Theorem I states that when the action is invariant under a finite continuous group characterized by N parameters the dynamics of the system is constrained to obey N corresponding conservation laws. For $N = 1$ an example is time translation symmetry. An isolated system whose action does not change from one day to the other or one hour to another has time translational symmetry; e.g., a pendulum fixed to a location on the earth's surface. The action of this system is the same regardless of when we choose the zero of time. This translational symmetry has $N = 1$, and there is precisely one conservation law that follows from it – energy conservation. There is a corresponding symmetry in the general theory of relativity but it is a subgroup of a much larger group, an infinite continuous group. In this case, Theorem II applies. It states that instead of conservation laws there are "dependencies" among the equations of motion and, possibly, their derivatives. These are identities which constrain the equations of motion. The Bianchi identities are examples in the general theory of relativity. The so-called Ward identities are examples in locally gauge invariant theories such as electrodynamics and chromodynamics. Historically the Bianchi identities were known but may not have been seen in this context before Noether's work.

As regards energy conservation, Noether's paper confirmed mathematically Hilbert's conjecture that the principle of local energy conservation is inconsistent with the basic symmetry of the general theory. From the physical point of view this can be seen as follows. The principle of local energy conservation requires that the energy in any arbitrary volume of space-time be well defined. This is not the case in general relativity. A basic principle of the theory is the principle of equivalence, which allows one to choose coordinates for a small neighborhood of any given point such that the gravitational field vanishes. Since the gravitational field carries energy, the total energy in that neighborhood will depend upon the choice of coordinates in that neighborhood and it is, therefore, not well defined. This does not mean that the principle of local energy conservation is wrong. It remains a powerful tool in the arsenal of physics, but doesn't apply in space-time regions where gravitational fields have to be taken into account. There is a large-scale principle of energy conservation for volumes whose surfaces are sufficiently far from gravitating sources and in asymptotically flat (field-free) regions. However, in general, in this case gravitational radiation needs to be taken into account.

The converses of Theorems I and II are proved along with the theorems in the 1918 paper. In the converse of Theorem I, Noether proved that a conservation law is evidence for an underlying symmetry. This fact played an important role in the discovery of the unified electroweak theory of elementary particles. R. P. Feynman and M. Gell-Mann in their famous paper on the conserved vector current in weak interactions concluded that "it may be fruitful . . . to study the meaning of the transformation groups involved" if it was indeed found.[9]

In this twenty-first century we may see theoretical physics advance with new ideas and new concepts that supersede the gauge theories of the Standard Model and general relativity, which now dominate our thinking. Nevertheless, Noether's Theorem, which gives the relation between conservation laws and symmetries, will remain an important contribution to physics. Formulated by Noether with such depth and generality, it has contributed very importantly to modern physics both in the discoveries of symmetries of fundamental interactions and in finding the dynamical consequences of symmetries. Her theorems have played a key role in the development of theoretical physics.

BIOGRAPHY

Emmy Noether was born March 23, 1882, in Erlangen, Bavaria, Germany, and died an untimely death on April 14, 1935, in Bryn Mawr, Pennsylvania, USA. Her father, Max Noether, was a distinguished mathematician and professor in the University of Erlangen. He is known as one of the leaders of nineteenth-century algebraic geometry.[10] Her brother, Fritz, was an applied mathematician, and there were people of known mathematical ability on her grandmother's side as well.

When Emmy was young, formal education of girls ended at age 14 or, in her case, 15. Women with intellectual interests born near the end of the nineteenth century worked as governesses and language teachers until the universities finally admitted women. After leaving school, Emmy studied French and English, and was certified to teach in a girls' school. In 1898, the Academic Senate of the University of Erlangen declared that the admission of women students would "overthrow all academic order." However, change was in the air. Emmy requested permission to audit lectures, and this was granted in 1900 with the consent of the lecturer. As her father and her father's friend, Paul Gordan, were lecturing on mathematics in the university, no doubt she was allowed to attend those lectures. The university registry for 1900 shows that two of the 986 students attending lectures were women. In the winter of 1903–4 she went to Göttingen and attended lectures given by the great mathematicians Schwarzschild, Minkowski, Klein and Hilbert. In 1904, after passing matura examinations at a nearby gymnasium, she was able to enroll in the University as a matriculating student.

She did a doctoral thesis under the supervision of Paul Gordan. The title of her thesis is *On Complete Systems of Invariants for Ternary Biquadratic Forms*. It contains a tabulation of 331 ternary quartic covariant forms! It was officially registered in 1908. She was Gordan's only doctoral student. She quickly moved on from this calculational phase to Hilbert's more abstract approach to the theory of invariants. In a famous paper of 1888, Hilbert gave a proof by contradiction of the existence of a finite basis for certain invariants. It was the solution to a problem Gordan had worked on for many years and Gordan, after reading it, exclaimed, *"Das ist nicht Mathematik; das ist Theologie."* Gordan was an algebraist of the old school.

After obtaining her doctorate, Emmy Noether stayed in Erlangen in an unpaid capacity doing her own research, supervising doctoral students and occasionally substituting for her father at his lectures until 1915, when Hilbert invited her to join his team in Göttingen. This was the most active and distinguished center of mathematical research in Europe. However, the mathematics faculty led by Hilbert and Klein found it impossible to obtain a University *Habilitation* for Emmy. Without that she could not teach or even give any University lectures. Her mathematical colleagues all supported her but at that time the *Habilitation* was awarded only to male candidates and Hilbert could not change this. He brought it up at Academic Senate meetings and was rebuffed by colleagues from other departments. Exasperated, at one such meeting he is reported to have exclaimed "I don't see why the sex of the candidate is relevant – this is after all an academic institution not a bath-house!"

From 1916 to 1919, when finally she was given *Habilitation*, she gave lectures that were formally advertised as Hilbert's; i.e., advertised as Mathematisch-Physikalisches Seminar, [title], Professor

Hilbert with the assistance of Frl. Dr. E. Noether. Finally awarded *Habilitation*, she could announce her own lectures. She remained, however, in an unpaid position, and it was not until 1923, when she was 41, that she was given a university position – but only that of *nicht-beamteter ausserordentlicher* Professor. The position carried with it no salary. However, Hilbert was able to arrange for her to have a *Lehrauftrag* for algebra, which carried a small stipend.

In 1933, when the Nazi Party came to power, Jews were forced out of their academic positions by decree. The Nazis didn't want "Jewish science" taught in the University. Emmy Noether was a Jewish woman and lost her position. At that time three of the four institutes of mathematics and physics were headed by Jews – Courant, Franck and Born. They all had to leave their teaching positions. Hermann Weyl took over from Courant for a while thinking he could hold things together, that this was a transitory bad patch and that reason would prevail. Before a year was out he saw otherwise and also left Göttingen. He says of that period:

> A stormy time of struggle like this one we spent in Göttingen in the summer of 1933 draws people closely together; thus I have a vivid recollection of these months. Emmy Noether – her courage, her frankness, her unconcern about her own fate, her conciliatory spirit – was in the midst of all the hatred and meanness, despair and sorrow surrounding us, a moral solace.

Otto Neugebauer's photo of her at the railroad station leaving Göttingen in 1933 is shown here. (See Fig. 8.2.)

There were only two positions offered Noether in 1933 when she had to flee the Nazis. One was a visiting professorship at Bryn Mawr supported, in part, by Rockefeller Foundation funds; and the other was in Somerville College, Oxford, where she was offered a stipend of fifty pounds, aside from living accommodations. She went to Bryn Mawr. While there, she was invited to give a weekly course of

FIGURE 8.2 Emmy Noether, standing at the Göttingen Railway Station, about to depart for the United States, 1933. Courtesy Otto Neugebauer.

two-hour lectures at the Institute for Advanced Study in Princeton, where she traveled by train each week. Nathan Jacobson, who later edited her collected papers, attended those lectures in 1935, and recollects that she announced a brief recess in her course because she had to undergo some surgery. Apparently, the operation was followed by a virulent infection and she died quite unexpectedly. According to Weyl, "she was at the summit of her mathematical creative power" when she died.

Many people have written about how helpful and influential she was in the work of others. She often tended to have her name not included as author on papers to which she had contributed, in order to promote the careers of younger people. She, apparently, was quite content with this and didn't feel a necessity to promote her own fame. She lived a very simple life and is reported to have been quite a happy person though she existed on meager funds. Einstein wrote this fitting epitaph for her in a letter to the Editor of the *New York Times*:

> The efforts of most human beings are consumed in the struggle for their daily bread, but most of those who are, either through fortune or some special gift, relieved of this struggle, are largely absorbed in further improving their worldly lot. Beneath the effort directed toward the accumulation of worldly goods lies all too frequently the illusion that this is the most substantial and desirable end to be achieved; but there is, fortunately, a minority composed of those who recognize early in their lives that the most beautiful and satisfying experiences open to humankind are not derived from the outside, but are bound up with the development of the individual's own feeling, thinking and acting. The genuine artists, investigators and thinkers have always been persons of this kind. However inconspicuously the life of these individuals runs its course, none the less the fruits of their endeavors are the most valuable contributions which one generation can make to its successors.

NOTES

1. Noether, E., *Collected Papers*, ed. N. Jacobson (Springer-Verlag, 1983).
2. Weyl, H., *Scripta Mathematica III.* **3** (1935) pp. 201–20. (Reprinted in Dick, A., *Emmy Noether (1882–1935)*, Birkhauser, 1981.)
3. Einstein, A., *New York Times*, May 5, 1935. (Reprinted in Dick, A., *Emmy Noether (1882–1935)*, Birkhauser, 1981.)
4. Invarianten beliebiger Differentialausdrücke. Nachr. v. d. Ges. d. Wiss. zu Göttingen 1918, pp. 37–44.
5. Bourbaki, N., *Elements of the History of Mathematics*, John Meldrum tr. (Berlin and Heidelberg: Springer-Verlag, 1994).
6. P. S. Alexandrov, *Proceedings of the Moscow Mathematical Society*, (1936) 2. [Reprinted in Dick, A., *Emmy Noether (1882–1935)*, Birkhauser, 1981.]
7. Documents of the Society of this period were lost and we have not been able to verify whether or not she was present at the meeting. Probably

only members could attend, and women were not eligible for membership. In 1918, women were not admitted to membership in the Royal Society of London, the Académie Française of Paris and other such societies and it seems very likely this was also the case in Göttingen.

8. Byers, N., Contributions of E. Noether to particle physics. In *History of Original Ideas and Basic Discoveries in Particle Physics*, eds. H. B. Newman and T. Ypsilantis (New York and London: Plenum Press, 1996). (This may be seen at http://xxx.lanl.gov/abs/hep-th/9411110.)

9. R. P. Feynman and M. Gell-Mann, *Phys. Rev.*, **109** (1958) 193.

10. Two well-known papers are M. Noether, *Math. Ann.*, **2** (1869) 293; *Math. Ann.*, **6** (1872) 351.

IMPORTANT PUBLICATIONS

Noether, E., Invariante Variationsprobleme. Nachr. v. d. Ges. d. Wiss. zu Göttingen 1918, pp. 235–57. (This may be seen at http://www.physics.ucla.edu/~cwp/articles/noether.trans/german/emmy235.html. An English translation, by M. A. Tavel, can be seen at http://cwp.library.ucla.edu/articles/noether.trans/english/ mort186.html.)

Noether, E. and Schmeidler, W., Moduln in nichtkommutativen Bereichen, insbesondere aus Differential-und Differen-zenaus-drucken. *Math. Zs.*, **8** (1920) 1–35.

Noether, E., Idealtheorie in Ringbereichen. *Math. Ann.*, **83** (1921) 24–66.

Hyperkomplexe Grossen und Darstellungstheorie. *Math. Zs.*, **30** (1929) 641–92.

Noether, E., Brauer, R. and Hasse, H., Beweis eines Hauptsatzes in der Theorie der Algebren. *Journal f. d. reine u. angew. Math.*, **167** (1932) 399–404.

Noether, E., Nichtkommutative Algebren. *Math. Zs.*, **37** (1933) 514–41.

FURTHER READING

Dick, A., *Emmy Noether (1882–1935)* (Birkhauser, 1981). (English translation by H. I. Blocher. An authoritative biography by a distinguished Viennese scholar.)

Noether, E., *Collected Papers*, ed. N. Jacobson (Springer-Verlag, 1983). (See Jacobson's long introduction.)

Brewer, J. W. and Smith, M. K., *Emmy Noether: A Tribute to Her Life and Work* (Marcel Dekker, Inc., 1986.)

Byers, N., E. Noether's discovery of the deep connection between symmetries and conservation laws, *Israel Mathematical Conference Proceedings*, Vol. 12 (1999). (This may be seen at http://www.physics.ucla.edu/~cwp/articles/noether.asg/noether.html.)

The life and times of Emmy Noether: contributions of E. Noether to particle physics. In *History of Original Ideas and Basic Discoveries in Particle Physics*, eds. H. B. Newman and T. Ypsilantis (New York and London: Plenum Press, 1996). (This may be seen at http://xxx.lanl.gov/abs/hep-th/9411110.)

Further information may be found at the following Internet sites:

http://www-history.mcs.st-and.ac.uk/history/Mathematicians/Noether_Emmy.html.

http://cwp.library.ucla.edu/Phase2/Noether,_Amalie_Emmy @861234567.html.

9 Inge Lehmann (1888–1993)

Bruce A. Bolt (deceased)
Department of Earth and Planetary Sciences, University of California, Berkeley

FIGURE 9.1 Photograph of Inge Lehmann.

DISCOVERY OF THE INNER CORE OF THE EARTH

The interior of the Earth is divided into four major regions: the rocky crust, the solid mantle, the fluid outer core and the solid inner core (see Fig. 9.3). The inner core, situated at the Earth's center, is now known to be solid and to be about the size of the moon. The 50th anniversary of the discovery of the Earth's inner core in 1936 by Inge Lehmann was marked by a symposium and special publications

Out of the Shadows: Contributions of Twentieth-Century Women to Physics,
eds. Nina Byers and Gary Williams. Published by Cambridge University Press.
© Cambridge University Press 2006.

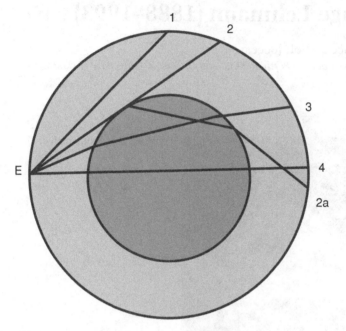

FIGURE 9.2 Schematic view of the Earth with a single inner core, showing different sound rays emitted from an earthquake in the crust at E.

(Bolt, 1987). This remarkable discovery would perhaps in atomic physics have earned a Nobel Prize! Her conclusion resolved difficulties that had arisen in interpreting the observed arrivals of seismic waves from earthquakes when only a single terrestrial core was assumed.

The earliest seismological studies used primitive seismographs at different points around the Earth to detect compressional sound waves emitted by an earthquake. Figure 9.2 shows the rays of sound emitted by an earthquake at E, where seismographs located at the ends of rays 1 and 2 would detect the waves, with a time delay depending on the distance the ray had traveled and the speed of sound in the Earth's mantle (about 10 km s^{-1}). However, seismographs between rays 2 and 3 found almost no wave arrivals, while waves were again observed at positions farther around the earth than ray 3 (see Box 9.1 for a more

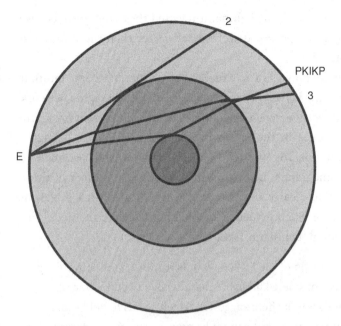

FIGURE 9.3 Simple two-shell earth model showing the PKIKP wave reflected at the inner core and arriving just inside the shadow zone defined by rays 2 and 3.

detailed discussion of this). This "shadow zone" was a puzzling feature of the measurements, and R. D. Oldham was the first to propose that this could be explained by the existence of a core region of the Earth with a different sound velocity from that of the surrounding mantle. As shown in Fig. 9.2 the rays that impinge on the core are refracted to different angles because of the change of sound velocity, and hence none of the sound can reach the region between rays 2 and 3. In 1914, an analysis by Gutenberg convincingly established the existence of the outer core and determined its properties.

The improved sensitivity of seismographs in the 1920s and 1930s enabled observers such as Lehmann to study more carefully the shadow zone region. They found that in fact there was a definite wave signal just inside the zone above ray 3 (the ray labeled PKIKP in Fig. 9.3), which could not be explained by the single-core model. There were attempts to explain the signal as simply diffraction effects from the single-core waves, but Lehmann rejected these theories because the amplitudes she observed were much larger than the diffraction models predicted.

Inge Lehmann was in an excellent situation to become familiar with these seismic waves, because her Danish seismographic network was almost directly across the diameter of the Earth from energetic earthquake sources in the South Pacific. For example, a damaging earthquake on June 16, 1929, in New Zealand was well recorded by European seismographs. Lehmann, in comparing a number of these recordings, could clearly see onsets of waves (PKIKP in Fig. 9.3) in the shadow zone. Such evidence enabled her to make the necessary imaginative jump (Lehmann 1936):

> An explanation of the PKIKP wave is required, since now it can
> hardly be considered probable that it is due to diffraction. A
> hypothesis will be here suggested which seems to hold some
> probability, although it cannot be proved from the data at hand.

Lehmann (1936) argued that the PKIKP wave was evidence of a second or "inner" core. The strength of her argument was buttressed by her ability to discard unessential information. She assumed a simple two-shell Earth model, as shown in Fig. 9.3, with constant seismic wave velocities in the mantle (10 $km s^{-1}$) and in the core (8 $km s^{-1}$. These were reasonable average values for both shells. She then introduced a small central core, again with a constant sound velocity. Such simplifications entailed straight seismic rays (chords) rather than curved paths, thus permitting the calculation of the wave travel times by elementary trigonometry. Lehmann then proceeded by successive adjustments to show that a reasonable velocity (10 $km s^{-1}$) and radius of the inner core (1400 km) could be found that predicted a travel-time curve close to the actual observations of the travel times of the anomalous core waves in question. In effect, Lehmann proved an existence theorem: namely, a plausible tripartite Earth structure could be found that explained the main features of the observed core waves. The final step was carried out two years later by

Gutenberg and Richter (1938), who calculated an inner core radius of about 1200 km and a mean inner-core wave velocity of 11.2 km s^{-1}.

In parallel research, Jeffreys (1939), after first demonstrating that the diffraction interpretation for the anomalous phases was unacceptable, adopted the Lehmann inner-core hypothesis and obtained a satisfactory numerical inversion of the wave travel-time measurements. Subsequently, K. E. Bullen established that the rapid increase in sound velocity at the inner core boundary entailed a transition from liquid to solid conditions, with a jump in shear wave velocity from zero to about 3.1 km s^{-1}. It was not until 1962 that direct new evidence supporting Lehmann's hypothetical sharp boundary was advanced, and not until 1970 that Engdahl *et al.* observed high-angle reflections of seismic waves incident on the inner core. Also, in 1970, Bolt and Qamar showed that the high-frequency reflections at steep incident angles on the inner-core boundary are evidence for jumps in both density and compressibility at the boundary. These inferences have been supported closely by recent Earth model inversions using higher modes of whole-Earth vibrations. Even today, the properties of the inner core remain a strong attractor of seismological research. Lehmann would no doubt have been delighted to assess the recent (and still controversial) evidence for significant anisotropy in the inner core and for its rotation relative to the outer core. An account of other studies of the inner core up to the late 1980s can be found in Bolt (1987).

The depression and World War II saw only the barest financial support for seismographic recording, and it became an unimportant sideline at astronomical observatories. Practical spectral analysis did not begin in the field until the late 1950s.

Lehmann's first important research after the war was a continuation of her studies of seismic body waves, working with one of her early seismological mentors, Professor Beno Gutenberg (at the California Institute of Technology at Pasadena) during four months in 1954. She made a careful study of travel times for small epicentral distances. She demonstrated that a low-velocity sound layer in the upper mantle, postulated by Gutenberg elsewhere, was not in evidence for

Europe and northeastern America. She endeavored to find a velocity structure that would fit revised travel times of Jeffreys by assuming the existence of a strong, possibly abrupt, velocity increase at a depth somewhat greater than 200 km. Her "220 km discontinuity" came to play an important role in subsequent modeling of the upper mantle by others.

Seismology was stimulated after 1960 by the problem of surveillance of clandestine underground nuclear explosions, with funding in the United States available through the VELA UNIFORM research program. This program and parallel efforts in other countries aimed first at improving standardized seismographs. These new instruments were installed in the early 1960s at about 200 existing seismographic stations around the world, including at Copenhagen, and formed the Worldwide Standardized Seismographic Network.

Although at that time over 70 years of age, Lehmann was greatly stimulated by the research opportunities of the VELA UNIFORM program. With that financial support she continued probing the structure of the upper mantle, using the seismic body waves from underground nuclear explosions recorded by the improved seismographs. Professor Jack Oliver, who was closely associated with Lehmann at Lamont Geological Observatory at Columbia University (now Lamont-Doherty Earth Observatory) has written:

> Her work with nuclear explosions was important, not so much because she found something new but because her stature and her integrity provided respectability and credibility for those whose studies she presented support for. This was not a trivial point in the partly-political, as opposed to purely-scientific, atmosphere during the time when the nuclear test-ban treaty was a hot political topic. Less-known seismologists could be suspected of warping results to support a particular political point of view.

In her centennial year, two colleagues and the author of this article raised the question of the most appropriate way to name a major structural discontinuity (Anderson *et al.* 1987) in the Earth after her: upper mantle or inner core? The inner-core boundary is one of the three first-order seismo-compositional discontinuities that divide the Earth's interior. The other two are well known by the names of their discoverers, Andrija Mohorovicic and Beno Gutenberg. In this tradition, the inner-core boundary is now called the Lehmann Discontinuity in honor of its discoverer.

Box 9.1   Waves from earthquakes

The compressional waves emitted by an earthquake at E in Figure 9.2 are known as P waves. In the single-core model, rays such as 2a and 3 will be refracted into the core region due to the change in the sound velocity in the liquid core, and these rays are known as PKP waves (K for *Kernwellen*), since they are refracted again from the core back into the mantle. The ray 2, which just misses the core, continues in a straight line, arriving at a point on the Earth's surface at 105° from the earthquake origin (with the angle measured from the Earth's center). The ray 2a, which just grazes the core, is very sharply refracted into the core, and winds up at an angle greater than 180° on the other side of the Earth. Rays between 2a and 3 hit the core at less steep incidence, and hence are refracted to smaller angles less than 180°, and the angles decrease smoothly until ray 3 reaches a minimum angle of 142°, where many rays converge to give an amplitude maximum. The receiving distance then begins to increase again for rays between 3 and 4, until the diameter ray 4 passes directly across the core at 180°. In this model there would be no signals arriving at angles between 105° and 142°.

Lehmann's observation of signals at less than 140° inside the shadow zone led her to propose the model of Figure 9.3, with a second smaller core region. There will now be a reflected wave from the inner core, denoted PKIKP, since it will be refracted at the outer core, reflected and then refracted again back into the mantle, arriving just inside the shadow zone defined by rays 2 and 3. By summing up the travel times over each segment Lehmann could show that this was consistent with the delay times she observed for these signals, confirming the existence of the inner core.

BIOGRAPHY

Born on May 13, 1888, at Osterbro by the Lakes in Denmark, Dr. Inge Lehmann spent much of her life there and at Kastelvej in Copenhagen. The origins of the Lehmann family were in Bohemia. The Danish branch had high-standing members, including lawyers, politicians and engineers. Inge Lehmann's paternal grandfather laid out the first Danish telegraph line (opened in 1854) and her great-grandfather was Governor of the National Bank. Her mother's father, Hans Jakob Torsleff, belonged to an old Danish family with a priest in every generation. A granddaughter, Anne Groes, was for a while Minister of Commerce. Inge's father, Alfred Lehmann, was a professor of psychology at the University of Copenhagen who pioneered experimental psychology in Denmark. (A fund endowed by her estate makes a travel award each year alternatively to a psychologist and a geophysicist.) She had a sister, Harriet.

Her childhood was a happy one in the peaceful atmosphere of the *fin de siècle*. Inge attended an enlightened co-educational school run by Hanna Adler, an aunt of Niels Bohr. She later remarked that she learnt there that boys and girls could be treated alike: "No difference between the intellect of boys and girls was recognized, a fact that brought some disappointment later in life when I had to recognize that this was not the general attitude."

She entered the University of Copenhagen in the autumn of 1907 and studied mathematics, physics, chemistry and astronomy. She passed the first part of the degree examination in 1910 and, in the autumn, was admitted for one year to Newnham College, Cambridge, England. She prolonged her stay because of the possibility of entering for the Mathematical Tripos, but serious overwork led her to return home in December, 1911. Her recovery was slow, and for a time she could not resume her university studies.

In the autumn of 1918 she re-entered the University of Copenhagen and graduated in the summer of 1920. In 1925, she was appointed assistant at "Gradmaalingen," which was responsible for Danish seismographic stations. In this fortuitous way, she entered the discipline of seismology, a rare profession in Denmark where earthquakes are nearly nonexistent.

> I began to do seismic work and had some extremely interesting years in which I and three young men who had never seen a seismograph before were actively installing seismographs in Copenhagen and also helping prepare the Greenland installations. I studied seismology at the same time unaided, but in the summer of 1927 I was sent abroad for three months. I spent one month with Professor Beno Gutenberg in Darmstadt. He gave me a great deal of his time and invaluable help.

In the summer of 1928, Lehmann passed an examination in geodesy at the University of Copenhagen, her thesis having a seismological subject, and obtained the degree magister scientarium. In 1928, she was appointed chief of the seismological department of the newly established Royal Danish Geodetic Institute, a post she held until retirement in 1953. She had to keep the instruments properly adjusted and to see to the maintenance of the establishments, including the buildings. She was free to do scientific work, but it was not considered a duty and was not encouraged. Sometimes she had assistants, but more often not.

Her cousin's son Nils Groes has commented:

> I remember Inge on Sunday in her beloved garden on Sobakkevej, it was in the summer, and she sat on the lawn at a big table, filled with cardboard oatmeal boxes. In the boxes were cards with information on earthquakes and the times for their registration all over the world. This was before computer processing was available, but the system was the same. With her cards and her oatmeal boxes, Inge registered the velocity of propagation of the earthquakes to all parts of the globe. By means of this information, she deduced new theories of the inner parts of the Earth.

Lehmann, like Gutenberg, brought to bear sharp observational insights in determining the times of arrival of seismic waves at her

Danish instruments. It was this work that led to her discovery of the inner core. Her work reading seismograms at Copenhagen continued through the dark years of World War II. As the postwar burden lifted, a fertile new turn in Lehmann's research occurred when Professor Maurice Ewing invited her to Lamont. Her consequent visit for several months in 1952 was followed after her retirement by additional visits to Lamont and other seismological research centers in the USA and Canada.

Lehmann retired from her post at the Geodetic Institute in 1953, five years before the mandatory retirement age of 70.

> This was probably the first time that she felt really free . . . she was first recognized abroad where she received several academic distinctions. But equally, Inge was pleased by the international academic friendships, which she nursed up to her death. The Danish recognition did not arrive until late. However, it was appreciated.
>
> *(Nils Groes)*

In 1996, the American Geophysical Union established the Inge Lehmann Medal for "outstanding contributions toward the understanding of the structure, composition, and/or dynamics of the Earth's mantle and core." It is of interest that the A.G.U. chose to associate the medal with a much broader geophysical research scope than Lehmann's own seminal seismological discovery of a major Earth structure.

There were not many Danish women who gained the international scientific celebrity status of Inge Lehmann:

> It was not easy for a woman to make her way into the mathematical and scientific establishment in the first half of the twentieth century. As she said, 'You should know how many incompetent men I had to compete with – in vain.' Inge was probably not always very diplomatic. Nevertheless, she obtained great scientific results.
>
> *(Nils Groes)*

Inge Lehmann always liked stimulating conversation and intelligent debate. She saw the need for constructive criticism and was impatient of the second-rate. She had great physical energy, enjoying many mountaineering trips to the Alps and the mountains of Norway. There can be no doubt that her almost 105 years were unusually stimulating and creative. Those of us who were fortunate to know her as a scientist and friend can only confirm her own assessment – "a life full of victories and good memories."

In writing this article, I have benefited from reminiscences, anecdotes and other help from many friends and colleagues of Inge Lehmann. I am especially grateful to Dr. Erik Hjortenberg who contributed greatly to an earlier memoir.

HONORS

1936–48   Member of the Executive Committee of the International Seismological Association.

1938   Tagea Brandt Award.

1957   Associate Royal Astronomical Society, London.

1959   Honorary Fellow Royal Society, Edinburgh.

1960   The Harry Oscar Wood Award in Seismology.

1963–67   Vice-President of the Executive Committee of the International Association of Seismology and Physics of The Earth's Interior.

1964   Doctor of Science (Sc.D) Columbia University in the City of New York.

1964   Deutsche Geophysikalische Gesellschaft, Emil Wiechert Medal.

1965   Kgl. Danske Videnskabarnes Selskab (Royal Danish Academy of Science and Letters) Gold Medal.

1967   Tagea Brandt Award.

1968   Doctor of Philosophy (Dr.Phil) h.c. University of Copenhagen.

1969   Foreign Member Royal Society, London.

1971   Bowie Medal, American Geophysical Union.

1973   Honorary Member, European Geophysical Society.

1977   Medal of the Seismological Society of America.

NOTE

See "Inge Lehmann" by Bruce Bolt in *Biographical Memoirs of the Royal Society* (1997) for a more extensive discussion of her life and work.

IMPORTANT PUBLICATIONS

Lehmann, I., P', *Publ. Bur. Cent. Seism. Int.* A, **14** (1936) 87–115. (One of the shortest scientific titles ever!)

Seismic time-curves and depth determination. *Mon. Not. R. Astron. Soc. Geophys. Suppl.,* **4** (1937) 250–71.

P and S at distances smaller than 25E. *Trans. Am. Geophys. Union,* **34** (1953) 477–83.

The times of P and S in northeastern America. *Annali Geofis.,* **X** (1955) 1–21.

S and the structure of the upper mantle. *Geophys. J. R. Astron. Soc.,* **4** (1961) *The Earth today,* dedicated to Sir Harold Jeffreys, 1961, 124–38.

On the travel times of P as obtained from the nuclear explosions. *Bull. Seism. Soc. Am.,* **54** (1964) 123–93.

Discontinuous velocity changes in the mantle as derived from seismic travel times and amplitudes. *Bur. Cent. Seism. Int.,* **24** (1968) 167–74.

The 400 km discontinuity. *Geophys. J. R. Astron. Soc.,* **21** (1970) 359–72.

Seismology in the days of old. *EOS Am. Geophys. Union,* **68** (1987) 33–5.

FURTHER READING

Anderson, D. L., Bolt, B. A. and Morse, S. A., The Lehmann discontinuity. *EOS Am. Geophys. Un.,* **68** (1987) 1593.

Bolt, B. A., 50 years of studies on the inner core. *EOS Am. Geophys. Union,* **68** (1987) 80.

Inge Lehmann. *Biographical Memoirs of the Royal Society 1997.*

Gutenberg, B. and Richter, C. F.. P' and the Earth's core. *Mon. Not. R. Astron. Soc. Geophys. Suppl.,* **4** (1938) 363.

Jeffreys, H., Times of core waves, *Mon. Not. R. Astron. Soc. Geophys. Suppl.,* **4** (1939) 594.

10 Marietta Blau (1894–1970)

Leopold Halpern and Maurice M. Shapiro
*Florida State University, Tallahassee, FL and
University of MD, College Park, MD*

FIGURE 10.1 Photograph of
Marietta Blau.

BLAU'S CONTRIBUTIONS TO TWENTIETH-CENTURY PHYSICS

Through most of her career, the nuclear physicist Marietta Blau suffered severe discrimination, enforced exile and other hardships. Yet she enriched the fields of nuclear physics, elementary-particle physics and cosmic-ray physics through her pioneering research. This she accomplished by investigating the registration and elucidation of particle tracks in nuclear emulsions, i.e., layers of photographic emulsion.

Out of the Shadows: Contributions of Twentieth-Century Women to Physics,
eds. Nina Byers and Gary Williams. Published by Cambridge University Press.
© Cambridge University Press 2006.

As early as 1911 it was known that an alpha particle striking a photographic film at glancing incidence alters the grains of silver bromide in its path, thereby producing a latent image of a microscopic track. However, the development of this effect into a powerful technique for nuclear physics, elementary particle physics and cosmic-ray physics awaited the pioneering research of Marietta Blau. In studies of elementary particle physics and cosmic rays as well as in nuclear physics, the photographic technique has some important advantages over other methods of particle detection. This is due, in large measure, to the high stopping power of the dense detector as compared to other instruments like cloud chambers or ionization chambers. The continuous sensitivity, the compact size, the high resolution and the low cost are further advantages. The emulsion technique is sometimes used in conjunction with other methods of particle detection.

An early review paper highlighted Blau's unique role in the development of the photo-emulsion technique.[1] This was up to 1939, when her work was interrupted by World War II. Prior to her development of photo-emulsion techniques, Blau had already published several papers on her research in nuclear physics.

An epitome of Blau's early contributions, in the 1920s and 1930s, to nuclear physics has been given by Oxford Professor R. H. Dalitz:

> Blau was the first physicist to show that proton tracks could be separated from alpha-particle tracks in emulsion. Of course, many physicists had studied alpha tracks from the decay of very heavy nuclei in the emulsion ... She went further, identifying proton tracks from (a) the elastic scattering of alphas by protons in the hydrogen in the emulsion, and (b) the reactions of alphas with the nuclei of the emulsion. She also exposed emulsion to neutron sources, and measured the energies of protons resulting from elastic scattering of the neutrons by hydrogen in the emulsion. In particular she used this method to determine the spectrum of neutrons resulting from specific nuclear reaction processes.[2]

TRACKS AND "STARS"

A major discovery by Blau working with the assistance of H. Wambacher was of cosmic-ray "stars", which registered in photographic emulsions exposed at mountain altitudes.[3] These stars were microscopic configurations of particle tracks radiating from a common center, which resulted from the breakup of an atomic nucleus in the emulsion hit by an incident cosmic ray. The phenomenon came to be known as "Blau–Wambacher stars." In 1937, Blau and Wambacher exposed photographic plates at an altitude of 2300 m on the Hafelekar Mountain near Innsbruck. They investigated the properties of cosmic-ray stars in these plates. They estimated that the total energy involved in these disintegrations often had to be at least 100 or 200 MeV. Another important result they obtained was that neutrons are a component of the cosmic rays in the atmosphere.

Special methods of dark-room processing are involved in developing uniformly the thick silver-bromide emulsions required for registration of cosmic-ray tracks. The earlier studies of Blau on the properties of the latent image and on dark-room techniques of development enabled her and other workers to devise methods for processing the photographic layers – a nontrivial, time-consuming ritual. (Enrico Fermi made extensive use of the photographic method to study particle tracks, and his students humorously entitled a paper, "How to Devil-up Nuclear Emulsions.")

Many of Blau's reports, and especially her earlier ones, appeared in the *Proceedings of the Viennese Academy of Sciences*. An early publication in 1925 was on "The Photographic Effects of Natural H-rays," i.e., protons.[4] In a 1927 paper, Blau studied the relative densities of the silver grains in the tracks of alpha particles and protons, finding that they were distinguishable, as the alphas had a higher grain density.[5]

Blau tried various types of photographic materials. Thus, in 1927, she used dental films made by Agfa, which seemed to be less sensitive to light than other materials. She concluded that the results on nuclear interactions obtained with these photographic plates "did not contradict findings made with the scintillation method." The latter method of studying alpha-particle interactions had been used by Rutherford and others in the early years of nuclear physics. In her comparative studies of various photographic media, Blau confirmed that finer-grained films gave better results than larger-grained ones. In 1931, she reported on the fading of the latent image during the interval between exposure and development.[6] This was to prove useful in later cosmic-ray studies, especially those at mountain altitudes, and, eventually, in extended space exposures.

NEUTRON STUDIES WITH RECOILING PROTONS

Soon after the discovery of the neutron by Chadwick in 1932 it was realized that, with a neutron mass approximating that of the proton, collisions of neutrons with hydrogen nuclei could provide a convenient means of investigating the spectra and angular distributions of neutrons emitted in various reactions.

Using the Ilford Halftone plates that Blau found to be superior for track registration, Blau and Wambacher studied the process of neutron emission from polonium-beryllium sources by photographic detection of the protons liberated by neutrons in the emulsion. As the gelatin in the emulsion is rich in hydrogen, they found that the recoil protons from neutron collisions could be utilized to get angular distributions and energy estimates. Although photographic films thicker than ordinary ones were provided by manufacturers, emulsion layers measuring some 70 microns were considered "thick"– a far cry from the 600-micron emulsions widely used in later decades for cosmic-ray studies and elementary particle physics. Particle trajectories, commonly oblique to the emulsion surface, are often incomplete due to the particle's escape from the sensitive layer. Eventually, it was possible to mitigate this limitation by using "stripped" layers, i.e., photographic emulsions free of glass backing, assembled in stacks. This permitted following a track from layer to layer without interruption in the intervening glass.[7]

SENSITIVITY OF NUCLEAR EMULSIONS

Until the end of World War II, nuclear emulsions had suffered from insensitivity to energetic charged particles owing to their low rate of

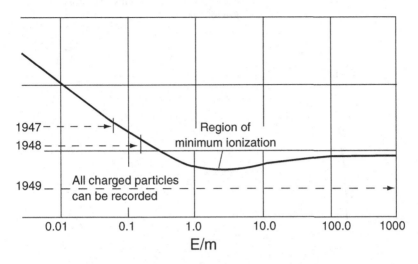

FIGURE 10.2 Rate of energy loss by ionization of a singly charged particle versus its kinetic energy in rest-mass units. The dashed lines show the gain in sensitivity of emulsions expressed in terms of the lowest rate of ionization loss that could produce visible tracks.

ionization loss. Blau had been aware of this, and had achieved some improvement in sensitivity that made it possible to register protons of higher energy. A decisive breakthrough was realized by R. W. Berriman – an emulsion sensitive to minimum ionization.[8] This expanded decisively the range of detectable particle energies in the photographic medium. (See Figure 10.2.)

It enabled the discovery of positively charged pions and their decay into muons, then these into electrons.[9] The development of cosmic-ray secondaries in the atmosphere, and the origin of cosmic-ray cascades could now be better understood.

The discovery of pions and their mode of decay earned a Nobel Prize for C. F. Powell. Arguably, Marietta Blau and G. P. S. Occhialini deserved to share in this recognition. In fact, Erwin Schrödinger twice recommended Blau for a Nobel Prize.

HEAVY PRIMARY NUCLEI IN COSMIC RAYS

In 1947, another major discovery, that of heavy primary nuclei in the cosmic rays, was made by deploying nuclear emulsions and a cloud

FIGURE 10.3 Example of pi-mu-e decay sequence recorded in an Ilford emulsion. (Courtesy of C. F. Powell's laboratory in Bristol, UK.)

chamber in a stratospheric balloon flight.[10] At the time, it was supposed that the primaries were protons. Energetic nuclei as heavy as iron were observed, although hydrogen remained the main component by far. Nuclear emulsions soon became the principal detectors in exploring the elemental composition of cosmic rays.[11] Studies of the breakup of heavy nuclei, and of their fragmentation products in the interstellar medium, became a veritable Rosetta Stone for the elucidation of various crucial parameters of cosmic-ray propagation and transformation in the galaxy.[12] (See Fig. 10.4.)

Among the parameters to be measured were the mean free path of the primaries, the path-length distribution, the residence time in the galaxy and – most important – the source composition.[13]

At this time before the GeV accelerators were built, cosmic-ray investigators were exploiting "nature's accelerator," which then provided energetic particles for producing the new menagerie of mesons and hyperons, unstable particles intermediate in mass between proton and deuteron. In the forties, cyclotrons were constructed at several universities – California, Chicago and Columbia – with proton beams of several hundred MeV. It was at the Nevis cyclotron of Columbia University that Blau and her colleagues exposed emulsion stacks to beams of pions and neutrons – the latter at 300 MeV.[14] Their aim was to distinguish between disintegrations produced in the light elements in the emulsion from those generated in the heavy elements.

PHYSICS WITH GEV ACCELERATORS

Soon the Cosmotron, and then the Bevatron, would produce intense beams of GeV protons, which enticed many cosmic-ray physicists to use these new, powerful accelerators for elementary particle research. In these investigations, too, nuclear-emulsion detectors played a significant role, and Marietta Blau used the newly available high-energy beams to good advantage.

By 1957, when one of us was invited to prepare a monograph on nuclear emulsions and their applications for the *Handbuch der Physik*,[15] it was not easy to decide which of the myriad publications

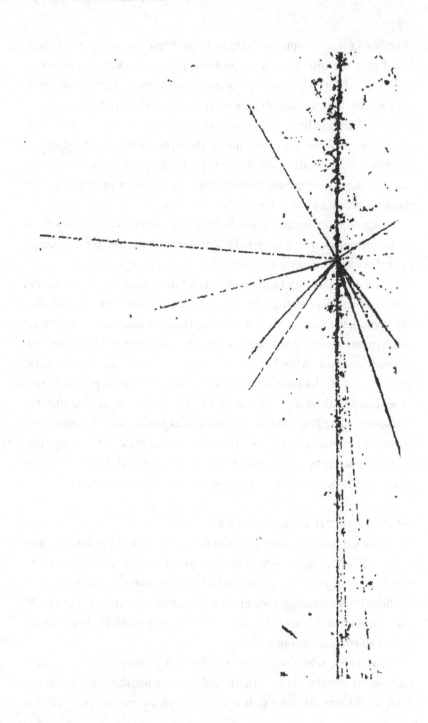

should be cited as references. As of this writing, four decades later, there have been literally thousands of scientific reports based on research that employed the photographic emulsion technique. Stacks of photo-emulsions are proving useful in some of the latest TeV-scale experiments at CERN and Fermilab, where gigantic and extremely complex detectors are required. This is due in part to the exquisite resolution obtainable with microscopic tracks at the site of an inter-action. The technique of track registration has far outlived the cloud chamber, which was used for so many of the early discoveries in twentieth-century physics. One can hardly overestimate the valuable contributions of the photographic emulsion methodology to nuclear physics, elementary particle physics and cosmic-ray astrophysics. Much of the credit for these achievements clearly belongs to the painstaking pioneering researches of Marietta Blau – performed under daunting circumstances that would have discouraged most scientific workers.[16]

Among Blau's researches in the 1950s and the succeeding years were studies of pion interactions, of hyperfragment and K-meson pro-duction by 3-GeV protons and of ionization parameters in nuclear emulsions.[17] The latter work involved a semiautomatic instrument for measurements of ionization devised by Blau and her co-workers.[18] At Columbia University, Brookhaven National Laboratory and finally at the University of Miami in Coral Gables, Florida, she enjoyed the collaboration of many colleagues who benefited from her matchless expertise in the methods of particle detection to which she had con-tributed so crucially.

←————————————————————————————

FIGURE 10.4 Breakup of a heavy cosmic-ray nucleus in collision with a nucleus in the photographic emulsion detector. The collimated stream of particles emerging beyond the collision point is mainly fragments of the incoming nucleus. The black tracks radiating in other directions from the collision point are mainly debris of the target nucleus. Collisions of a similar type with interstellar matter give rise to fast nuclei of lithium, beryllium and boron, which are extremely rare among the general abundances of elements in the universe. (Laboratory for Cosmic-ray Physics, NRL.)

In assessing the legacy of Marietta Blau and her influence on twentieth-century physics, no short summary can do her justice. However, among the many notable discoveries made with the photographic emulsion technique, the following seem especially important: pi-mu-e decay (Fig. 10.3); the presence in galactic cosmic rays of nuclei up to iron; evidence for the fragmentation of complex cosmic-ray nuclei in the interstellar medium,[19] providing a "Rosetta Stone" for elucidating their propagation, source composition and path-length distribution in the galaxy; the lifetime of the neutral pi meson;[20] and the production of hyperfragments,[21] K-mesons[22] and hyperons in high-energy interactions.[23]

BIOGRAPHY

Marietta Blau's story requires insight into the times when her achievements were accomplished. These were the times of Nazi fascism when Goebbels realized that the repetition of propaganda slogans makes people believe them, and when Hitler sought political solutions by working German crowds into a frenzy against Jews and other non-Aryan people. Physicists were not all beyond such persuasions. For example, Philip Lenard proclaimed in his "German Physics" that scientific discoveries were made by Aryans of German descent. Student life was under the influence of politically biased fraternity-like organizations with powerful protectors in the faculty and government. Alcoholism and anti-Semitism were fairly common among these groups. Jews frequently suffered social discrimination. Access to university positions required approval of the faculty council, which rejected some brilliant physicists recommended by physics professors.

After a decade of groundbreaking research in Vienna without salary or status, Blau asked for a regular paid position. The answer was, "You are a woman and a Jew – this together is too much! There is no chance." Ironically, Marietta Blau's character and personality were as opposite to the anti-Semitic stereotype of a Jew as the photonegative of her emulsions. She was modest, quiet, disinterested in material gains and recognition, striving for genuine achievements and exceedingly helpful and caring.

When one of us (Leopold Halpern) was introduced to Dr. Blau in 1953, he met a tender little person whose soft-spoken kindness and humor could win every heart. The palms of her hands showed creased lines and her eyes were clouded by cataracts – results of her tireless work under exposure to radiation.

THE EARLY YEARS

Marietta Blau was born in 1894, the daughter of Florentine Goldenzweig-Blau and Mayer Blau, a lawyer and publisher of music literature. She began her study of physics and mathematics at the University of Vienna in 1914. Her doctoral dissertation, on the absorption of gamma rays, was sponsored by S. Meyer and F. Exner. After graduating *summa cum laude* in 1919, she studied and worked at a Viennese medical X-ray laboratory, striving to qualify for research in medical physics.

Then, following a short period of industrially oriented activity in Berlin, she was appointed in 1921 as assistant at the Institute of Medical Physics at the University of Frankfurt-am-Main. There she lectured to medical students and performed experiments on X-ray absorption. She also elaborated a theory on the biological effects of radiation. She tried to obtain a suitable position, but felt constrained to interrupt these efforts in order to return to Vienna because of the ailing health of her widowed mother. She realized that she could not find any suitable position with income or status in Vienna. Her living expenses had to be covered by the family.

START WITH PHOTO-EMULSIONS

In Vienna she pursued her own research, working as a volunteer at the Institute f. Radium Research (IRR) and the II. Physical Institute of the University. H. Peterson, in a newly established nuclear physics group at the IRR, suggested that she investigate the possibility of using photographic emulsions to detect recoil protons from nuclear processes, e.g., from alpha particles incident on paraffin. After fruitless experimentation and collaboration with E. Rona, she finally succeeded and the method became useful for investigating the disintegration of nuclei with alpha particles.

Sometime around 1931 Blau was approached by a student, Hertha Wambacher, who had studied law but was undecided about her future course. Helpful as ever, Blau suggested that Wambacher work with her on the development of her emulsions. Thereafter, Wambacher appeared as co-author of many of Blau's publications. Blau received a grant from the Austrian Association of Woman Academicians to spend two years abroad. In 1933, she went to Göttingen and worked with R. W. Pohl. They published a paper on crystal physics but this collaboration was interrupted when Hitler came to power. She moved to the Institut Curie in Paris. There she met the Norwegian nuclear physicist Ellen Gleditsch, and experimented on neutrons produced in beryllium by alpha particles. Blau returned to Vienna in 1934, and continued her work with Wambacher to improve the emulsions. Meanwhile, a productive collaboration with photographic firms in Britain had been established.

PHYSICISTS AND NAZIS

In 1934, Chancellor Dolfuss was murdered in a failed Nazi putsch. The Nazi party was outlawed but had won many supporters in Austria, among them members of the nuclear physics group in which Blau worked. G. Stetter, an ardent supporter, led these activities in the group. Wambacher was also a member of the illegal Nazi party. It was an open secret that Stetter, who was married, had relations with Wambacher. He tried to discourage Blau, a Jew, from continuing her work at the Institute. The Aryan Wambacher (who had perforce repeated her final exams and then finally passed with the lowest admissible grade) would take over everything, including credit for their successful work.

In these difficult circumstances the collaboration of the two women continued. In 1936, they applied the photographic method to the investigation of cosmic radiation at an altitude of 7000 ft. In

1937, they received the Lieben Prize from the Austrian Academy of Sciences for their nuclear research with photographic plates. Their fruitful research provided information about neutrons in the cosmic radiation, and about cosmic-ray induced nuclear disintegrations. They then received a grant from the Austrian Academy of Sciences to extend their work to higher altitudes with balloons.

However, Blau's work in Vienna was interrupted in 1938 when Hitler annexed Austria. She was then in Oslo working with E. Gleditsch. The refugee quota for Scandinavia was filled, so Blau could not remain there. Moreover, she also sought a refuge for her mother. Albert Einstein had visited Vienna earlier in 1938. Recognizing Blau's difficult situation, he recommended her for a secure position in Mexico City and she was appointed professor at the Polytechnic Institute. Fearing the outbreak of war, Blau chose the quickest transportation. During a stopover in Hamburg, officials came on board and confiscated her scientific papers, which included her future research plans. She was then allowed to fly on. Later, Stetter and Wambacher exposed photoplates in balloon flights, as envisaged in Blau's confiscated research plans!

Blau liked Mexico despite a heavy teaching load and limited equipment – some of which turned up in pawn shops when faculty salaries were overdue. Attempting to help a biologist emigrant find a position, she visited the bacteriological laboratory. Learning that a disgruntled graduate student had left a typhus culture unconfined, both visitors were immediately vaccinated. However, Blau contracted the disease by inhalation and remained for days helpless in her flat with a high fever, worrying between faintings about the fate of the other woman. Her attending physician claimed to have a diploma from Vienna, but a subsequent glance at the diploma showed that it had been issued for a special Viennese hair styling.

A NEW LIFE IN THE UNITED STATES

In 1944, Blau obtained a visa to enter the United States. At first she worked at the Canadian Radium and Uranium Co., and then at

Columbia University where she had support from the US Atomic Energy Agency. There she helped in the development of photomultipliers as another technique of particle detection. She also improved emulsion techniques for use at the high-energy accelerators, and collaborated on construction of an instrument for measuring multiple Coulomb scattering along particle tracks. In 1950, Blau moved to Brookhaven National Laboratory, where she worked as associate physicist in the group of E. O. Salant, collaborating in the development of semiautomatic devices for measurements of ionization and scattering. She investigated interactions of high-energy protons and mesons with the nuclei in emulsions, and demonstrated that secondary mesons could be generated in collisions of incident mesons.

In 1955, Dr. Blau left Brookhaven to become Associate Professor at the University of Miami, Coral Gables. An Air Force grant allowed her to construct improved devices for microscopic track measurements, and to study mesons and hyperons. She was invited by C. S. Wu (see Chapter 24) and Y. Yuan to contribute to their important volume on "Methods of Experimental Physics."

RECURRING HEALTH PROBLEMS AND ELUSIVE AWARDS
The cataracts resulting from her lifelong work with radioactivity had worsened. She had to give up her position and needed an operation. She could not afford this in the US, but living in Austria was still cheap, and social medicine made an operation affordable. The surgeon whom she consulted found her health too weak to incur the risk of operation until she improved. Her recovery took years. She returned to Austria in 1960, determined to work again in the USA after recovery. She had never had a paid position in Austria, hence had no rights to an Austrian pension. Austria has declared itself a victim of the Nazis and made no restitution to Nazi victims after the occupation ended. Germany would have done so for Austrian victims had Austria not confiscated the private property of German residents. In this situation both sides

felt themselves relieved of any effort to compensate victims of the Nazis.

Blau once again worked at the Radium Institute in Vienna and helped research students – unpaid. She could not feel at home. Stetter, her old nemesis, had been reinstated despite his Nazi past as Director of the Institute. In 1960, K. Prizbarn nominated Blau for membership in the Austrian Academy of Sciences. Had this succeeded, this membership would have alleviated her situation considerably. She did receive the Schrödinger Prize posthumously with Wambacher, who had died in 1950. Schrödinger had recommended only Blau. In the reinstatement of active Nazis, he saw a blow to the feelings of their victims.

Once Blau's eye operation was successful, she was able to work again. But inflation had shrunk the purchasing power of her meager income from the US Social Security System. She refused to accept financial help from anyone, even from her brother, who would have assisted her. Attempts by friends to intervene on her behalf with various institutions, such as the Institute in Copenhagen, CERN in Geneva and the International Atomic Energy Agency proved fruitless. When one of us (Leopold Halpern) next visited Dr. Blau in Vienna, she said that her serious heart condition would preclude any more systematic work. He then suggested to S. Wouthuysen in Amsterdam that the Nobel Committee be notified of Blau's failing health and a last chance to award her a Nobel Prize. The answer to him was that Blau indeed deserved it, but the prize had already been given to C. F. Powell for application of the photographic method. Erwin Schrödinger had written to Max Born, expressing frustration: twice he had nominated Blau for the Nobel award to no avail. Dr. Blau had twice been offered the prestigious Leibniz medal. Unfortunately, she could not accept it, as she was then working for the US Atomic Energy Commission, at a time when this prevented her from visiting the Academy in East Berlin that selected the award winners.

Marietta Blau died in Vienna in 1970.

NOTES
The authors are grateful to many colleagues who have shared valuable information with them, and especially to the following: N. Byers, R. Chorcherr, R. H. Dalitz, P. A. M. Dirac, O. Frisch, P. Galison, P. Havas, O. Klein, B. Laurent, H. Leng, C. Quigg, M. Rentizi and S. Rosenthal. Interesting new material was also provided by the Central Library for Physics in Vienna and its director, Dr. Kerber.

1. Shapiro, M. M., Tracks of nuclear particles in photographic emulsions. *Rev. Mod. Phys.*, **13** (1941) 58.

2. http://www.physics.ucla.edu/~cwp/Phase2/Blau,_Marietta @843727247.html.

3. Blau, M. and Wambacher, H., Range of H-rays by the photographic method. *Acad. Wiss. Wien*, **146** (1937) 259.

4. Blau, M., The photographic effects of natural H-rays. *Akad. Wiss. Wien*, **134** (1925) 427.

5. Blau, M., Relative grain densities in tracks of alpha particles and protons. *Akad. Wiss. Wien*, **136** (1927) 469.

6. Blau, M., Fading of the latent image in tracks. *Akad. Wiss. Wien*, **140** (1931) 623.

7. Demers, P., Cosmic-ray investigations with special photographic emulsions. *Canad. J. Research*, **28** (1950) 628. Stiller, B., Shapiro, M. M. and O'Dell, F. W., *Phys. Rev. A*, **85** (1952) 712. Stiller, B., Shapiro, M. M. and O'Dell, F. W., Techniques for processing thick nuclear emulsions *Rev. Sci. Inst.*, **25** (1954) 340. Powell, C. F., *Phil. Mag.*, **44** (1953) 219.

8. Berriman, R. W., A photographic emulsion sensitive to minimum ionization, *Nature*, **162** (1948) 992.

9. Lattes, C. M. G., Occhialini, G. P. S. and Powell, C. F., *Nature*, **160** (1947) 453, 486.

10. Freier, P. S., *et al.*, Observation of heavy nuclei in the cosmic radiation. *Phys. Rev.*, **74** (1948) 213.

11. Shapiro, M. M. and Silberberg, R., Heavy cosmic-ray nuclei. *Ann. Rev. Nucl. Sci.*, **20** (1970) 323.

12. Shapiro, M. M., Silberberg, R. and Tsao, C. H., Diffusion of cosmic rays and their source composition. In *Cosmology, Fusion and Other Matters*, ed. F. Reines (Boulder, CO: Univ. of Colorado Press, 1972) p. 124 ff.

13. Shapiro, M. M., Silberberg, R. and Tsao, C. H., series of papers in *Acta Physica Hungarica* 29 Supplement (1970): (1) Transformation of

cosmic-ray nuclei in space, p. 463 ff (2) The distribution function of cosmic-ray path length, p. 471 ff (3) Lifetime of cosmic rays, p. 479 ff (4) Relative abundances of cosmic rays at their sources, p. 485 ff.

14. Blau, M. and Wambacher, H., Disintegration processes by cosmic rays with the simultaneous emission of several heavy particles. *Nature*, **140** (1937) 585, and Blau, M. and Wambacher, H., Report on photographic investigation of the heavy particles in the cosmic radiation. *Akad. Wiss. Wien*, **146** (1937) 623.

15. Shapiro, M. M., Nuclear Emulsions, in Vol. 45, *Encyclopedia of Physics* (*Handb. d. Physik*), eds. S. Flugge and E. Creutz (Berlin: Springer-Verlag, 1958) p. 342 ff.

16. One of us (Maurice M. Shapiro) acknowledges his debt to the early publications of Dr. Blau; these were the starting point for his work in all three of the aforementioned disciplines.

17. Blau, M., Oliver, A. R. and Smith, J. E. Neutron and meson stars induced in the light elements of the emulsion. *Phys. Rev.*, **91** (1953) 949. Blau, M., Hyperfragments and slow K mesons in stars produced by 3-BeV protons. *Phys. Rev.*, **102** (1956) 495.

18. Blau, M., Bloch, S. C., Carter, C. F. and Perlmutter, A., Studies of ionization parameters in nuclear emulsions. *Rev. Sci. Inst.*, **31** (1959) 289.

19. O'Dell, F. W., Shapiro, M. M. and Stiller, B., Relative abundances of the heavy nuclei of the galactic cosmic radiation. *J. Phys. Soc. Japan*, **17** Suppl. A-III 23 (1962).

20. Glasser, R. G., Seeman, N. and Stiller, B., Lifetime of the neutral pion. *Phys. Rev.*, **123** (1961) 1014.

21. Danysz, M. and Pniewski, J., Hyperfragments. *Philos. Mag.*, **44** (1953) 348.

22. O'Ceallaigh, C., Tau mesons. *Philos. Mag.*, **42** (1951) 1032.

23. King, D. T., Seeman, N. and Shapiro, M. M., Charged particles of mass intermediate between proton and deuteron. *Phys. Rev.*, **92** (1953) 838. Lal, D., Pal, Y. and Peters, B., Bagneres Conf. Report (3rd Intl. Conf. On Cosmic Radiation, Bagneres-di-Bigorre), 146 (1953). Levi-Setti, R., Bagneres Conf. Report (1953). Shapiro, M. M., The International Conference on cosmic radiation at Bagneres di Bigorre. *Science*, **118** (1953) 701. Shapiro, M. M., Mesons and hyperons. *Am. J. Phys.*, **24** (1956) 196.

IMPORTANT PUBLICATIONS

Blau, M. and Wambacher, H., Photographic detection of protons liberated by neutrons. II. *Akad. Wiss. Wien*, **141** (1932) 617.

Blau, M. and Oliver, A. R., Interaction of 750 MeV mesons with emulsion nuclei. *Phys. Rev.*, **102** (1955) 489.

FURTHER READING

Blau, M., in *Methods of Experimental Physics*, eds. L. C. L. Yuan and C. S. Wu, Vol. 5, Nuclear Physics, Part A (1961), and Part B (1963).

Possibilities and limitations of the photographic method in nuclear physics and cosmic radiation. *Acta Physica Austriaca*, **3** (1950) 384.

Friedlander, M. W., *Cosmic Rays* (Cambridge: Harvard Univ. Press, 1989).

Galison, P. L., Marietta Blau: Between Nazis and nuclei. *Physics Today*, Nov. (1997), p. 42 ff.

Halpern, L., Marietta Blau (1894–1970). In *Women in Chemistry and Physics*, eds. L. S. Grinstein, R. K. Rose and M. H. Rafailovich. (1993) p. 56 ff. (This contains a fairly complete bibliography of Blau's publications.)

Powell, C. F., Fowler, P. H. and Perkins, D. H., *The Study of Elementary Particles by the Photographic Method* (Pergamon Press, 1959). (An account of the main techniques and discoveries, and an atlas of beautiful and instructive photomicrographs.)

Shapiro, M. M., A brief introduction to the cosmic radiation. In *Chemistry in Space*, eds. J. M. Greenberg and V. Pirronello (Dordrecht: Kluwer Acad. Publ., 1989).

Tracks of nuclear particles in photographic emulsions. *Rev. Mod. Phys.*, **13** (1941) 58.

Nuclear emulsions. In Vol. 45, *Encyclopedia of Physics* (Handb. d. Physik), eds. S. Flügge and E. Creutz (Berlin: Springer-Verlag, 1958) p. 342 ff.

Stiller, B., Shapiro, M. M. and O'Dell, F. W., Techniques for processing thick nuclear emulsions. *Rev. Sci. Instr.*, **25** (1954) 340.

Tsao, C. H., Shapiro, M. M. and Silberberg, R., Cosmic-ray isotopes at energies >2 GeV/amu. *Proc. 13th ICRC, IUPAP Conf. Papers* (Denver CO, Univ. of Denver) **1** (1973) 107.

11 Hertha Sponer (1895–1968)

Helmut Rechenberg
Werner-Heisenberg-Institute, Max-Planck-Institut für Physik, Munich, Germany

FIGURE 11.1 Photograph of Hertha Sponer.

SIGNIFICANT CONTRIBUTIONS

Hertha Sponer occupied professorships in Germany and the United States, and published over 80 scientific papers in a field then ruled by high standards and great competition. James Franck called her the second female physicist (in Germany) next to Lise Meitner (see Chapter 7), and Friedrich Hund wrote: "She realized that she did something extraordinary for a woman, but she also possessed the capacity, toughness and abilities to defeat the difficulties involved." The talk

Out of the Shadows: Contributions of Twentieth-Century Women to Physics, eds. Nina Byers and Gary Williams. Published by Cambridge University Press.
© Cambridge University Press 2006.

is about Hertha Dorothea Elisabeth Sponer, who was born on September 1, 1895, in Neisse, then German Silesia, and died on February 17, 1968, in Ilten near Hanover.

Hertha Sponer began her scientific career in March, 1920, with a Ph.D. thesis entitled (in English translation) "On the infrared absorption of diatomic gases." According to her supervisor in Göttingen, Peter Debye, she "solved the task of clarifying the situation (i.e., to describe the molecular spectra) on the basis of the present quantum concepts." Hund judged later: "Had this dissertation been published, the name of Sponer might have been added, besides those of Torsten Heurlinger, Wilhelm Lenz (his paper appeared simultaneously) and Adolf Kratzer (he got his Ph.D. six months later), to mark the beginnings of the theory of band spectra." After this theoretical work, she then took up experimental research in James Franck's group at the Berlin Kaiser Wilhelm Institute, determining "the frequency of inelastic collisions of electrons with mercury atoms" (1921), a follow-up to the famous Franck–Hertz experiments, which had proved the existence of stationary states in Niels Bohr's atomic theory.

During the following years Sponer, then in Göttingen, investigated further electron–atom collision phenomena, notably the anomalous behavior of slow electrons in noble gases, the so-called Ramsauer effect. A review article for the *Ergebnisse der exakten Naturwissenschaften*, written jointly with Rudolf Minkowski (1924), summarized the experimental and theoretical status before the advent of quantum mechanics.

In the mid twenties, Göttingen was one of the leading centers in atomic physics. Since 1922, Max Born, professor of theoretical physics, had begun, with his brilliant students and collaborators Wolfgang Pauli, Werner Heisenberg and Pascual Jordan, to examine systematically the defects of the old Bohr–Sommerfeld quantum theory. By carefully analyzing the data of atomic spectroscopy (to which Sponer contributed in 1924 and 1925) they established in 1925 the first formulation of quantum mechanics (Heisenberg, Born and Jordan) and the foundation of electron spin (Pauli). The Göttingen theoreticians also had great interest in the theory of molecules. Notably Born's assistant Friedrich Hund became a pioneer applying wave mechanics to many-atom systems. From 1925–7 he developed the theory of molecular energy states based on a combination of dynamical and symmetry considerations, and he introduced for the first time the "tunnel effect" in connection with isomeric complex molecules. Heisenberg, Hund and their successor Walter Heitler prepared the basis of quantum chemistry. Max Born and distinguished guests of his institute, such as J. Robert Oppenheimer, Eugene Wigner, Gerhard Herzberg and Edward Teller, contributed substantially to this.

On the experimental side, Franck and his assistants, from Otto Oldenberg to Günther Cario, worked intensely on molecular problems. Hertha Sponer joined this bandwagon in 1926, and from that point the study of molecular spectra remained the sole theme of all her future research. Starting from an experimental determination of the excitation energy of nitrogen via the electron-collision method, with a theoretical analysis of the data, she clarified (with Raymond Birge in California) the concept of "the heat of dissociation" in the case of non-polar diatomic molecules (1926, see her review in the *Ergebnisse* of 1927). Supported by the German institutions supporting science *Helmholtz-Gesellschaft* and the *Notgemeinschaft*, she became an expert at vacuum spectroscopy, which was particularly suited to probe the constitution of polyatomic molecules (see Box 11.1). She used her own measurements and those of other authors to investigate carefully essential theoretical concepts like the "Franck–Condon principle" and the hypothesis of "predissociation" (see Box 11.2). In these fundamental studies she collaborated with (besides Franck), excellent younger colleagues, e.g., the experimentalist Heinz Maier-Leibnitz and the theoreticians Gerhard Herzberg (1934) and Edward Teller (1932). "*Von der Theorie wird man nicht heller, Gott geb' uns täglich unsern Teller* (when theory does not enlighten us, may God give us our daily Teller") she rhymed jokingly in an appreciation of the latter's stimulating ideas.

Box 11.1 Molecular spectroscopy

Molecules consist of atoms bound by ionic or covalent chemical forces (the former being explained by electric attraction of opposite charged ions, the latter as quantum-mechanical exchange of electrons between different atoms). Structure and forces give rise to (essentially discrete) energy states with spectra often having a characteristic band type, being created by transitions between these states. Molecular spectra are generally classified as fine-structure spectra (e.g., due to paramagnetic resonance), rotation spectra, vibration-rotation spectra and electronic spectra. Rotation spectra (arising from the rotation of the atoms in the molecules) are observed in the far infrared to microwave region, the mixed vibration-rotation spectra (the atoms can also vibrate in the molecule) in the infrared region, while the electronic spectra (originating from changes in the electronic state of the molecule) occur in the infrared, visible and vacuum ultraviolet regions. Depending on the frequencies observed, spectroscopic methods in the microwave, infrared, visible and ultraviolet regions are used. Since some frequencies are absorbed by air or the prisms of the spectroscopic apparatus, one needs special materials (quartz or fluorspar) and often evacuated chambers to observe specific molecular spectra. The electronic spectra of molecules are frequently emitted by certain atomic groups, which can be better detected (intensity!) when the corresponding substances are condensed in the solid state. Hertha Sponer concerned herself especially with spectra in the near vacuum-ultraviolet region, the most difficult spectral region to work in.

Box 11.2 Principles of molecular theory

Franck–Condon principle: For transitions involving changes of the electronic and vibrational state of a molecule James Franck suggested in 1925 that, because of the smallness of the electron mass as compared with that of atomic nuclei, every change of the electronic state occurs so fast that the position and velocity of the nuclei are not altered appreciably (hence the separation of the nuclei remains constant). As a consequence, most transitions of that kind are associated with electrons assembling at the points where the molecular potential shows a flex point. This principle Edward U. Condon later (in 1928) formulated wave-mechanically, obtaining the result that the probability is large for such transitions, where the eigenfunctions of the two combining states possess maxima at the same distance between the nuclei.

Predissociation: Especially in the spectra of many-atom molecules often after a sharp band region there occur diffuse bands (without any rotational structure) before the continuum starts. This property, discovered by Victor Henri in 1923 could be easily explained by wave mechanics as an allowed transition of the electrons into the continuum. Often the probability of predissociation is large where also the maximum probability for Frank–Condon transitions is expected; hence, the separation of the two effects causes some difficulties in the analysis of molecular structure.

The insecure situation between 1934 and 1936 (see the biography later) greatly interfered with Sponer's scientific activities; still she completed and published during this time her only book, the two volumes on molecular spectra in the well-known Springer series *"Struktur der Materie"*(Structure of Matter) (1935, 1936). Then, as an established professor at Duke University, she plunged into work on a systematic program, which she titled "Many-atom molecules: structure determination, relation between binding moments and absorption, general comparison of sequences of compounds," and which she executed in detail with numerous students and collaborators. Again, she mainly used vacuum spectroscopy as an experimental tool and associated herself with Teller and other theoreticians in pioneering the analysis of the electronic spectra of organic substances, notably of benzene, methane and their derivatives. At the Chicago "Conference on Spectroscopy" in June, 1942, Sponer presented the results of this particular program, which involved experiments with another German refugee, Hedwig Kohn (details were published only in 1948). After World War II she turned her attention to fluorescence phenomena, and she developed in particular the field of "low temperature spectroscopy of aromatic molecules": the experimental procedure allowed one to observe the electronic spectra of complex molecules (more accurately, of characteristic parts of the molecule) in the solid state and led to the discovery of several subtle effects, e.g., the influence of excitons on phosphorescence (1960). In the last two decades of her career, as an accomplished member of the worldwide scientific establishment, she published amply, both alone and in cooperation with students or senior colleagues – e.g., Karl Herzfeld (1952), Per-Olof Löwdin (1954) or James Franck (1956). Her excellent mastery of the literature, both theoretical and experimental, was displayed particularly in the article on "Electronic Spectroscopy" (1955 for *Annual Review of Physical Chemistry*).

BIOGRAPHY

Hertha Sponer was the first of five children of Robert Sponer, who ran a stationery shop, and his wife Elisabeth née Heerde, the daughter of a state official. After receiving higher education at various secondary schools, she went to a female teachers' seminar. From 1915–16 she substituted for a teacher at an elementary school, but then she decided to get her *Abitur* (in Spring, 1917) in order to be able to study physics. The first year she spent at the University of Tubingen (participating in Friedrich Paschen's very good laboratory course); the following years in Göttingen, attending there lectures of Peter Debye and Woldemar Voigt in physics, Erich Hecke in mathematics and Adolf Windaus in chemistry. Already in March, 1920, she completed her Ph.D. with a thesis supervised by Debye, and an excellent examination with Debye, Windaus and the mathematician David Hilbert.

When Debye left Göttingen, he recommended that Sponer go to Berlin, where she enjoyed the most stimulating atmosphere provided by the world-famous scholars there, from Max Planck, Albert Einstein and Walther Nernst to Otto Hahn, and she worked at Fritz Haber's

Kaiser-Wilhelm-Institut für physikalische und Elektrochemie in the group of James Franck. When the latter in 1921 became *Ordinarius* at Göttingen, he proposed that she come along as an assistant. Soon she took over the responsibility for Franck's laboratory courses, and she was remembered as "a cheerful, amiable and always interested colleague who stimulated also a private theoretical seminar for younger students (at her home), where even Heisenberg presented first his quantum mechanics." Further "she kindly helped the frequent visitors from abroad as well as young doctoral students" (Hund). In October, 1925, she got her *Habilitation* (and became in Göttingen the second female *Privatdozent* after Emmy Noether [see Chapter 8] in 1919). Five years later, she was appointed *Oberassistentin* and as such ran Franck's institute in the professor's absence, and in January, 1932, finally she obtained the title of an extraordinary professor. In the meanwhile Sponer had distinguished herself by research centering on Franck's interests, namely molecular dissociation, atomic collisions and fluorescence, and she developed her own special research method, vacuum spectroscopy. She had also spent the academic year 1925/6 on a Rockefeller grant at Berkeley, collaborating there with Raymond Birge, a specialist on molecular spectra.

The extremely happy and fruitful Göttingen period ended in 1933 when the Nazis came to power in Germany. Franck resigned voluntarily to protest the actions of the new government, and Sponer – although she was neither Jewish nor politically active (her sister, Margot, would later die in a concentration camp) – lost her paid position in 1934 (also because the remaining experimentalist in Göttingen, Robert Pohl, did not appreciate female colleagues). After considerable hesitation to leave her home country, she accepted again in the fall of that year a Rockefeller fellowship to join Lars Vegard's institute in Oslo. Franck, who had in 1935 gone to John Hopkins University, Baltimore, helped to secure his protegée a position at the not-too-distant Duke University in Durham, North Carolina.

Thus, in February, 1936, Sponer became a full American professor, and she stayed at Durham until she retired in March, 1965.

She cared quite a lot about building up the physics institute, won old and new collaborators, e.g., Edward Teller and Lothar Nordheim – the latter joined Duke University in 1937 as professor of theoretical physics after Max Born and Teller had turned down the offer – and Ida Bruch-Willstätter (daughter of the German chemist and Nobel Prize winner Richard Willstätter). She educated a large number of Master's and Ph.D. students (altogether 23 Ph.D. theses were completed between 1938 and 1966 under her supervision, a considerable number by female candidates). During World War II she participated in a special three-year teaching program for the Navy. Together with her mentor James Franck she cooperated before the war to assist emigrated scientists from Germany, e.g., Hedwig Kohn, Peter Pringsheim and Fritz Reiche, get appropriate positions in America. After the war the couple married on 29 June, 1946, following the death of Franck's first wife, and they organized care-packet actions and other help for friends in Germany. Franck had moved in 1938 to the University of Chicago. While separated professionally, they continued the exchange of scientific ideas. They spent summer vacations together at Cape Cod, and they also traveled together to conferences or institutes abroad. Sometimes Hertha went alone, e.g., she went alone to Uppsala in 1953/4 and to Japan and India in 1962. In spite of her original fears of leaving Germany, she became at home in the USA and was a respected member of the American physics community. She was elected to the New York and North Carolina Academies of Science. She also served as associate editor of the *Journal of Chemical Physics*. While at Duke University she received many grants to support her scientific programs and to provide the salaries of visitors, including Per-Olof Löwdin from Uppsala. Since 1949 Sponer, and after 1953 also Franck, returned for visits to Germany. On 21 May, 1964, Franck died in Göttingen. Soon afterwards his widow fell ill with Alzheimer's disease. Christoph Schonbein took his aunt to Celle in 1966, but in the following year she had to be taken to a mental hospital in nearby Ilten, where she died in 1968.

Hertha Sponer's work as an original physicist was definitely acknowledged by her colleagues and quoted appropriately in the literature, e.g., in the review article on band spectra of Walter Weizel in the *Handbuch der Experimentalphysik* (1931) and the one of Karl Herzfeld on molecules in the Springer *Handbuch der Physik* (1933), or in the later monography of John Slater on *Quantum Theory of Molecules and Solids* (Vol. 1, 1963). One might argue that she was overshadowed by Franck, since they collaborated closely over her entire career; thus, it may appear that she lacked the originality and independence of research as shown by the "German Madame Curie" – as Lise Meitner (see Chapter 7) was often called. In addition, Sponer investigated areas that after 1930 were considered as applied rather than fundamental physics. And finally, her molecular research, which combined the two disciplines, physics and chemistry, never gave rise either to ardent controversies or to spectacular discoveries that attracted the attention of a wider public. On the other hand, one has to admit that she possessed several decisive characteristics of a professor and organizer of research. For example, already in Göttingen the young Wilhelm Hanle highly respected the "severe *(gestrenge)* *Oberassistentin* Fräulein Sponer" and characterized her as follows: "Franck was only little interested in administration, and perhaps he occasionally showed less knowledge of human nature. [She] had a better overview of the difficulties in the institute and took care to remedy the abuses." In summary, Sponer really showed quite early the qualities of an efficient and successful institute director, being able to organize and conduct research and to educate a large number of collaborators. Perhaps she was the first woman who equaled, or even surpassed, many of her male colleagues in these important characteristics.

NOTE

Rossiter, M. W., *Women Scientists in America: Struggles and Strategies to 1940* (Baltimore and London: Johns Hopkins University Press, 1982).

IMPORTANT PUBLICATIONS

Sponer, H., Frequency of inelastic collisions of electrons with mercury atoms. *Z. Phys.*, **7** (1921) 185–200.

Free paths of slow electrons in noble gases. *Z. Phys.*, **18** (1923) 249–57.

Remark on the series spectra of lead and tin. *Z. Phys.*, **32** (1925) 19–26.

Excitation potential of the band spectrum of nitrogen. *Z. Phys.*, **34** (1925) 622–33 .

Sponer, H. and Birge, R., The heat of dissociation of non-polar molecules. *Phys. Rev.*, **28** (1926) 259–83.

Sponer, H. and Franck, J., Determination of the dissociation energy of molecules from band spectra. *Nachr. Ges. Wiss. Göttingen* (1928) 241–53.

Sponer, H. and Cordes, H., Molecular absorption bands of chlorine, bromine, . . . in the extreme U.V. I; II. *Z. Phys.*, **63** (1930) 334.
Z. Phys., **79** (1932) 170–85.

Sponer, H. Franck, J. and Teller, E., Predissociation spectra of triatomic molecules. *Z. Phys. Chem. B*, **18** (1932) 88–101.

Sponer, H., Nordheim, G., Sklar, A. L. and Teller, E., Analysis of near U.V. electronic transition of benzene. *J. Chem. Phys.*, **7** (1939) 207–20.

Sponer, H. and Teller, E., Electronic spectra of polyatomic molecules. *Rev. Mod. Phys.*, **13** (1941) 75–123.

Sponer, H., Remarks on radiationless transitions in complex molecules. *J. Phys. Chem.*, **25** (1956) 172.

Molekülspektren und ihre Anwendung auf chemische Probleme, 2 volumes, Springer, Berlin, 1935 and 1936.

FURTHER READING

Franck, J., Sponer, H. and Goeppert-Mayer, M. Interview with T. Kuhn, July 12, 1962. Archive for the History of Quantum Mechanics.

Hund, F., Hertha Sponer-Franck 1.9.1895–17.2.1968, *Phys. Bl.*, **29** (1968) 166.

Hanle, W., *Memoiren* (Giessen: Physikalisches Institut der Universität, 1989).

Maushart, M.-A., *"Um nicht zu vergessen" – Hertha Sponer, Ein Frauenleben für die Physik im 20.Jahrhundert* (Bassum: Verlag für Geschichte der Naturwissenschaften und Technik, 1997).

12 Irène Joliot-Curie (1897–1956)

Hélène Langevin-Joliot and Pierre Radvanyi
Institut de Physique Nucléaire, Orsay, France

FIGURE 12.1 Irène
Joliot-Curie (1897–1956) in
front of her Wilson cloud
chamber. (Archives Curie et
Joliot-Curie, photo Robert
Doisneau.)

SOME IMPORTANT CONTRIBUTIONS

Irène Curie learned the practice of separating, measuring and using
radioactive substances working along with her mother, Marie Curie,
in the newly established Institut du Radium in Paris. This work consisted of both chemistry and physics. Her first important scientific
work dealt with the characteristics (range, ionization power and

Out of the Shadows: Contributions of Twentieth-Century Women to Physics,
eds. Nina Byers and Gary Williams. Published by Cambridge University Press.
© Cambridge University Press 2006.

energy loss) of alpha particles in their passage through matter. She prepared very strong polonium sources and, among various apparatuses, she used to good purpose was a Wilson cloud chamber. This careful study became the subject of her Dr.Sc. thesis. It cleared up discrepancies between previous experiments and gave additional firm support to current theories. For a few more years she studied the decay properties of several radioelements.

Irène Curie and Frédéric Joliot started working together in 1928, using a variety of measuring apparatuses. Their most important joint experiments began with investigations on very low intensity radiation accompanying the alpha rays of polonium, and the preparation of very intense and localized polonium sources. The polonium resulted from the decay of long lived radium D (an isotope of lead) deposited in sealed glass tubes having contained radon gas. These glass tubes had been patiently collected by Marie Curie.

Toward the end of 1931, they used these strong sources to study the mysterious and very penetrating radiation observed the previous year by W. Bothe and H. Becker in Berlin. Bothe and Becker had been bombarding two light elements, beryllium and boron, with alpha particles. On January 18, 1932, Irène Curie and Frédéric Joliot announced that they had discovered, by a delicate experiment, that this penetrating radiation – which they believed to consist of very energetic gamma rays – was able to eject fast protons out of hydrogeneous screens. This publication prompted J. Chadwick, in Cambridge, to perform further experiments, which led him, on February 17, 1932, to the discovery of the neutron, a particle having about the same mass as the proton but no electric charge.

After this dramatic episode and various experiments on the production of neutrons in nuclear reactions, Frédéric and Irène Joliot-Curie carried on their study of the action of intense alpha radiation. C. Anderson in California had just discovered the positron (a positive electron) and the two French scientists were among the first to study positrons resulting from the materialization of energetic gamma rays into electron pairs. They also observed that the bombardment of light target materials – in particular aluminum foil – with alpha particles gives rise to the emission of positrons. They believed first that these positrons were "transmutation positrons" produced directly together with neutrons in a nuclear reaction. This hypothesis was challenged at the Solvay meeting in Brussels in October, 1933. Back in Paris, they looked for confirmation by examining the threshold of such a reaction (i.e., the minimum energy of alpha rays producing positrons and neutrons) using various measuring instruments. In the course of these experiments, in January, 1934, they observed, using a Geiger–Müller counter, that the emission of positrons, contrary to that of neutrons, is not instantaneous, but takes place only after several minutes of irradiation, and continues for a short time after the end of the irradiation, following an exponential decrease. They concluded that the transmutation (a nuclear reaction), induced by alpha particles in aluminum nuclei, had produced, not a stable nucleus as in all previous cases, but a radioactive nucleus of phosphorus. This as yet unknown isotope of phosphorus then decayed, with a half-life of about three minutes, into a stable silicon.

They had thus discovered a new type of radioactivity – by positron emission. Similar phenomena were observed for boron and magnesium. Two weeks later the Joliot-Curies were able to separate chemically the new radioactive species. The necessary chemical reactions had to be completed in a few minutes before the activity disappeared. From then on, radioactivity was recognized as a general property of all elements.

In 1935, Frédéric and Irène Joliot-Curie received jointly the Nobel Prize in chemistry for their discovery of "artificial radioactivity." They suggested that the natural radio-elements were only the rare survivors of the very numerous radio-elements, which must have been formed at the beginning of the Earth's history. New applications were predicted for medicine. A considerable development of the use of these new radioactive nuclei could be anticipated: e.g., as indicators to study the behavior of their stable inactive isotopes, in particular in biological phenomena. Very soon thereafter G. Hevesy and O. Chiewitz, in September, 1935, in Copenhagen, used radioactive phosphorus-32 to study the metabolism of phosphorus in rats. This became a key method in modern biology. Many other novel applications have been developed, e.g., Nobel laureate Rosalind Yalow's method of radioactive immunoassay (see Chapter 27).

Within a few years, radioactive isotopes were obtained of all chemical elements, even of yet unknown ones. Considerable success was achieved, already in 1934, by E. Fermi and his team in Rome, using neutrons as projectiles especially suited for the heavier elements. When bombarding uranium (atomic number 92), the heaviest of all known elements, they observed several new radioactive nuclei. They thought that they had obtained at least one transuranic element, of atomic number 93.

These difficult experiments were taken up in stricter conditions, in Berlin, by O. Hahn and L. Meitner, joined later by F. Strassmann. Their efforts lasted for almost four years. Using fast and slow neutrons to bombard uranium, they found nine different activities, which they believed to correspond to at least four transuranic elements. Everybody at that time believed that a nuclear reaction with slow neutrons on uranium could only lead to isotopes of uranium or of an element in the neighborhood of uranium. Irène Curie entered the competition in 1937, studying first thorium and then uranium. Together with P. Savitch, she confirmed several of the Berlin group's results.

Curie and Savitch then adopted a new approach to the problem. Using copper screens thick enough to get rid of the other activities, they selected the most energetic beta emitters. A new activity of 3.5 hours half-life separated out clearly from all others in rather long irradiation times. Irène Curie decided to focus on the identification of the chemical element responsible for that activity; its chemical properties were systematically checked against those expected for the elements close to uranium (including the supposed transuranic elements) without success. In their paper written in July, 1938, and published in October, Curie and Savitch concluded that this new radioisotope "has, on the whole, the properties of lanthanum from which it seems it could only be separated by fractionating methods" (lanthanum is a rare-earth metal).

The Paris results were first dismissed in Berlin. However, after the forced emigration of Lise Meitner, O. Hahn and F. Strassmann resumed their experiments. They thought that the 3.5 hours half-life activity was a mixture and they decided to look for radium isotopes, using barium as carrier; they observed what they believed to be new radium activities; but they came finally to the conclusion that they had actually obtained barium isotopes. This was the discovery of fission, the splitting of uranium nuclei into two lighter nuclei. Among these were barium and lanthanum nuclei! Irène Curie had come very close to the truth. It happened that the 3.5 hours half-life substance contained also a small amount of another fission product (an isotope of yttrium) in addition to lanthanum. True transuranic elements were discovered later in Berkeley.

War and illness nearly interrupted the scientific work of Irène Joliot-Curie. From 1944 on, she resumed research and gave still several valuable scientific contributions: in particular, in 1946, the setting up of a method of estimating very small uranium and thorium contents in rocks by observing alpha particle tracks in photographic emulsions, and, in 1953, the realization of a method for detecting and proportioning carbon in steel using artificial radioactivity.

BIOGRAPHY

Irène Curie, the eldest daughter of Pierre and Marie Curie, was born on September 12, 1897, in Paris, one year before her parents discovered radium. She was only nine years old when Pierre Curie died in a street accident. Her grandfather, a retired physician of rather radical opinions, greatly helped Marie Curie in taking care of Irène and her younger sister Eve. His influence on Irène's interests and general ideas has been especially important, together with that of Marie. The young Irène benefited from a quite unconventional education, especially during two years (1907–8) when Marie Curie and several of her friends organized a special school. The children learned doing simple experiments and thinking for themselves on every subject. They spent time in outdoor activities. This teaching method contributed to Irène's taste for mathematics and scientific matters. She then spent four years at the College Sevigne preparing for her Baccalaureate degree.

She was 17 years old when World War I broke out in 1914. She started taking university courses in physics to get a degree. In parallel, she contributed to the operation of radiological cars near the battle-front in the framework of Marie Curie's program. She ensured later the practical training needed for handling correctly X-ray apparatuses in complement to Marie Curie's lectures to groups of nurses. After the war, she was appointed Preparateur (1918), then Assistant (1923) at the Radium Institute. She would perform all her research work there. The fact that Irène grew up with radioactivity certainly played a role in her choice to devote her entire working life to that research field. She was not afraid of her mother's fame. Her collaboration with her as a radiographer during the war reinforced her natural confidence in her own capability.

In October, 1926, Irène Curie married Frédéric Joliot, a young engineer appointed the year before as a personal assistant to Marie Curie. Their personalities were complementary. Irène appeared to many as a very distant person. Frédéric Joliot discovered "underneath the surface" a quite different young woman. As he explained later "I observed her . . . . I found a human being with extreme sensitiveness and poetic character . . . She had the same purity, the practical sense and quietness as her father Pierre Curie." She had also a real sense of humor. Irène Curie shared with Frédéric her passion for research, rather radical views about changes needed in social organization for achieving social progress, as well as the love of sports and active holidays. Irène continued using her birth name in her scientific papers after her marriage. She was Mrs. Joliot otherwise, until the joint name Joliot-Curie became more and more used after the Nobel Prize. Irène and Frédéric Joliot-Curie had two children, Hélène, born in 1927, who became a nuclear physicist, and Pierre, born in 1932, who became a biophysicist. In spite of her ardour for scientific research, Irène always found time to meet the education and affection needs of her children. After the age of 30, tuberculosis attacks periodically obliged her to slow down or even interrupt her activities.

Irène Curie was appointed *Chef de travaux* in 1932, *Maître de recherches* in 1935, professor at the Paris Faculty of Sciences in 1936 and full professor for general physics and radioactivity in 1946. After her thesis defense in 1925, she brought an increasingly precious help to Marie Curie in managing the laboratory. After Marie Curie's death in July, 1934, she directed a large part of the laboratory activity even before becoming the Director in 1946.

Irène Joliot-Curie served in a number of national and international committees. She assumed important responsibilities and, in 1936, was among the three first women ever to participate in a French government. She was appointed the first Under Secretary of State for Scientific Research in the government headed by the socialist L. Blum, following the success of the *Front Populaire* movement. She accepted the job from June until September, 1936, in support of French women's fight for equal political rights. It was also a turning point in the recognition of the importance of science.

Little research could be performed at the Radium Institute during the German occupation of France. Irène's health was especially poor during these years. She left Paris secretly in May, 1944, and moved to Switzerland with her children. She was able to resume her activity at the laboratory soon after the liberation of Paris.

From late 1945 until 1951, Irène Joliot-Curie shared her time between the Radium Institute and the newly created Atomic Energy Project, headed by Frédéric Joliot and Raoul Dautry. She was appointed scientific commissioner together with Pierre Auger and Francis Perrin. She was mainly interested in uranium prospecting and chemistry problems. After Frédéric Joliot's dismissal for political reasons during the cold war, she was not renewed in her position at the end of her five-year term.

She returned full time to the Radium Institute with a major concern about the future. The laboratory inside Paris, too small anyway, could not fulfill the development needs of modern nuclear physics and radiochemistry. She had searched for many years for a place to build a new laboratory and finally discovered one at Orsay, only 30 kilometers south of Paris. In 1954, she obtained the necessary funding for a well-equipped laboratory, including a synchrocyclotron. Her initiative later led to the creation of the Orsay Scientific Center of the University Paris-Sud. Irène Joliot-Curie died on March 17, 1956, from leukemia.

Irène Joliot-Curie was active in antifascist movements. Deeply concerned with the risk of a nuclear war, she supported the French and international peace movements. In 1950, she signed the Stockholm Appeal against the use of atomic bombs. She was a member of women's organizations and she expressed herself in many occasions for woman rights. She often emphasized the role of scientific research for the benefit of mankind.

She was Officer of the Legion of Honour. Though she was a member of several foreign Academies, and Dr. honoris causa of several Universities, she was not elected to the French Academy of Sciences. She received major scientific prizes and medals, together with F. Joliot (Nobel Prize in Chemistry, 1935; Mateucci Medal, 1932; Barnard Medal, 1940), and alone (Lavoisier Medal, 1954).

From the obituary by Sir James Chadwick (in *Nature*, 1956):

> ... She was born in the stirring days of radioactivity when her parents were making their great discoveries, she grew up with radioactivity, and all her working life was devoted to its study. She bore an honoured name, to which she added lustre by many contributions of great importance in radioactivity and in the development of nuclear physics ...

During these years . . . she continued at the same time to publish work on various aspects of radioactivity, for her ardour for scientific research was such that neither administrative duties nor failing health could keep her from her laboratory.

Her parents were both persons of strong and independent mind, and Mme. Joliot-Curie inherited much of their character as well as their scientific genius. She had a powerful personality, simple, direct and self-reliant. She knew her mind and spoke it, sometimes perhaps with devastating frankness; but her remarks were informed with such regard for scientific truth and with such conspicuous sincerity that they commanded the greatest respect in all circumstances. In all her work, whether in the laboratory, in discussion, or in committee, she set herself the highest standards and she was most conscientious in the fulfillment of any duties she undertook.

Some quotes by Irène Joliot-Curie:

Among the goals of the feminist movement, none is more important than the right of a woman to obtain jobs for which she is qualified by her knowledge and her aptitudes . . .

(In a paper published by La France vivante, *1937.)*

Scientific research is a comforting activity from the moral point of view, for the pleasure of discovery even if it is of weak significance, for the pleasure of having met and overcome difficulties, for being aware that every new knowledge is definitively acquired to Mankind.

(In a broadcast talk to students, 1938.)

Box 12.1   glossary

Alpha, beta and gamma rays Rays emitted during radioactive decay, the alpha rays are positively charged and consist of helium nuclei; the beta rays are electrons (positive or negative); the gamma rays are high-energy electromagnetic radiation.

Carrier In order to separate a radioactive substance, present only in minute quantity, one can add a sizeable quantity of a carrier, a stable substance or element having the same or similar chemical properties.

Geiger–Müller counter This detector, a gas-filled cylindrical tube with an insulated wire along its axis, allows counting of alpha, beta and (with a lower efficiency) gamma rays.

Half-life The characteristic time lapse during which a radioactive substance loses one half of its radioactivity.

Indicator method The addition of a small quantity of a radioactive isotope allows one to follow the chemical or biological behaviour (metabolism) of a stable element.

Isotopes Atoms having the same chemical properties (i.e., the same number of protons and therefore the same atomic number), but different masses (i.e., different numbers of neutrons).

Nucleus, proton, electron The atom consists of a tiny positively charged nucleus, where almost all the mass of the atom is concentrated, and peripheral negatively charged electrons. The mass of the proton is about 1836 times larger than the mass of the electron. The simplest atom is the atom of hydrogen, consisting of a single proton and a single electron. Heavier nuclei are composed of protons and neutrons.

Polonium, radium Heavy, rare, radioactive elements, discovered in 1898 by Pierre and Marie Curie.

Radioactivity General property of many elements and isotopes which transmute spontaneously into other chemical elements.

Radon An inert gas discovered by E. Dorn in 1900; it is the daughter element of radium.

Transuranic elements Elements heavier than uranium; they have therefore an atomic number larger than 92 and do not exist normally in nature.

Wilson cloud chamber Fog is artificially produced in a closed glass vessel by the expansion of a gas (e.g., air) saturated with the vapour

> of a liquid (e.g., water). The resulting liquid droplets are preferentially
> formed along the tracks of ionizing charged particles, which can thus
> be seen, identified and measured.

IMPORTANT PUBLICATIONS

Curie, I., Recherches sur les rayons alpha du polonium, oscillation de parcours, vitesse d'émission, pouvoir ionisant, thèse de Doctorat, 27 Mars 1925. *Ann. de Phys.*, **11** (1925) 403. (Research on polonium alpha rays, range straggling, emission speed, ionizing power.)

Curie, I. and Joliot, F., Emission de protons de grande vitesse par les substances hydrogènées sous l'influence des rayons gamma très pénétrants. *C. R. Acad. Sci.*, **194** (1932) 273. (Emission of high-speed protons from hydrogeneous substances due to very penetrating gamma rays.)

Un nouveau type de radioactivité. *C. R. Acad. Sci.*, **198** (1934) 254. (A new type of radioactivity.)

Séparation chimique de nouveaux radioéléments émetteurs d'électrons positifs. *C. R. Acad. Sci.*, **198** (1934) 559. (Chemical separation of radioelements emitting positive electrons.)

Curie, I. and Savitch, P., "Sur les radioéléments formés dans l'uranium irradié par les neutrons, II". *J. Phys. et Rad.*, **9** (1938) 355. (On radioelements produced in uranium irradiated by neutrons.)

Joliot, F. and Curie, I. Artificially produced radioéléments. *Papers and discussion of the joint conference of the International Union of Pure and Applied Physics and the Physical Society, London*, vol 1 (Cambridge, UK: Cambridge University Press, 1935).

Joliot-Curie, I., *Les Radioéléments Naturels, Propriétés Chimiques, Préparation, Dosage* (Paris: Editions Hermann, 1946). (Natural Radioelements, Chemical Properties, Preparation, Measurement.)

Frédéric et Irène Joliot-Curie, Œuvres Scientifiques Complètes (Paris: Presses Universitaires de France, 1961).

Marie Curie, ma mère. *Revue Europe*, **108** (1954). (Marie Curie, my mother.)

Correspondance Marie Curie – Irène Curie (Paris: Les Editeurs Francais Réunis, 1974).

FURTHER READING

Curie, E., *Madame Curie, a Biography* (New York: Doubleday, Doran and Co., 1937).

Cotton, E., *Les Curies et la Radioactivité* (Paris: Editions Pierre Seghers, 1963).

Radvanyi, P. and Bordry, M., *La Radioactivité Artificielle et Son Histoire* (Paris: Le Seuil, 1984). (Artificial Radioactivity and its History.)

Guinier, A., *La Radioactivité Artificielle a 50 Ans* (Paris: Editions de Physique-CNRS, 1985). (Artificial Radioactivity is 50 Years Old.)

Pflaum, R., *Madame Curie and Her World* (New York: Doubleday and Dell Publishing Group, 1989).

Loriot, N., Irène Joliot-Curie (Paris: Presses de la Renaissance, 1991).

Bertch McGraine, S., *Nobel Prize Women in Science*, Birch Lane Press (Carol publishing group, 1993).

Radvanyi, P., *Les Curie, Pionniers de l'Atome* (Paris: Berlin Pour la Science, 2005).

13 Katharine Burr Blodgett (1898–1979)

Gary A. Williams

Department of Physics and Astronomy, University of California, Los Angeles

FIGURE 13.1 Photograph of Katharine Burr Blodgett. (General Electric Research laboratory, courtesy AIP Emilio Segrè Visual Archives.)

IMPORTANT CONTRIBUTIONS

Katharine Burr Blodgett spent nearly her entire professional career at the General Electric (GE) Laboratories in Schenectady, New York, the first woman employed as a research scientist there. She joined the lab in 1918, just after receiving her Master's degree from the University of Chicago, where her thesis research involved the surface adsorption of gases on activated charcoal (a topic important during World War I for the design of gas masks). Her initial assignment at GE was assisting the chemist Irving Langmuir in measurements of molecular films floating on the surface of water. Langmuir had developed an improved version

Out of the Shadows: Contributions of Twentieth-Century Women to Physics,
eds. Nina Byers and Gary Williams. Published by Cambridge University Press.
© Cambridge University Press 2006.

of a surface-tension trough first used by Agnes Pockels (see Chapter 3), and this device is still widely used today, known as a Langmuir trough. He was able to show that the floating oil and soap films could be diluted until they formed a single molecular layer on the water surface, and that by further lowering the density of the molecules they approached the characteristics of a nearly ideal two-dimensional gas. This was a major finding, and was the beginning of what is known today as the field of surface science. As the assistant who had not yet earned her Ph.D., Blodgett was seldom included as an author of the scientific publications reporting the work, but it was clear that Langmuir greatly valued her help. In one of these papers he acknowledges,[1] "The writer is much indebted to Miss Katharine Blodgett for carrying out most of the experimental work." Langmuir received much acclaim for his studies of surface films, and in 1932 was awarded the Nobel Prize in chemistry.

Langmuir encouraged Blodgett to pursue her Ph.D., and on his recommendation she was admitted in 1924 to Cambridge University. Her adviser there was the famous physicist Ernest Rutherford, who had discovered the atomic nucleus. Her thesis topic was in the area of gaseous electronics, studying the motion of electrons through ionized mercury vapor. As was the custom at Cambridge, her work was carried out nearly independently of Rutherford, and Blodgett was the sole author of the resulting publication.[2] In 1926, she became the first woman to receive a Ph.D. in physics from Cambridge.

On returning to the GE lab, Blodgett worked again with Langmuir, but now on a project to improve the properties of light bulbs, one of GE's principal products. This involved basic studies of space-charge currents and thermionic emission of electrons from filaments, as well as applied topics such as changes in the surface chemistry of different types of filaments. In the publication of this work Blodgett was now an equal author with Langmuir.[3] She was particularly known for the clarity and precision of her scientific writing.

In 1933, after he had won the Nobel Prize, Langmuir asked Blodgett to collaborate with him again on surface films. The chemistry of the monolayer films was now fairly well understood, and they set about to study thicker films. They developed a process by which monolayer molecular films such as barium stearate (a soap film) could be transferred from the water surface to coat a solid surface with the same film. By repeating the process Blodgett found that very uniform thick films could be built up layer by layer,[4] to thicknesses of as many as 3000 molecular layers.[5] These films on solid surfaces are to this day an important subject of study, and are universally known as Langmuir–Blodgett films. This was Blodgett's most important discovery; hundreds of research papers on this topic are still published every year.

The physics and chemistry of molecular films on the surface of water is a complex and interesting field that continues to be intensively researched, driven by many practical applications. Molecules that are able to float on the surface generally have an elongated structure where hydrocarbon $-CH_3$ complexes are found at one end of the molecule, while an acidic group such as $-COOH$ is found at the other end. The hydrocarbons tend to repel water molecules ("hydrophobic"), while the $-OH$ of the acid group attracts water ("hydrophilic"). When placed on the water surface these opposing tendencies lead to a stable film state where the acidic end of the molecule is at the surface of the water, and the hydrocarbon end protrudes away from the water.

The properties of the surface molecules can be monitored in the trough device that Langmuir developed. A sliding barrier is used to compress or expand the molecules, changing the surface density, and by using a balance technique similar to that of Agnes Pockels (Chapter 3) the changing surface tension can be measured. The surface pressure that the molecules exert can also be measured by observing the tiny deflection of a floating barrier holding the molecules on the surface. This allows studies of the thermodynamics of the two-dimensional gas of molecules. A rich variety of transitions between gas, liquid and solid phases of the molecules are observed.

Blodgett and Langmuir discovered that the surface monolayers could be transferred from water onto a solid surface such as a glass plate, and that multiple layers could be built up. First, the glass plate is dipped into the water in the Langmuir trough, and then the surface monolayer is introduced, with its surface pressure monitored with the balance to keep the molecular density constant. The glass is then pulled slowly out of the trough, and as the plate surface comes out of the water the hydrophilic acid groups at the end of the molecules are attracted to the glass and attach themselves there. The hydrocarbons at the other end repel the water, and so the glass emerges from the water completely dry. The glass is then dipped back into the water, and this time the hydrocarbon groups on the water are attracted to the hydrocarbon groups on the plate, forming a second molecular layer with reversed direction of the molecules. This is now repeated: every time the plate is pulled up and dipped back two more molecular layers are added to the film on the plate. It was found that thicknesses of as many as 3000 layers of molecules such as barium stearate could be formed.

Blodgett began to pursue applications of the thick molecular films, whose thicknesses could be controlled with unprecedented accuracy, at the atomic scale of Ångströms. In this later work, she operated nearly independently of Langmuir, and began to mentor and collaborate with other researchers in the lab. One famous application of the films was their utilization as optical coatings on glass surfaces.[6] By precisely controlling the thickness to one-fourth of the average wavelength of visible light in the film, the light reflected from the glass surface interferes with the light reflected from the film surface, canceling out any net reflected light. Blodgett was able to demonstrate that non-reflective glass was a practical reality, marking the beginning of the science of optical coatings. Today, anti-reflection coatings are used pervasively on optical glass to cut down unwanted reflections, and are commonly found on such consumer items as eyeglasses, picture frames, camera lenses and TV and computer screens.

During World War II, Blodgett worked on several projects to support the war effort. One involved devising a more efficient means of de-icing airplane wings in sub-zero weather conditions. Blodgett and Langmuir built the first large-scale analog computer at the GE lab to do simulations of the growth of ice crystals on fiber surfaces. She also participated in a project to improve smoke screens that were used by invading infantry in North Africa and Italy. In 1947, she worked on applying her thin films to measure the air humidity on military weather balloons.

In her later years, before her retirement in 1963, Blodgett remained active in the laboratory. She continued her work on films, and took up other areas of research. One involved a study of electrically conducting glass, and another was the use of electrical discharges in gases to clean solid surfaces of impurity atoms. Both of these topics are of considerable importance today in the fabrication of complex semiconductor devices.

Box 13.1   Non-reflective glass coatings

Blodgett exploited the fact that her molecular films could be layered to very precise thicknesses in her development of non-reflecting glass. By coating a glass surface with exactly 44 layers of cadmium arachidate (a type of soap film) the thickness of the film is 1388 Ångströms, one-fourth of the average wavelength of visible light. With the film on the surface of the glass, there will be interference between the light reflected from the film surface and that reflected from the glass surface, as shown in Fig. 13.2a. The ray traveling through the film and reflecting from the glass will have traveled a total additional distance of one-half of a wavelength farther than the ray reflected from the top surface of the film. The two rays are then exactly 180° out of phase, and if their amplitudes are the same they will exactly cancel each other. The amplitudes will be the same if the index of refraction of the surface film is equal to the square root of the index of refraction of the glass. Since glass has an index of 1.5, this means that the film index must be equal to 1.23. Blodgett found that she was able to vary the index of refraction of the films by dipping them into an organic solvent such as acetone for a brief period of time. The solvent removes some of the fatty acids on the molecules, changing the index of refraction, but does

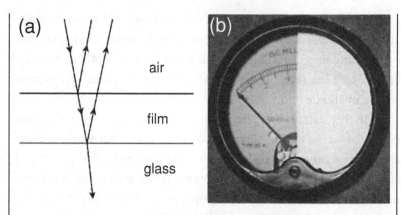

FIGURE 13.2 (a) A light ray incident on the film surface produces two reflected rays, one from the film surface and one from the glass surface, and they interfere since one has traveled an extra half-wavelength. (b) The left-hand half of the glass in front of the meter has been dipped to form a non-reflecting film on it, while the right half is uncoated glass. Illuminated from the front with a flood light, the difference is quite apparent.

not affect the thickness of the films. By monitoring the brightness of the reflected ray as the film is repeatedly dipped into the solvent, Blodgett could change the index of the film to the point where nearly all of the reflected light is canceled, at least for light incident normally on the glass, as shown in Figure 13.2b (light incident at grazing angles to the glass will travel a longer distance in the film and the reflected rays will not cancel as effectively, leading to a partial reflection).

Although Blodgett's technique worked very successfully in demonstrating the principle of non-reflecting glass, the main practical problem with her method was the fragility of the film coating the glass. Since it is basically a soap film, it is easily wiped from the glass with just a gentle touch. To be used in consumer devices the films needed to be much more resistant to damage. Within days of Blodgett's announcement of the discovery in December, 1938, a second team of researchers at the Massachusetts Institute of Technology announced a different technique for applying a uniform coating to glass, using the vacuum evaporation of calcium fluoride films onto the glass surface.

This approach also forms a non-reflective coating, following the same ideas that Blodgett had used, but had the advantage that the coating is less fragile than her films. Vacuum evaporation is now the standard technique for making optical coatings, and these are often very sophisticated structures involving multiple layers of different materials, with a final protective coating that resists scratches.

BIOGRAPHY

Katharine Burr Blodgett was born in 1898 in Schenectady, New York, where her father had been a patent lawyer for the General Electric Company. Unfortunately, he died just weeks before she was born, and her mother moved the family (including her older brother) first to New York City for three years, and then to France for four years so the children could experience European culture and languages. Returning to New York City, Blodgett was educated at the Rayson School, an exclusive private school. In 1913, at the young age of 15, she enrolled in Bryn Mawr College near Philadelphia, having won a scholarship in a competitive exam. There she excelled in mathematics and physics, encountering several inspiring teachers, and graduated second in her class.

Because of her father's connection with GE, Blodgett was able to visit the GE laboratory in Schenectady during the Christmas vacation of her senior year at Bryn Mawr. Her guide for the tour was Irving Langmuir, who advised her to pursue further training in physics. After graduating from Bryn Mawr in 1917, she enrolled in a Master's program at the University of Chicago, and received her M.Sc. degree in 1918. She then went to work at GE, serving as Langmuir's assistant. After six years of collaboration, Langmuir helped her gain admission to the Ph.D. program in physics at Cambridge. Finishing the degree in two years, Blodgett returned to Schenectady in 1926 to work again with Langmuir, but now as a scientist in her own right.

Blodgett lived a quiet life in Schenectady, never marrying. She was active in social organizations and served as a leader of the GE employees' club. She enjoyed cooking, gardening and the outdoors, and had a summer retreat at Lake George, New York.

The announcement of non-reflecting glass in December, 1938, caused quite a stir in the popular media, and Blodgett gained much attention. *Time, Look* and *Life* magazines all had feature articles on it,[7] although they were filled with misconceptions about the effect, calling it "invisible glass." Part of the novelty was that the discovery had been made by a woman scientist; women were still quite rare in technical fields then. The notoriety led to honorary degrees from Elmira College (1939), Brown University (1942), Western College (1942) and Russell Sage College (1944). In 1945, she was presented with the Annual Achievement Award of the American Association of University Women, and in 1951 the Garvan Medal of the American Chemical Society, intended "to honor the American woman for distinguished service in chemistry." She was made a Fellow of the American Physical Society, and in 1972 she received the Progress Medal of the Photographic Society of America for her contributions to optical coatings.

In spite of the awards and honors and the distinction Blodgett brought to the GE laboratory, most research enterprises in those years were still quite male dominated. A glimpse of the prejudice facing Blodgett can be seen in an article published in the journal *Science* in 1953,[8] which celebrated the 75th anniversary of the founding of the GE laboratory. Although the accomplishments of the male scientists in the lab are reviewed in detail, Blodgett is not even mentioned in the article. She apparently did not let instances like this bother her, and was content to take pleasure from the creative work of her research.

Blodgett retired from the lab in 1963, and died at home in 1979 at the age of 81. To honor her, an entire issue of the journal *Thin Solid Films* was dedicated to her and featured papers inspired by her work on films. Vincent Schaefer, a co-worker whom she had mentored, eulogized her in an obituary:[9] "The methods she developed have become the classic tools of the science and technology of thin films.

She will be long- and rightly-hailed for their simplicity, elegance, and the definitive way in which she presented them to the world."

NOTES

1. Langmuir, I., The mechanism of the surface phenomena of flotation. *Trans. Far. Soc.*, **15** (1920) 62.
2. Blodgett, K. B., A method of measuring the mean free path of electrons in ionized mercury vapor. *Phil. Mag.*, **4** (1927) 165.
3. Langmuir, I., MacLane, S. and Blodgett, K. B., The effect of end losses on the characteristics of filaments of tungsten and other materials. *Phys. Rev.*, **35** (1935) 478.
4. Blodgett, K. B., Films built by depositing successive monomolecular layers on a solid surface. *J. Am. Chem. Soc.*, **57** (1935) 1007.
5. Blodgett, K. B. and Langmuir, I., Built-up films of barium stearate and their optical properties. *Phys. Rev.*, **51** (1937) 964.
6. Blodgett, K. B., Use of interference to extinguish reflection of light from glass. *Phys. Rev.*, **55** (1939) 391.
7. New inventions. *Time*, **33**, Jan. 9, 1939; Woman makes glass invisible by use of film one molecule thick. *Life*, **6**, Jan. 23, 1939; Look applauds, Dr. Katharine Burr Blodgett. *Look*, **15**, Aug. 28, 1951.
8. Suits, C. G., Seventy-five years of research in General Electric. *Science*, **118** (1953) 451.
9. Schaefer, V. J., Obituary: Katharine Burr Blodgett 1898–1979. *J. Coll. Int. Sci.*, **76** (1980) 269.

FURTHER READING

Finley, K. T. and Siegel, P. J., Katharine Burr Blodgett (1898–1979). In *Women in Chemistry and Physics*, eds. L. S. Grinstein, R. K. Rose and M. H. Rafailovich (Westport, Connecticut: Greenwood Press, 1993) p. 65.

Holman, J., Katharine Burr Blodgett. In *Notable Women in the Physical Sciences*, eds. B. F. Shearer, and B. S. Shearer (Westport, Connecticut: Greenwood Press, 1997) p. 20.

14 Cecilia Payne-Gaposchkin (1900–1979)

Vera C. Rubin
Department of Terrestrial Magnetism,
Carnegie Institution of Washington

FIGURE 14.1 Photograph of Cecilia Payne-Gaposchkin. (Credit: AIP Emilio Segrè Visual Archives, Physics Today Collection.)

SOME IMPORTANT CONTRIBUTIONS

Cecilia Payne arrived at Harvard in 1923, a time when physics and astronomy were starting to merge into the rewarding field of astrophysics. Classical astronomers had long measured star positions and identified stellar constituents of the universe; astrophysicists now wished to use the newly understood properties of atoms to determine the composition and evolution of stars.

For over 30 years, Harvard astronomers had been taking spectra of stars, spectra that displayed the starlight as a band of light, with

Out of the Shadows: Contributions of Twentieth-Century Women to Physics,
eds. Nina Byers and Gary Williams. Published by Cambridge University Press.
© Cambridge University Press 2006.

superposed lines, which were characteristic of the atoms producing the light. Three of the "Harvard women," Williamina Fleming, Antonia Maury and Annie Jump Cannon, had examined the spectral properties of over 300 000 stars. By the presence or absence of particular atomic lines, they placed the stars in about seven broad classes, but the underlying physics that produced those different spectra was not understood.

In studies during the early 1920s, the Indian astrophysicist M. N. Saha showed that the temperature and pressure in a stellar atmosphere determine the degree of ionization of the atoms. By noting the conditions under which a given spectral line just becomes visible, Saha was able to assign a temperature to each spectral class. Payne believed that it should be possible to use the Saha theory to establish the chemical composition and temperatures of the stars, and hence to understand the systematic differences between the stars of different spectral classes.

Exacting technical studies of the large collection of stellar spectra at Harvard led to "months, almost a year as I remember it, of utter bewilderment," but ultimately Payne discovered that stellar atmospheres were composed principally of atoms of hydrogen and helium, with only traces of other elements. She calculated that hydrogen was one million times more abundant in the sun than on earth.

The great Princeton authority Henry Norris Russell disbelieved this startling result, although in a letter to Shapley he called Payne's thesis the best doctoral thesis he had ever read, except perhaps Shapley's. It was undoubtedly Russell's reaction to an early draft of her thesis that caused the young Payne to modify her published statement (*Stellar Atmospheres*, p. 188), and conclude "The enormous abundance derived for these elements (hydrogen and helium) in the stellar atmosphere is almost certainly not real."

Yet four years after the publication of Payne's thesis, Russell published in the *Astrophysical Journal* an elaborate quantitative study "On the Composition of the Sun's Atmosphere." In the abstract, he noted that "the abundance of hydrogen is difficult to estimate."

Finally, on the 55th published page, he compared his results with those of Payne, and noted the fine agreement. Following the calculations that showed that hydrogen is 80 times as abundant as all the metals together, he concluded "The numerical values should not be stressed; but the great overabundance of hydrogen can hardly be doubted ... The calculated abundance of hydrogen in the sun's atmosphere is almost incredibly great."

It is unfortunate that there is no record of Payne's reaction to Russell's disbelief of her startling result. At least she must have had some satisfaction in seeing Russell's belated agreement with her work. In the introduction to Payne-Gaposchkin's autobiography (1979), Kidwell wrote: "Neither the original draft of Payne's paper nor any account of her reaction to Russell's letter has survived." But Payne-Gaposchkin did publish (1977) a laudatory account of Russell's work on the composition of stellar atmospheres in which she never alluded to her own work. I suspect that her opening sentence, "Henry Norris Russell knew a good thing when he saw it," as well as in her closing one, "How he would delight in the complex detail that is now invading the fundamental simplicity that he was one of the first to recognize," applied to herself as well.

Eddington, in commenting to Payne on her work, called her ideas "not so wild as might at first appear." It is to the credit of (most of) the astronomical community that history has awarded to Payne the distinction of having discovered the enormous abundance of hydrogen in the universe. Her thesis work led to several publications and also to her first book, *Stellar Atmospheres*. In their history of twentieth-century astronomy (1962), Otto Struve and Velta Zebergs called Payne's dissertation "undoubtedly the most brilliant Ph.D. thesis ever written in astronomy." By the early 1940s, Payne-Gaposchkin had moved to the study of variable stars, formerly called "pathological" stars by Shapley. Here too she found a gold mine in the Harvard collection of sky images. Believing that it would be possible to arrange the variable stars into a coherent scheme, her husband studied the eclipsing variables (double stars whose orbits are aligned so that the light varies as one star eclipses the second); she tackled all others. This work led to their enduring interest in novae, dwarf novae and cataclysmic variables.

At the start of Payne-Gaposchkin's career, these stars had been identified by their intensity variations (see Box 14.1). By 1948, she and her husband had completed analysis of about 1500 variables of 5 types: long period, semi-regular, Cepheids, cluster variables and eclipsing variables. Not only had they learned the chemistry and physics of these stars, but they knew their orbits in the galaxy. While the Cepheid orbits resembled the circular orbit of the sun about the Galactic Center, the cluster variables were interlopers in the solar neighborhood, with eccentric orbits that just happened to pass near the sun.

These studies helped to bring order to the complex behavior of the variable stars. Today, with the understanding of the causes of these variations, astronomers know not only the present physical condition of such stars, but also their past and their future. For many stars, "pathology" is just a phase in their evolution.

Having conquered variable stars in the solar neighborhood, Payne-Gaposchkin's next step was the study of variables in the Small Magellanic Cloud (SMC), one of the two satellite galaxies of our own Milky Way visible in the southern hemisphere. The differences that Payne-Gaposchkin enumerated between SMC variables and those in our galaxy are now understood to arise from the smaller fraction of metal atoms in the SMC interstellar gas, from which those stars formed.

The recent construction of large telescopes has rejuvenated the study of variable stars. Variability is detected in fainter, and hence more distant, stars. Variable stars are found in more distant regions of our own galaxy, and also in more distant galaxies in the universe. Cepheid variables are basic to determining distances in the universe; supernovae are just starting to tell us about the geometry of the universe. Payne-Gaposchkin would not be surprised to learn that her field of inquiry is today a thriving industry, fundamental to understanding our place in the cosmos.

Box 14.1 Light curves of variable stars

Some stars, when observed repeatedly over long periods, show systematic variations in their light intensity. Such stars can be identified by comparing their images on photographic plates with images of stars known to have constant brightness. Cecilia Payne-Gaposchkin and her

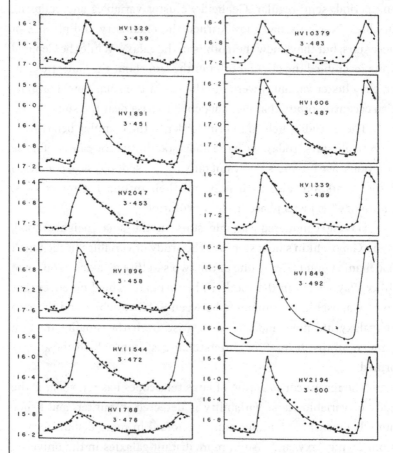

FIGURE 14.2 Typical light curves of periodic variable stars in the Small Magellanic Cloud, as determined by Cecilia Payne-Gaposchkin and Sergei Gaposchkin (Variable Stars in the Small Magellanic Cloud, Smithsonian Contributions to Astrophysics, Vol. 9, 89, 1966). Each point is the mean of many successive periods. Over 500 photographs, taken since 1898, were searched in order to establish periods for almost 1600 stars. Horizontal axis: Time. Vertical axis: Brightness.

husband Sergei Gaposchkin determined light curves for thousands of variable stars, and classified their properties. Cepheid variables, one of these classes, are supergiant stars, with brightnesses up to 30 000 times that of the sun. The atmospheres of these stars undergo radial pulsations, with periods that may be as long as 50 days. The light curve for a given star is very regular, and its period is related to the true brightness of the star. Thus, once the period of a Cepheid is established, its true brightness is known. This relationship, discovered by Henrietta S. Leavitt (see Chapter 5) at Harvard in 1912, is used to obtain the distance of a Cepheid, by comparing its apparent brightness with the absolute brightness inferred from its period. Cepheids in nearby galaxies are used to determine the distances of the galaxies, and hence are valuable for establishing the distance scale of the universe.

BIOGRAPHY

Cecilia Payne-Gaposchkin was one of the great astronomers of the twentieth century, a woman astronomer at a time when the male scientific establishment was just beginning to permit women entrance into their world.

Born in Wendover, England, in 1900, she was a student of natural history from childhood. She entered Newnham College, Cambridge University, in 1919, to study botany, chemistry and physics. Early in her studies, the brilliant astronomer A. S. Eddington lectured on the results of his solar eclipse expedition to Brazil in 1918. His observations detected the apparent shift in positions of stars seen near the sun, as predicted by Einstein's Theory of Relativity. "The result was a complete transformation of my world picture" she would write; she determined to become an astronomer.

She received her B.A. degree in Natural Sciences in 1923; she was generally the only woman in the physics courses. She then left England to accept a National Research Fellowship at Harvard University to study with Harlow Shapley, the newly appointed Director of the Harvard College Observatory. For her doctoral thesis, she applied contemporary theories of atoms (Saha ionization theory) to determine the chemical compositions and the temperatures of stars.

Payne remained at the Harvard Observatory for her entire career. Many of her studies were made in collaboration with Sergei I. Gaposchkin, whom she married in 1934. She published over 300 papers and notes, many on individual stars of interest. After years of having an ambiguous relation with Harvard College, she was made professor of astronomy in 1956, the first female professor in the University's history. She later told a group of young women astronomy students that her salary was then increased by a factor of four! She retired from the astronomy department in 1966, but continued doing research almost until her death in 1979. She and her husband had three children.

Her many awards include the first Annie J. Cannon prize of the American Astronomical Society (AAS) for distinguished contributions to astronomy by a woman, honorary D.Sc. degrees from Wilson College (1942), Smith College (1943), Western College (1951), Cambridge University (1952), Colby College (1958) and Women's Medical College of Philadelphia (1961). In 1958, she presented the 28th Joseph Henry Lecture of the Philosophical Society of Washington, DC. In 1976, she gave the Henry Norris Russell Prize Lecture of the American Astronomical Society, the first woman to do so. This, the highest award of the AAS, is presented for a lifetime of distinction in astronomy. And in another historical first, the prize scroll was awarded by the president of the Society, Dr. E. Margaret Burbidge (Chapter 25), the first woman president of the AAS.

It is generally known that Payne's Ph.D. was the first astronomy degree awarded by Harvard or Radcliffe, although the reason is rarely stated. In fact, astronomy students at Harvard formerly had received degrees in physics, but Theodore Lyman, Chairman of the Physics Department, refused to accept a woman student. Hence, Shapley arranged for her degree to be in Astronomy.

At the Harvard Observatory, Payne had the support and friendship of the many "Harvard women" (see Figure 14.3) employed to

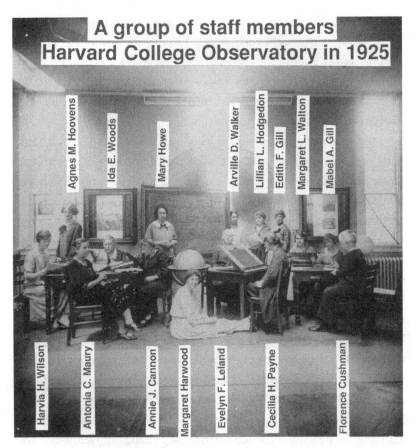

A group of staff members Harvard College Observatory in 1925

Agnes M. Hoovens · Ida E. Woods · Mary Howe · Arville D. Walker · Lillian L. Hodgedon · Edith F. Gill · Margaret L. Walton · Mabel A. Gill

Harvia H. Wilson · Antonia C. Maury · Annie J. Cannon · Margaret Harwood · Evelyn F. Leland · Cecilia H. Payne · Florence Cushman

FIGURE 14.3 A group of "Harvard women" in the observatory, 1925. Cecilia Payne is 5th from the right, sitting in front of the photographic plate holder. Annie J. Cannon, who classified spectra of more than 30 000 stars, is 5th from left (sitting).

measure the photographic plates, classify stars and spectra and catalog the results. Even before the turn of the century, Director Pickering learned that the female assistants were "capable of doing as much routine work as astronomers who would receive much larger salaries. Three or four times as many assistants can thus be employed, and the work done correspondingly increased for a given expenditure."

Although at times Payne-Gaposchkin denied that being a woman had made a difference in her career, her own statements and the facts show otherwise. Having been required by regulations at Cambridge to sit alone (as the only woman) in the first row of the physics lectures and to endure the applause and foot stampings of the "boys" as the professor began "Ladies and Gentlemen," she wrote (1979) "at every lecture I wished I could sink into the earth. To this day I instinctively take my place as far back as possible in a lecture room."

At Harvard, her position was minimized and ambiguous. Although she was first research fellow and later a Ph.D. researcher, Payne had little more freedom than the "Harvard women." She never asked what her salary would be. She was not paid as an instructor or professor, but as Shapley's "technical assistant." As such, she was not free to choose her research projects; these were specified by Shapley. Following her brilliant early work on stellar astrophysics, she agreed to Shapley's suggestions that ". . . I should turn my attention to the determination of photographic magnitudes and standards. Alas for my beloved spectra." Thus, at a time when astronomers elsewhere had begun to obtain accurate photoelectric brightnesses, Payne was set to work estimating magnitudes of questionable accuracy from old Harvard photographic plates. Discussing this work (1979) that occupied several years in the late 1920s, she wrote "I wasted much time on this account. This was another lesson: loyalty and devotion are not sufficient to qualify one to do science. My change in field made the end of the decade a sad one."

Her autobiography, *The Dyer's Hand* (1979), is replete with contradictory statements about insults she ignored. For most of her career, she did not have an appointment from the Harvard Corporation. Courses she taught were not listed in the catalog until 1945. Payne learned from Shapley that Harvard President Lowell told him "Miss Payne should never have a position in the University while he was alive."

In describing her position, she wrote (1979) "I had no official status, as little as that of the students who provided the 'girl-hours' in which Shapley counted his research expenditures. I was paid so little that I was ashamed to admit it to my relations in England."

Opportunities to improve her status were rare, at Harvard and elsewhere. When Shapley left in 1959, she was not considered for the Director's position, and she recalled: "Had I been a man, perhaps he would indeed have wanted to hand over the throne to me. But it would have been a great mistake." Why a mistake? Perhaps she was only denying her hurt.

Late in her life, I asked her many questions about her treatment as a woman in the field of astronomy. "Oh no, it had made no difference" she replied. The next day she approached me and said, "You know those questions you asked me last night? I decided I had given you all the wrong answers." She then proceeded to detail the many injustices that she had ignored, in order to do her astronomy. I puzzled for years over this turn-about. Now I understand that these indignities bothered her only at times. More likely, throughout her life she realized that she was taking steps never before open to a woman, and accepted them as part of the difficulties of being first. As a pioneer, she suffered from many of the prejudices of her male colleagues. But she also must have recognized that her achievements not only advanced science, but also opened doors for succeeding generations of women astronomers.

It is an enormous injustice that Payne-Gaposchkin was never elected to the US National Academy of Sciences, whose membership until the 1970s excluded other distinguished women scientists. During a week-long visit to Harvard during the 1970s, I asked the Chairman of the Harvard Astronomy Department if he would nominate her for membership. This was a brazen act, for someone not (and never expecting to be) a member. He declined, saying he was "really a geophysicist," and a member of the geophysics NAS section. Some time later, Dr. Leo Goldberg, a former Director of the Harvard Observatory and then Director of the Kitt Peak National Observatory, agreed to do so, and Payne-Gaposchkin was nominated for membership. Unfortunately, it was very late in her life, and it is not certain that she would have been elected; she died before the election took place.

IMPORTANT PUBLICATIONS

Books

Payne, C. H., *Stellar Atmospheres* (Cambridge, UK: W. Heffer and Sons, 1925). (This book is her Ph.D. thesis, submitted to Radcliffe College in 1925.)

The Stars of High Luminosity (New York: McGraw-Hill, 1930).

Payne-Gaposchkin, C., 1957. *The Galactic Novae* (New York: Interscience, 1957). (Reprinted in 1964, New York: Dover Publ. Inc.)

Stars and Clusters (Cambridge, MA: Harvard University Press, 1979).

The Dyer's Hand (1979). (Her privately printed autobiography. See Further reading.)

Research papers

Payne, C. H., On the spectra and temperatures of the B stars. *Nature*, **113** (1924) 783.

Why do galaxies have a spiral form? *Sci. Am.*, **189** (1953) 34.

Payne-Gaposchkin, C., Russell and the composition of stellar atmospheres. *Dudley Observatory Report*, **13** (1977) 15.

Fifty years of novae. *Astron. J.*, **82** (1977) 665. (This is her Henry Norris Russell Prize Lecture before the American Astronomical Society.)

The development of our knowledge of Variable Stars. In *Annual Review of Astronomy and Astrophysics 16*, ed. G. Burbidge (Palo Alto, CA: Annual Reviews Inc., 1978).

FURTHER READING

Haramundanis, K., *Cecilia Payne-Gaposchkin: An Autobiography and Other Recollections* (Cambridge, UK: Cambridge University Press). (This contains Payne-Gaposchkin's autobiography, as well as valuable introductions by her daughter, K. Haramundanis, Jesse Greenstein, an astronomer, and Peggy Kidwell, an historian of science.)

Smith, E. V. P., Obituary of Cecilia Payne-Gaposchkin. *Physics Today*, **33** (1980) 64.

15 Mary Lucy Cartwright (1900–1998)

Freeman J. Dyson
Institute for Advanced Study, Princeton

FIGURE 15.1 Photograph of Mary Lucy Cartwright, 1950. Courtesy of the Mistress and Fellows, Girton College, Cambridge.

SOME IMPORTANT CONTRIBUTIONS

Mary Cartwright was a great pure mathematician who devoted only a small fraction of her long and productive life to applied problems. She became an applied mathematician in 1938, when Britain was belatedly preparing to defend itself in the impending war against Hitler. The most important technical instrument for the defense of Britain was radar. Radar, as it turned out, played a decisive role in winning the Battle of Britain two years later. But in 1938, the deployment of radar was greatly hampered by the lack of reliable high-powered radio

Out of the Shadows: Contributions of Twentieth-Century Women to Physics,
eds. Nina Byers and Gary Williams. Published by Cambridge University Press.
© Cambridge University Press 2006.

amplifiers. Amplifiers worked well so long as they remained within the linear range, with the output signal proportional to the input. But at the high powers needed for an effective radar system, the output became nonlinear and the transmitted signals became garbled. Instead of putting out a steady signal with a single frequency, the transmitters were putting out signals with a mixture of several frequencies, or with no fixed frequency at all.

The Department of Scientific and Industrial Research, responsible for overseeing the development of radar, issued a call for help. The department appealed to pure mathematicians for "really expert guidance," to diagnose the problem and understand what was going wrong. Cartwright responded to the appeal. She worked on the problem, on and off, for seven years, from 1938–45, sometimes alone and sometimes in collaboration with the mathematician, John Littlewood. With characteristic modesty, she said later, "Although I myself have helped to develop the general theory and settle certain theoretical problems, I do not think that I have ever produced a result useful for any specific practical problem when it was needed." This disclaimer may well be true. The radio-engineers in 1938 could not afford to wait for a general theory. Cartwright wrote to me in a recent letter, "We made very slow progress and they [the engineers] found that trial and error was the quickest method." By trial and error, the engineers had the radars working reliably in 1940 when they were needed. Meanwhile, Cartwright was systematically exploring a new world of mathematical physics, the world that later became known as "chaos."

The standard equation describing the dynamics of a nonlinear amplifier is the Van der Pol equation. Balthasar Van der Pol was a Dutch electrical engineer who experimented with electrical circuits and wrote down his equation to describe them. He explored the solutions of his equation numerically, and found mysterious irregularities that he was unable to explain. Cartwright was able to make sense of these irregularities by looking at them from a new and more abstract point of view. She studied the work of Henri Poincaré, the French mathematician who had analyzed the nonlinear equations of motion of planets and satellites in the solar system and found similar irregularities there. To understand the intricacies of celestial mechanics, Poincaré invented a new branch of mathematics, which later became known as topology. The topological viewpoint gave him a simple abstract language to describe the complicated behavior of planetary orbits. Cartwright could speak Poincaré's language and used it to understand the sequence of periodic and aperiodic solutions of the Van der Pol equation. The aperiodic solutions had a disastrous effect on the radars, but had a beautifully intricate topological structure, which Cartwright and Littlewood continued to elucidate long after the radars had been fixed.

Littlewood had a famously difficult temperament, and working with him was not easy. Here is Cartwright's description of the way they worked:

> We communicate mainly by letter or postcard or in brief encounters out of doors, occasionally by long walks or telephone but hardly ever sitting indoors, seldom in his room, never in mine, never on a blackboard. Several important points were settled at accidental meetings out of doors. I remember drawing curves with my finger on the stone at the back of the Guildhall and meeting near Girton College when he was returning or starting one of his favourite walks from the bus at the corner.

In letters written to me in 1993, Cartwright (then 92 years old) reminisced more frankly about the difficulties of the collaboration:

> Littlewood was always changing his mind and very slow . . . He was dogged by bouts of depression from 1931 until about 1955 when Beresford-Davies cured him. Pills to counteract it, I think. I got fed up with his pernickerty alterations and he agreed that I could finish sundry things that he had left half finished . . . All that time, 1947 approx. to 1955, Littlewood fiddled about generalizing the functions in the equation in a not (to me) interesting

way . . . I think that this gives you some idea of how things came about. I kept on tempting Littlewood with steady state results and he sent me sketches of proofs. Yours ever, Mary.

Littlewood described the collaboration in more poetic language: "Two rats fell into a can of milk. After swimming for a time one of them realised his hopeless fate and drowned. The other persisted, and at last the milk was turned to butter and he could get out." Although Littlewood does not explicitly say so, we can be sure that the rat who did not drown was Cartwright.

The end result of the collaboration was a deep understanding of the intricacies of the solutions of the Van der Pol equation. In Littlewood's words: "We went on and on at the thing with no earthly prospect of results: suddenly the entire vista of the dramatic fine structure of solutions stared us in the face." The vista included the phenomena of period-doubling, bifurcation, abrupt transitions from periodic to aperiodic motion and basins of attraction of different periodic solutions separated by boundaries with a strange structure known to mathematicians as indecomposable continua. These are the phenomena that became known as "chaos" after they were rediscovered 20 years later by the meteorologist Edward Lorenz. Indecomposable continua seemed less formidable and easier to grasp when they were given the friendlier name, "strange attractors." The dramatic fine structure of solutions became more intelligible when it was displayed in brightly colored pictures and given the name "chaos."

A solution of a nonlinear equation is said to be chaotic if the effect of a small change of initial conditions increases exponentially with time. Chaotic solutions are unpredictable, because the initial conditions can never be known with zero error, and the effects of any finite error in the measurement of initial conditions will become large as time goes on. The equations that Lorenz studied were only a toy model of the equations of real meteorology, but his discovery of chaotic solutions made it immediately plausible that the solutions of equations describing a real atmosphere would be likewise unpredictable. The discovery that weather is in a precise mathematical sense unpredictable marked a turning-point in the history of science.

After the work of Lorenz, chaos became fashionable. Mathematicians such as Levinson, Smale, Yorke and Li worked out general theories of chaos, in which the earlier discoveries of Cartwright and Littlewood, and the new discoveries of Lorenz appeared as exemplary special cases. The work of Cartwright and Littlewood was published in 1945, the work of Lorenz in 1963. Why did the work of Lorenz precipitate an immediate public response while the work of Cartwright and Littlewood 20 years earlier did not? There were probably two main reasons for the delayed response. First, Cartwright and Littlewood had found chaotic behavior in radar transmitters that were hidden behind walls of military secrecy, while Lorenz found it in the weather that was familiar to everybody. Second, Cartwright and Littlewood were working with pencil and paper before the age of computers, while Lorenz had computers and graphical displays to make his chaotic solutions visible. Public interest in chaos grew explosively as a result of the artistry of Mandelbrot and others whose brilliant images of chaos graced the covers of magazines all over the world.

Before chaos became fashionable, Cartwright put the Van der Pol equation aside and continued her distinguished career as a pure mathematician, working mainly in analytic function theory. She wrote to me in 1993: "I heard that you were extolling me as one of the pioneers of work on chaos. I do not know what is meant by chaos. My nephew lent me a large book on chaos and there was no mathematics in it." Presumably the book that her nephew lent her was *Chaos: Making a New Science* by James Gleick, an excellent popular exposition of the luxuriant growth of ideas and images that grew out of her work. She remained a pure mathematician at heart, unimpressed by the revolution that she had helped to bring about. It often happens that scientists who lead the way to a scientific revolution are dissatisfied with the revolution when it comes. Examples of revolutionaries who became dissatisfied after the revolution are Planck, Einstein and Schrödinger. Cartwright belongs in their company.

The story of the Van der Pol equation is worth telling because it illustrates vividly the blindness of scientists to discoveries in unfashionable fields. I was myself as blind as everybody else to the importance of Cartwright's work. I had the good fortune to hear her lecture about it when I was a young student in 1942, before any of it had been published. I was delighted with the beauty of her results. She had disentangled what Littlewood called "the dramatic fine structure of solutions." I saw the beauty of her work, but I did not see its importance. I said to myself, "This is a lovely piece of work. Too bad it is only a practical wartime problem and not real science." I did not say, "This is the birth of a new field of science that I could spend the rest of my life exploring." In 1942, the study of chaotic behavior in classical dynamical systems was profoundly unfashionable. I missed the opportunity to be a pioneer in a new field, because I shared the tastes and prejudices of my contemporaries.

BIOGRAPHY

Mary Lucy Cartwright was born in 1900 in the village of Aynho in the English county of Northamptonshire. Her father was rector of the village church, and her mother was daughter of another local clergyman. In those days the clergy belonged to the educated aristocracy of the county. Cartwright's family tree contained many names distinguished in the history of the British army, church and parliament. One of her ancestors was John Theophilus Desaguiliers, a Huguenot who escaped from France, allegedly concealed in a barrel, in 1685. He afterwards wrote books on physics, invented the planetarium, and became curator of instruments and demonstrations at the Royal Society of London. Another ancestor was the poet John Donne.

Cartwright was educated by governesses at home until age 12, then went to local schools, and finally went to the Godolphin

School, one of England's elite boarding-schools for girls, at age 16. At Godolphin she had a remarkable self-taught mathematics teacher, Miss Hancock, who taught her calculus, analysis and analytic geometry. At 19 she went to Oxford and took a first-class degree in mathematics. At Oxford she attended evening classes run by G. H. Hardy, at that time the leading mathematician of England, and decided to devote her life to mathematics. The evening classes started after dinner on Mondays. They consisted of a talk followed by mathematical discussions that went on until late in the evening. These sessions set the course for her career. She picked up from Hardy her impeccable taste in choosing mathematical problems to work on, and her elegant style as a writer. After four years of teaching mathematics in schools, she had saved enough money to return to Oxford as a research student, with Hardy as her supervisor. Hardy encouraged her to work on the summability of divergent series. She quickly moved ahead from where Hardy had stopped, published two substantial papers and obtained an Oxford D.Phil. degree. In 1930, she moved to Cambridge with a research fellowship at Girton College, and for the rest of her life she was a professional mathematician.

At Cambridge during the 1930s, Cartwright published a series of important papers on analytic function theory, which established her reputation as a world-class analyst. Her most famous theorem says that under suitable conditions any analytic function $f(z)$ that is bounded for all integer z is also bounded for all real z. Then, in 1938, she turned her attention to the Van der Pol equation and established a second reputation as a world-class expert on nonlinear differential equations. During the 1950s and 1960s she continued to make discoveries in function-theory and wrote a book on entire functions. Through these years she was also heavily involved with teaching and administration. From 1936–49 she was Director of Studies for Mathematics, and from 1949–68 she was Mistress of Girton College, a position that in America would have been called President. She ran the college during a period of rapid expansion and reform, foreseeing the need for changes before they were demanded by the increasingly radical students of the 1960s. During her tenure, Girton College changed from an isolated and austere community on the outskirts of Cambridge to a full participant in the life of the university, sharing with the traditionally male colleges their uninhibited style of thinking and living. She said that she felt she had really arrived when she became the first female member of the *Septemviri* (Latin for seven men), an ancient judicial body empowered to adjudicate disputes between office-holders in the university.

In addition to her work as teacher and administrator at Cambridge, Cartwright led an active life outside. She became a Fellow of the Royal Society in 1947 and the first female member of the Royal Society Council in 1955. She became president of the London Mathematical Society in 1961, and became Dame Mary Cartwright (the female equivalent of a knighthood) in 1969. She traveled to Russia in 1956 to establish closer relations between the Royal Society and the Soviet Academy of Sciences. She traveled extensively in Europe, Africa and America, both as a Royal Society delegate and as a private citizen. After she retired as Mistress of Girton College in 1968, she spent a year at Brown University in Providence, and was given an honorary degree, walking proudly in the academic procession with Duke Ellington at her side.

Cartwright grew increasingly frail after she passed the age of 90, but never lost her brains or her character. At the age of 96 she took part in a television documentary, "Our Brilliant Careers," that captured the sharp sparkle of her wit. A year later she died, full of years and honors, recognized as a first-rate mathematician and as one of the ablest women of the twentieth century. Among the many achievements of her long career, the one that received the least attention was perhaps the greatest, her exploration of the new field of science that later became known as chaos.

In her obituary of Cartwright in the British newspaper, *The Guardian*, Caroline Series concludes:

I was once present at a heated debate among some very eminent
mathematicians about which women would have been worth a
chair at one of the top U.S. mathematics departments. Emmy
Noether? Yes. Sonya Kovalevskaya? Perhaps. The only other name
on the table was Cartwright's. Generations of other women have
been inspired by Mary Cartwright's achievements.

IMPORTANT PUBLICATIONS

Cartwright, M. L., On certain integral functions of order one. *Q. J. Math.*, **2**:7 (1936) 46–55.
(This contains the proof of her theorem mentioned in the text.)

Cartwright, M. L. and Littlewood, J. E., On non-linear differential equations of the second order,
I. *J. Lond. Math. Soc.*, **20** (1945) 180–9. (This is the classic paper describing the pathology
of solutions of the Van der Pol equation.)

Cartwright, M. L., Integral Functions. *Cambridge Tracts in Mathematics and Mathematical
Physics*, **44** (1956). (A monograph summarizing the discoveries of Cartwright and others
concerning the behavior of entire functions [functions of a complex variable analytic over
the entire complex plane].)

From non-linear oscillations to topological dynamics, Presidential address to the London
Mathematical Society, 1963. *J. Lond. Math. Soc.*, **39** (1964) 193–201. (A survey for non-
experts of the topological viewpoint that emerged from Cartwright's study of the Van der
Pol equation.)

FURTHER READING

Hayman, W. K., *Biographical Memoirs of Fellows of the Royal Society*, Vol. 46 (2000) 19–
35. (Most of the information in this article is derived from the excellent memoir of Mary
Cartwright written by her student Walter Hayman, himself a distinguished mathematician.)

McMurran, S. L. and Tattersall, J. J., The mathematical collaboration of M. L. Cartwright and
J. E. Littlewood. *Am. Math. Mthly*, **103** (1996) 833–45. (A good technical account of the
Cartwright–Littlewood collaboration.)

Gleick, J., *Chaos: Making a New Science* (New York: Viking Penguin, Inc. 1987). (A good
non-technical account of the history of chaos. This account begins with the discoveries of
Edward Lorenz in the 1960s and barely mentions the earlier discoveries of Cartwright and
Littlewood. Apart from this lapse, the book is historically accurate and well written. As
Cartwright observed in her letter to me in 1993, it contains no mathematics.)

16 Bertha Swirles Jeffreys (1903–1999)

Ruth M. Williams

Girton College and Department of Applied Mathematics and Theoretical Physics, Cambridge

FIGURE 16.1 Photograph of Bertha Swirles Jeffreys.

SOME IMPORTANT CONTRIBUTIONS

A substantial part of the research of Bertha Swirles Jeffreys was done in the early days of quantum mechanics when she was Bertha Swirles, and the new ideas were being applied to calculations in atomic physics. In fact, her first paper on the polarizability of atomic cores, a problem given to her by Douglas Hartree, was pre-quantum mechanics. Soon after that while still a research student, she became interested in the photoelectric effect and set about trying to explain observed data on electron emission from atoms when gamma ray

Out of the Shadows: Contributions of Twentieth-Century Women to Physics,
eds. Nina Byers and Gary Williams. Published by Cambridge University Press.
© Cambridge University Press 2006.

FIGURE 16.2 Photograph of Bertha Swirles Jeffreys, 1903–1999, with Harold Jeffreys.

photons emitted from the nucleus are absorbed by electrons in the atom (internal conversion of gamma rays). Following R. H. Fowler's suggestion she modeled the process by placing a dipole at the centre of the atom and including quantum mechanical effects. A weakness of the work was that it did not treat the problem relativistically, but it led to calculations by H. R. Hulme, and by Nevill Mott and Harold Taylor, in which this was put right.

When Swirles moved to Manchester as an Assistant Lecturer after completing her Ph.D., E. A. Milne was Head of Department there. Under his influence, as she put it, she "strayed" into astrophysics and worked on the absorption of radiation by a gas, both by free electrons and by electrons bound to nuclei. S. Chandrasekhar was also investigating this and their results were published independently. Swirles worked first on a highly degenerate gas and then a partially degenerate gas. Her results are very important in models of stellar structure when the central core of a star is a degenerate gas and there is necessarily a transition zone where the gas is only partially degenerate.[1]

Returning to her earlier area of research, Swirles then followed up a suggestion made by Douglas Hartree as they chatted on Euston station while attending a conference in London in 1934. (The women at the conference appeared in the press as the "atom women"!) She extended Hartree's self-consistent field method to the relativistic case, which was clearly a very important step in the study of electron interactions.[2] (Her calculations were reproduced years later using modern facilities, by Ian Grant in Oxford.) The work was especially significant as it was the very first formulation of the Dirac–Fock method for atoms, and it was used subsequently in the first numerical Dirac–Fock calculation on copper (Williams, A. O., 1940, *Phys. Rev.* **58**, 723). She then improved the results on two electrons by taking into account both the interaction of the spins and the retardation, applying this to the example of helium. The accuracy of the results was limited by the approximate hydrogenic relativistic wave functions used, which were too simple. She tried another approach in joint work with Hartree, where "superpositions" or linear combinations of wave functions, a typical feature of quantum mechanics, were used.[3] Although the wave functions were non-relativistic, this was very important work as it was one of the first *ab initio* configuration interaction calculations. Configuration interactions had been used earlier (Bacher, R. F., 1933, *Phys. Rev.* **43**, 264) but this calculation used hydrogenic wave functions derived using Slater's screening rules. Finally, in a classic paper with Douglas Hartree and his father, who did the numerical work using a Brunsviga hand calculating machine, both exchange interactions and superposition were used to obtain approximate wave functions for various ions of oxygen.[4] This was the very first introduction to, and application of, this important method, using two configurations for the oxygen atom. Swirles's main contribution to this paper was the extensive algebraic calculations, an aspect of research that she particularly enjoyed. When she described her research, Swirles was always careful to give credit to others who had suggested ideas to her or stimulated her to think in particular directions. In her modesty, she often gave the impression that she had just worked out other people's ideas, but her contributions were undoubtedly much more than that.

Box 16.1   Multi-electron wave functions

In an atom with just one electron, it is relatively simple to solve the Schrödinger equation to find the electron's wave function, the square of the modulus of which gives the probability distribution for its position. However, most atoms have more than one electron so it is important to be able to solve the problem in the more general case. If the electrons had no interaction with each other, then the multi-electron wave function would just be the product of single-electron ones, but not only do the electrons repel each other electrostatically as they have like charges, but there are also quantum mechanical restrictions, as we shall see.

The self-consistent field method

One way of finding the wave function for an electron while taking the effect of the charges of the other electrons into account, is called the *self-consistent field method*. The idea amounts to imagining the total charge due to the other electrons to be spread out in a spherically symmetric negatively charged cloud, the effect of which can then be calculated from the classical potential. This method was first developed by Hartree and is described very clearly in his book *The Calculation of Atomic Structures* (New York: Wiley, 1957).

As a first approximation to the wave function for an atom with n electrons, one can take the separable form:

$$\Phi = \psi_{\alpha 1} \psi_{\beta 2} \cdots \psi_{\eta n}$$

where the Greek letters label the one-electron wave functions and the numbers label the electrons. Each wave function ψ_α could be determined from the Schrödinger equation for the electron in the field of the nucleus and of the total average charge distribution of the other electrons. Thus, the field of the average electron distribution derived from these wave functions $\psi_\alpha, \ldots, \psi_\eta$ is the same as the field used in evaluating them, hence the name "self-consistent field."

In practice, the wave functions are calculated using the variational principle, by finding stationary values of the expression:

$$E = \int \Phi^* \, H \, \Phi \, dV \bigg/ \int \Phi^* \, \Phi \, dV$$

evaluated using the approximate wave function Φ, with the Hamiltonian H given by a sum of the kinetic terms, the potential energy of the electrons in the field of the nucleus and the mutual potential energy of pairs of electrons. Variation leads to a set of Schrödinger-type equations, called Fock equations, for each ψ, which can be solved numerically by successive approximation.

The theory described so far is a non-relativistic one, appropriate to low-energy situations. The next step is to extend this to the relativistic case, in which a single electron is described by the Dirac equation and has an intrinsic spin of $\frac{1}{2}$. The relativistic self-consistent field method also takes into account the interaction between the spins of the electrons, and the retardation, which enters through an exponential factor involving the energy difference between each pair of electrons and the distance between them.

Exchange forces

Electrons, having spin $\frac{1}{2}$ are fermions, so the overall wave function describing a collection of them must be antisymmetric under the exchange of any pair. An elegant way of achieving this is to use, instead of the simple product wave function above, a determinant wave function, with rows made up of the different possible assignments of the electrons to the available wave functions. Such a wave function gives rise to integral terms in the Fock equations for the individual wave functions, coming from the mutual potential energy terms. These are called *exchange* terms because they can be thought of as coming from the "exchange" of electrons between the different one-electron wave functions. The exchange terms make the equations much more difficult to solve, but they are clearly important for accurate results.

Superposition

In quantum mechanics, the most general state of a system can be written as a superposition of eigenstates of some appropriate operator. This leads to the idea that another way of improving wave functions calculated from the self-consistent field method is to consider a linear combination of the wave function for the state under consideration, and that for an excited state with the same parity, orbital angular momentum and spin, introducing the concept of *configuration interaction*. The variational method can then be used either to find the coefficients in the linear combination if the wave functions are given, or to find both coefficients and wave functions.

The self-consistent field method with the various refinements described here has led to tabulations of wave functions for many multi-electron atoms. These tables and the method itself are still used today by atomic physicists and theoretical chemists.

The next major paper was published in 1942 when she was already Bertha Jeffreys, having married Harold Jeffreys in 1940. This was the study of the transmission of waves through a potential barrier, done in the context of quantum mechanics for solutions of the Schrödinger equation, but applicable more widely.[5] She had corrected George Gamow's English in his book *Atomic Nuclei and Radioactivity* in 1930 but had not noticed then that his mathematical treatment also needed some correction; this she then did. The transparency or transmission coefficient was checked for the well-known case of a rectangular barrier, and approximations were found for triangular and trapezoidal barriers. In these last two cases, the exact solutions involve Airy functions, and these solutions were derived in a paper published in 1956, where numerical values were tabulated for the transmission coefficient.

Between the two papers just mentioned, Jeffreys produced, in collaboration with Harold, what has turned out to be her most influential contribution, the book *Methods of Mathematical Physics*, first published in 1946 and reprinted many times.[6] (A reissue of the third edition has appeared recently in the Cambridge Mathematics Classics series.) This is an extremely comprehensive account of the techniques that form the essential mathematical equipment of a theoretical physicist or applied mathematician, including complex analysis, differential equations and special functions. It is written in a very clear and scholarly way, with appropriate and amusing quotations taken from sources ranging from Ibsen to Lewis Carroll, at the start of each chapter. Affectionately known as *"J&J,"* it has instructed generations of students, and it is still a recommended textbook for courses on mathematical methods in universities all over the world. When Jeffreys wrote recently to her Member of Parliament about the inadequacy of the extra 25p of pension for the over 80s, her letter was passed on to the Minister for Social Security, whose reply to her mentioned that he had very much enjoyed using *"J&J"* as a student!

Work on *"J&J"* was a major preoccupation for a number of years, and this, combined with her teaching, meant that Jeffreys had less time for research. Remembering that Harold Taylor had asked her at the Royal Society in 1932 how much she knew about multipoles, she decided much later to find out about them and the result was a detailed treatment of multipole radiation, expanding on what was outlined in *"J&J."*[7] She calculated the exact form of the radiation from a charge-current distribution confined to a small region and vibrating with a single frequency, and the approximation for large distances was derived. Although these calculations were classical, they were in a form suitable for the calculation of quantum mechanical transition probabilities if the multipole moment were replaced by the corresponding quantum theoretical matrix element. Jeffreys said that she enjoyed this work most. When she was in her mid-80s, she revisited the topic in a short paper on axisymmetric multipoles.

In a book on quantum theory edited by D. R. Bates and published in 1961, Jeffreys gave a very clear description of an important use of the asymptotic approximation method (or AA method, as she wished to rename the technique often known as the WKB method, but more accurately the JWKB method, to acknowledge the contribution of Harold Jeffreys).[8] The problem considered was one in quantum mechanics, the solution of the Schrödinger equation in a potential that was not piecewise constant. This includes general potential well and potential barrier problems, and also the solution of the radial equation for a hydrogen-like atom. By looking at cases where the exact solution was known, she was able to demonstrate the validity and power of the AA method.

It seemed inevitable that Jeffreys should be drawn, at least to some extent, into Sir Harold's research interests. In 1965, at his instigation, she wrote a paper clarifying the transformation of spherical harmonics under rotation between reference frames, which is important in the theory of satellites. This was partly to answer Sydney Goldstein's criticism that the treatment in "J&J" was unintelligible. In the 1970s, she worked with Sir Harold on editing his collected papers; she was jointly responsible for Volumes 3–6 (Volumes 1 and 2 were on seismology). These were very important and influential pieces of work, for Sir Harold was a pioneer not only in geophysics and seismology but also in hydrodynamics, celestial mechanics and probability. (His 1939 book *The Theory of Probability* has just been republished by OUP.) In 1983, while he was busy on other matters, she wrote a short paper updating a section of his book *The Earth*, first published in 1924, using more recent and reliable data on the motion of the Earth and the Moon.

Typical of Jeffreys's sense of humour and her untiring desire to understand the origins of scientific knowledge, down to the matter of notation, were two papers she wrote in her eighties, "A Q-rious tale . . ." about the origins of the parameter Q used in electromagnetism.

Summary

With publications spanning seven decades, Bertha Swirles Jeffreys certainly left her mark on twentieth-century physics. The book *Methods of Mathematical Physics* has been a central part of the education of generations of mathematicians and physicists. Her papers on atomic physics in the early days of quantum mechanics were pioneering and exciting, and stimulated many further avenues of research. Some are still regarded as world classics.

BIOGRAPHY

The life of Bertha Swirles Jeffreys spanned almost the whole of the twentieth century. Bertha Swirles was born on May 22, 1903, in Northampton, a county town in the English East Midlands. Her father, who was a leather salesman, died while she was still a toddler. She was involved in the world of education from her earliest days; her mother and seven of her nine aunts were all teachers. From the primary school where her mother taught and an aunt was headmistress, she went at the age of 12 to the brand new Northampton School for Girls, where she was taught mathematics by three Cambridge women graduates. A major excitement was being invited to hear J. J. Thompson when he came to a Speech Day at the neighboring boys' school. As a girl, she read *Scientific American*, to which her mother subscribed for her, and in 1997 she wrote "As I was brought up almost entirely by women, it did not occur to me that there was anything strange in wanting to become a mathematician or physicist."

In 1921, Swirles went as a Major Scholar to Girton College, Cambridge, to read mathematics, graduating with first class honours in 1924. She was encouraged to become a physicist by a research student of E. V. Appleton, Mary Taylor, with whom she studied James Jeans's *Electricity and Magnetism*, and she spent the next academic year on the Part II Physics course in the Natural Sciences Tripos, attending lectures by J. J. Thompson and Ernest Rutherford, amongst others. She did not particularly enjoy the practical work; she was the only woman in the class and she thought that perhaps her shyness made her somewhat clumsy, resulting in some rather disastrous experiments.

Swirles's imagination had been caught by what she had read about the Bohr theory of spectra, and having found the financial support she needed to be a research student, she was sent to see Rutherford, who said "They tell me that you are not much good at experiment; you had better go and see Fowler." In 1925, she did indeed become a research student of Fowler, joining a distinguished company of students, which included Nobel Prize winners such as P. A. M. Dirac and Chandrasekhar. Her first research problem actually came from D. R. Hartree, himself still a research student. Fowler sent her to see him at his private house, and when she was shown into a room with two pianos (Hartree's upright and a grand piano belonging to his wife who was a professional musician), Swirles thought to herself "This is the place for me!" This was the beginning of a lifelong friendship with the Hartrees.

During the next two years in Cambridge, Swirles attended Dirac's lectures on "Quantum Theory (Recent Developments)," which described his and Werner Heisenberg's amazing pioneering work. She encountered many eminent visitors from abroad, but it was difficult to meet and talk informally with other students and more senior physicists. The societies that met in the evenings, the $\nabla^2 V$ and Kapitza Clubs, were not open to her as a woman.

With the encouragement of Mary Taylor, who was an important role model for her, and of Fowler, Swirles spent the winter semester of 1927–8 in Göttingen where she studied under Max Born and Heisenberg, and interacted with many of the other leading continental workers in the very exciting new field of quantum mechanics. She returned to Cambridge to finish her thesis, and by the time she was awarded her Ph.D. in 1929, she was already an Assistant Lecturer in Manchester. This was followed by similar appointments at Bristol and Imperial College, London. While in London, she attended meetings of the discussion group on atomic physics at the Royal Society, and was delighted when Dirac turned round to her after an election of members and told her that she had been made an official member. She was the only woman. In 1933, she returned to Manchester as a Lecturer in Applied Mathematics. D. R. Hartree at Manchester was extremely sorry to lose "such a valued colleague" when she returned to Cambridge in 1938, to an Official Fellowship and Lectureship in Mathematics at Girton.

For more than 30 years, until her retirement in 1969, Swirles taught mathematics (which included theoretical physics) at Girton, and also took part in university lectures and examining. Between 1938 and 1948, she shared the College teaching with Mary Cartwright (see Chapter 15), who became Mistress in 1949; she then took over from her as Director of Studies in Mathematics at Girton. This involved selecting, teaching and advising generations of women mathematicians (Girton did not admit men until the 1970s) who have themselves gone on to propagate her influence to ever-widening circles.

In 1940, Swirles married Harold Jeffreys, a distinguished Cambridge mathematician and geophysicist, whom she had known for some time. (To quote Rosa Goldstein, Sydney's wife: "It took a major war to get them together!") Harold was knighted in 1953 for his services to science, and Jeffreys became Lady Jeffreys (or Lady J. as she was affectionately known). Theirs was a very long and happy marriage; Sir Harold died in 1989 when he was almost 98.

After her retirement, Jeffreys continued to be involved in Girton as a Life Fellow. She kept in touch with large numbers of her ex-pupils, who, with their own families, constituted an enormous extended family for her. She maintained a very active correspondence with colleagues and friends all over the world, often sorting out details of scientific history, which she remembered better than most did! For example, when she heard about Michael Frayn's play "Copenhagen," which is based on the historic meeting between Bohr and Heisenberg in 1941, she was dubious about whether a playwright could possibly have it right, so she wrote to Hans Bethe in Cornell, who was very enthusiastic about the play. Thus reassured, she wrote to Frayn, enclosing a copy of Bethe's letter, and received a delighted reply.

Jeffreys was never a narrow specialist. She was a skilful linguist and had a wide knowledge of literature. Music was a very important part of her life; she was an accomplished pianist and cellist, and still played piano duets with a friend in her early nineties. She directed studies in music at Girton between 1939 and 1947. In all aspects of her life, she always set herself the highest standards and expected others to do the same. Her advice was never stereotyped: she approached each problem with an open mind and an enormous amount of common sense.

In recognition of her leading role in education in the twentieth century, Jeffreys was awarded honorary doctorates by the University of Saskatchewan in 1995 and the Open University in 1996. Although the people who have benefited first-hand from her teaching have been mainly women, because Girton was a women's college during her teaching days, her influence has spread through them worldwide to both men and women, as she wished. She said quite emphatically that she would prefer to appear in a book entitled "Contributions of *People* to Twentieth Century Physics," reflecting her misgivings about classifying people specifically as "*women* physicists" or "*women* mathematicians." However, her lack of sympathy for scientific activities arranged for women never stopped her from being extremely supportive of individuals.

After a few months of failing health, Jeffreys died at home (as she wished) on December 18, 1999, at the age of 96.

NOTES

The author thanks Professor N. C. Handy, F. R. S., Dr. N. Pyper and, most of all, Lady Jeffreys, for help with the preparation of this chapter.

1. Swirles, B., The coefficients of absorption and opacity for a partially degenerate gas. *Proc. R. Soc. A*, **141** (1933) 554–66.
2. Swirles, B., The relativistic self-consistent field. *Proc. R. Soc. A*, **152** (1935) 625–49.
3. Hartree, D. R. and Swirles, B., The effect of configuration interaction on the low terms in the spectra of oxygen. *Proc. Camb. Philos. Soc.*, **33** (1937) 240–9.
4. Hartree, D. R., Hartree, W. and Swirles, B., Self-consistent field, including exchange and superposition of configurations, with some results for oxygen. *Philos. Trans. R. Soc. A*, **238** (1939) 229–47.

5. Jeffreys, B., Note on the transparency of a potential barrier. *Proc. Camb. Philos. Soc.*, **38** (1942) 401–5.
6. Jeffreys, H. and Jeffreys, B. S., *Methods of Mathematical Physics* (Cambridge, UK: Cambridge University Press, 1946).
7. Jeffreys, B., The classification of multipole radiation. *Proc. Camb. Philos. Soc.*, **48** (1952) 470–81.
8. Jeffreys, B. S., The asymptotic approximation (AA) method. In *Quantum Theory*, Vol. I, ed. D. R. Bates (Academic Press, 1961).

17 Kathleen Yardley Lonsdale (1903–1971)

Judith Milledge
Department of Earth Sciences, University College London

FIGURE 17.1 Photograph of Kathleen Yardley Lonsdale.

SOME IMPORTANT CONTRIBUTIONS

Kathleen Yardley began to study mathematics at Bedford College, but changed over to physics at the end of her first year, partly because she enjoyed experimental physics. However, the first of her major contributions was mathematical and theoretical, done while she was waiting for experimental components to be made in a workshop. This work was a major contribution to crystallography,[1] which she made

Out of the Shadows: Contributions of Twentieth-Century Women to Physics,
eds. Nina Byers and Gary Williams. Published by Cambridge University Press.
© Cambridge University Press 2006.

with W. T. Astbury after she joined W. H. Bragg at University College London. Lonsdale (then Yardley) and Astbury related geometrical crystallography and the theory of space groups to the diffraction patterns obtainable from crystals using H. H. Hilton's book. (See Box 17.1.)

Box 17.1   Effect of crystal symmetry on diffraction patterns

Three-dimensional repeating patterns (crystals) are characterized by the combination of two types of symmetry elements, point symmetry elements and translational symmetry elements. The latter result in systematic absences among certain types of reflections in X-ray diffraction patterns, and indicate which of the allowed 230 symmetry combinations, or space groups, are possible for the crystal. The symmetry elements give the result that a number of positions (x, y, z) in the unit cell are equivalent. They determine the relative phases of all the structure factors in the form {hkl}. (See Box 17.2).

Astbury and Yardley produced tabulated data and devised diagrams illustrating the equivalent positions for all the 230 space groups. This was published in 1924. Kathleen also provided the formulae for the calculation of structure factors (see Box 17.2), which were included in the *International Tables for the Determination of Crystal Structures* (1935) and later (1936) published in a facsimile edition by the Royal Institution (RI). These tables were also published in Volume I of the *International Tables for X-ray Crystallography* (1952), of which she was General Editor, and are basic to all structure determination.

In 1923, Kathleen Yardley went to the RI with W. H. Bragg, where she determined the structures of a number of simple organic and inorganic compounds. In 1927, she married and moved to Leeds with her husband Thomas Jackson Lonsdale.

KathleenYardley Lonsdale determined the structure of hexamethyl benzene in Leeds. This was the first structure of an aromatic compound to be determined by X-ray diffraction.[2] She considered it to be her most important scientific contribution because it established

Box 17.2   Structure functions and crystal structure determination

Crystal structures are determined from the observed intensities of X-ray reflections $I(hkl)$. These are used to obtain structure factors $F(hkl)$, which are then used to compute a three-dimensional Fourier series that shows the distribution of electron density (and hence the positions of the atoms) in the unit cell of the crystal. In order to do this, the relative phases of the structure factors must be known. This information cannot be obtained experimentally except by utilizing anomalous dispersion for some atoms in the structure, which also enables Absolute Configuration to be determined. This is possible because although in general $I(hkl) = I(-h, -k, -l)$, i.e., the reflection intensity is the same for reflection from the back or the front of any set of planes (hkl), resonance effects, which occur when the X-ray wavelength lies close to an absorption edge, result in differences between $F(hkl)$ and $F(-h, -k, -l)$. When differences are large enough to be measured they can be used to establish the absolute configuration of optically active molecules, and to help to determine the relative phases of structure factors in any crystal.

When this cannot be done, two other approaches can be used. The first, the Patterson Synthesis, uses the values of $|F(hkl)|^2$ (for which the phases are all positive) in a Fourier series, which shows vectors between the atoms all displaced to the origin, and Patterson Harker sections, which utilize translational symmetry elements to simplify the interpretation of the resulting maps.

The second approach involves Intensity Statistics if the values of $F(hkl)/F_{max}(hkl) = e(hkl)$, i.e., value of a structure factor as a fraction of its maximum possible value, can be obtained. It can be shown that there are definite relationships between the phases of structure factors with high values of $e(hkl)$, and statistically significant relationships between those with moderately high values. Hence, the relative phases of many of the strongest $F(hkl)$ values can be obtained after assigning arbitrary phases to three $F(hkl)$ terms (which must be correctly chosen). This leads to approximate electron density maps in

> which the structure can be recognized. The simplest example of such relationships for a centrosymmetric crystal is that if e(hkl) = 1, then F(2h, 2k, 2l) is positive.

the structure of the benzene ring. This important achievement was followed by the determination of the structure of hexachlorobenzene, the first Fourier analysis of an organic crystal, and one used by Patterson in the development of the Patterson Synthesis (see Box 17.2) in 1935.

When she returned to the R.I. in 1934, Kathleen was offered the use of a large electromagnet, as all available X-ray apparatus was then in use. She was working in Faraday's old room, and reading his notebooks, as well as papers by K. S. Krishnan on the diamagnetic anisotropy (Box 17.3) of aromatic molecules and crystals. She was able to adapt the electromagnet to make measurements on single crystals by fitting it with a torsion head from which a crystal could be suspended on a thin glass fibre between the poles of the electromagnet. In collaboration with Krishnan, she produced many new results that aroused the interest of a number of theoreticians, including Linus Pauling and Fritz London.

Box 17.3 Magnetic anisotropy of crystals

The measured molecular diamagnetic susceptibility of aromatic crystals is much greater normal to the plane of the molecule than parallel to it. This can be explained as being due to the orbits of the electrons, in particular to the relatively large radii of pi-electrons in the plane of the molecule. This diamagnetic anisotropy can be measured by suspending a crystal in a magnetic field and measuring the torsion required to deflect it from its preferred orientation from suspension about different crystallographic axes.

Measured diamagnetic anisotropy provided the first direct experimental confirmation of the existence of molecular orbitals, and when combined with a knowledge of the crystal structure led to improved calculations of molecular susceptibility. This in turn allowed molecular orientations in unknown structures to be inferred from measurements of diamagnetic anisotropy. Calculations are also possible for other types of organic molecules.

Optical anisotropy presents a problem for crystallographers because the absolute configuration of optically active molecules cannot be determined from diffraction patterns unless anomalous dispersion (see Box 17.2) can be utilized. In 1938, Kathleen became fascinated by the diffuse reflections seen on Laue patterns of benzil, a crystal with very high optical anisotropy being studied at the RI by Ellie Knaggs, for whom she made diamagnetic anisotropy measurements. Diffuse scattering (see Box 17.4), first observed for KCl in 1913, and by 1935 attributed to thermal motion by Faxen and Waller, had recently been studied by Laval in France before he became a prisoner of war. Kathleen persuaded H. A. Jahn, a theoretician then working at the RI, to determine the shape of diffuse reflections from cubic crystals to be expected on the Faxen–Waller theory, and these were found to agree quantitatively with Laval's data. She then began an intensive experimental investigation of diffuse scattering in many types of crystals, involving, amongst other things, making measurements at different temperatures. Such studies were to remain a major interest for the rest of her life, and developed into work on thermal expansion, order–disorder phenomena and solid-state reactions.

Box 17.4 Diffuse scattering

The appearance of sharp reflections on X-ray photographs of a stationary crystal (Laue photographs) is a consequence of the existence of a large array of identical unit cells – the "perfect" crystal. Anything that reduces the dimensions of this array in one, two or three directions results in corresponding extensions of reflecting power in reciprocal space, i.e., in the observed diffraction pattern. There are two kinds of deviations from perfection that can affect the diffraction pattern. The first is crystal texture: i.e., when the neighboring crystallites are not strictly continuous or completely parallel. The second is static or dynamic disorder on a molecular scale. The frequencies of atomic vibrations are much lower than those of the incident X-rays, and consequently the atoms appear randomly displaced about their mean positions (dynamic disorder). This gives rise to relatively weak diffuse scattering in the vicinity of reflections produced by characteristic radiation. In some crystals the atoms or molecules are not strictly ordered (static disorder), and this also results in diffuse scattering. It is not possible to distinguish between static and dynamic disorder by making measurements at one temperature only. The strength and geometry of the diffuse scattering can be related to the elastic constants of the crystal at the temperature used. Its intensity increases as the melting point is approached, and it is thus usually much stronger for organic than for inorganic crystals measured at RTP.

Among the diffuse phenomena Kathleen was studying lurked another that stayed with her for the rest of her life: surrounding the rather weak diffuse spot observed near the 111 reflection on Laue photographs of some (but not all) diamonds was a curious triplet of sharper spots, which were shown to be due to extensions of reflecting power ("horns") along the cube axes. Such an effect could not be related to crystallite size, because all three horns did not occur for other reflections, and was eventually shown to be related to small precipitates . . . thin plates normal to the cube axes. Their precise structure remains controversial to this day, but they aroused Kathleen's interest in diamond, and she used a demountable X-ray tube at the RI to obtain divergent-beam photographs of several diamonds and of a number of other crystals. Such photographs consist of a number of black (reflection) and white (absorption) conics whose strength and sharpness are related to crystal texture (see Box 17.4) and whose positions are related to unit cell parameters. She showed that very accurate cell parameters (errors $\sim 1 \times 10^{-6}$) and ratios of X-ray wavelengths could be obtained from such patterns, and that cell parameters for individual diamonds differed by small but detectable amounts.

In 1945, Kathleen had become one of the first two women to be elected Fellows of the Royal Society. In 1946, she left the RI to become Reader to establish crystallography in the Department of Chemistry of University College London (UCL), brought the electromagnet and other pieces of equipment with her and established a research group there. In 1949, she became the first woman professor at UCL. She and her research group continued to develop the themes of her earlier work – diffuse scattering, thermal expansion, diamagnetic anisotropy and studies on diamond. However, the International Union of Crystallography was formed soon after the end of World War II, and from 1946 she was involved in the preparation of the new edition of the International Tables, of which she was Editor-in-Chief, and did little further experimental work herself, although she interpreted the photographs taken for her by Eric Nave, her experimental officer.

Her expertise with Laue photographs, essential for studies of diffuse scattering, enabled her to realize in 1954 that absolute configuration could be determined from Laue photographs by positioning selected reflections (see Box 17.2), so that anomalous dispersion could be utilized, a procedure now superseded by appropriate choice of synchrotron wavelengths. Her predilection for systematizing space-group information led to the derivation of rules for choosing the three independent reflections necessary to develop phase relationships using intensity statistics (see Box 17.2).[3]

Various studies – of disorder and short-range order in organic solid solutions; of photochemical reactions; of spontaneous solid-state reactions (failed explosives because the reactions were too slow, but fast enough for systematic X-ray studies); of the composition of human stones; of n-methomium compounds of pharmacological interest – were undertaken, but these did not introduce new physical methods or equipment of the same innovative importance as much of her previous research. She was, however, involved in the early work on synthetic diamonds, and the hexagonal form of diamond found in Canyon

Diablo and in other meteorites was named lonsdaleite in recognition of her contributions to diamond research.

BIOGRAPHY

Kathleen Yardley was born in Newbridge, Southern Ireland, on January 28, 1903, the youngest of ten children of Jessie Cameron and Harry Frederick Yardley, who was then the local postmaster. The family was poor, and four of her brothers died in infancy. Her mother brought the children to England in 1908 and settled in Essex. Here Kathleen had a very successful school career, winning scholarships to the County High School for Girls, and attending classes (unavailable in the girls' school) in physics, chemistry and higher mathematics at the County High School for Boys . . . the only girl to do so. She refused the chance to remain at school to try for Cambridge and, aged 16, entered Bedford College for Women to study mathematics, but changed to physics at the end of her first year, partly because she did not want a career in teaching, the only option likely to be available. She was warned that she had little chance of distinguishing herself in physics, but in fact she headed the University list in the honours B.Sc. examination in 1922, for which W. H. Bragg was one of the examiners, and gladly accepted his invitation to join his research team at University College London (UCL). In 1923, Bragg took her with him to the RI, where he had a particularly successful research school. She gained M.Sc. (1924) and D.Sc. (1936) degrees from UCL.

Kathleen had been brought up by her mother, a strict fundamentalist Baptist who had persuaded the children to become teetotal (her father had a drinking problem, possibly due to unrecognized diabetes), but she found fundamentalism increasingly unacceptable, and with a fellow physics student, Thomas Jackson Lonsdale, whom she married in 1927, attended various other churches before finally joining the Society of Friends (Quakers) in Leeds. The Lonsdales had three children, Jane (1929), Nancy (1931) and, after they had returned to London, Stephen (1934). In 1931, W. H. Bragg obtained funding for her to return to the RI, where she remained until 1946, when she went to University College London, where she remained until she retired in 1968. Here she developed a 30-hour practical course in crystal structure analysis, which was copied in many colleges throughout the world.

She was deeply involved in Quaker activities, becoming a pacifist, and during World War II spent a month in Holloway women's prison for refusing to register for firewatching, although she was actually doing so, and in any case was exempt as a mother of young children. This was one of the most formative experiences of her life, and led to a lifelong interest in prison conditions; she became a member of the Board of Visitors at Aylesbury Prison for Women, and later at Bulwood Hall Borstal Institution.

She traveled widely both for scientific purposes and for the Society of Friends, wrote pamphlets and gave public lectures, was a member of the Atomic Scientists Association and involved in the formation of the Pugwash movement, was a member of the Quaker delegation that visited Russia in 1951 and President of the British section of the Women's International League for Peace and Freedom. She was also the first woman President of the British Association for the Advancement of Science, and of the International Union of Crystallography in 1966, and was a Vice-President of the Royal Society at the time of the tercentenary celebrations in 1960.

LIST OF HONORS

Fellow of The Royal Society (London) (1945). (One of the first two women elected.)

Dame Commander of the British Empire (1956).

Davy Medal of The Royal Society (1957).

Honorary Doctorates:

Universities of Wales, Leicester, Manchester, Lancaster, Leeds, Dundee, Oxford and Bath.

Hexagonal diamond found in some meteorites named lonsdaleite.

Professor Emeritus and Fellow of University College London.

NOTES

1. Astbury, W. T. and Yardley, K.,Tabulated data for the examination of the 230 space-groups by homogeneous X-rays. *Philos. Trans. R. Soc. A,* **224** (1924) 221.

2. Lonsdale, K., The Structure of the benzene ring in hexamethylbenzene. *Proc. Roy. Soc. A*, **123** (1929) 494.
3. Grenville-Wells, H. J. and Lonsdale, K., Determination of absolute configuration by Laue photographs. *Nature*, **173** (1954) 490.

IMPORTANT PUBLICATIONS

Lonsdale, K., An X-ray analysis of the structure of hexachlorobenzene, using the Fourier method. *Proc. Roy. Soc. A*, **133** (1931) 552.

Magnetic anisotropy and electron structure of aromatic molecules. *Proc. R. Soc. A*, **159** (1937) 149.

Lonsdale, K. and Smith, H., An experimental study of diffuse X-ray reflection by single crystals. *Proc. Roy. Soc. A*, **179** (1941) 8.

Lonsdale, K., Extra reflexions from the two types of diamond. *Proc. R. Soc. A*, **179** (1941) 375.

Divergent beam X-ray photography of crystals. *Philos. Trans. R. Soc. A*, **240** (1947) 219.

Lonsdale, K., Milledge, H. J. and Nave, E., X-ray studies of synthetic diamonds. *Mineralog. Mag.*, **32** (1959) 185.

Lonsdale, K., Experimental studies of atomic vibrations in crystals and of their relationship to thermal expansion. *Z. Kristallogr.*, **112** (1959) 188.

Crystal structure analysis for undergraduates. *J. Chem. Educ.*, **41** (1964) 240.

Lonsdale, K., Nave, E. and Stephens, J. F., X-ray studies of a single crystal chemical reaction: photo-oxide of anthracene to (anthraquinone, anthrone). *Philos. Trans. R. Soc. A*, **261** (1966) 1.

FURTHER READING

Hodgkin, D. M. C., Kathleen Lonsdale. *Biographical Memoirs of Fellows of the Royal Society*, Vol. 21 (1975).

Milledge, H. J., Kathleen Lonsdale. *Acta Cryst. A* **31** (1975) 705–8.

Lonsdale, K., X-ray diffraction & its impact on physics. In *Fifty years of X-Ray Diffraction*, ed. P. P. Ewald (Netherlands: International Union of Crystallography, 1962) pp. 221–49.

Early work at University College London. In *Fifty years of X-Ray Diffraction*, ed. P. P. Ewald (Netherlands: International Union of Crystallography, 1962) pp. 408–10.

Crystallography at the Royal Institution. In *Fifty years of X-Ray Diffraction*, ed. P. P. Ewald (Netherlands: International Union of Crystallography, 1962) pp. 410–20.

Personal Reminiscences: K. Lonsdale. In *Fifty years of X-Ray Diffraction*, ed. P. P. Ewald (Netherlands: International Union of Crystallography, 1962) pp. 595–602.

(This book is worth reading for an overview of the development of crystallography.)

"Kathleen Lonsdale" by H. Judith Milledge. *Dictionary of National Biography*.

"Dame Kathleen Lonsdale". In Series Women Scientists. *Shell Times* issue No. 69 (1989).

Crystals and X-rays (London: G. Bell & Sons, 1949). (This book is based on a course of public lectures given at University College London in 1946 and was "designed to interest those who do not now use X-ray crystallography, but who might well do so; and to instruct those who do use X-ray crystallographic methods without altogether understanding the tool that has been put into their hands." It is an excellent introduction for those who want an essentially nonmathematical text.)

18 Maria Goeppert Mayer (1906–1972)

Steven A. Moszkowski

Department of Physics and Astronomy, University of California, Los Angeles

FIGURE 18.1 Photograph of Maria Goeppert Mayer. Courtesy of Peter C. Mayer. An archive of Maria Goeppert Mayer papers may be found in the Mandeville Special Collection Library at the University of California at San Diego.

SOME IMPORTANT CONTRIBUTIONS

The nuclear shell model

The discovery and application of the nuclear shell model was one of the most important developments in nuclear physics. For this Maria Goeppert Mayer received the Nobel Prize in 1963, together with Hans Jensen who had made a similar discovery independently.[1] Maria Goeppert Mayer was the first woman to win the Nobel Prize for theoretical physics.

Out of the Shadows: Contributions of Twentieth-Century Women to Physics,
eds. Nina Byers and Gary Williams. Published by Cambridge University Press.
© Cambridge University Press 2006.

Her contribution to the nuclear shell model can be roughly divided into three parts:

(1) discovery of the magic numbers,
(2) explanation of the magic numbers,
(3) nuclear pairing.

We discuss here her role in the discovery and explanation of the magic numbers.

Discovery and explanation of the magic numbers

It was known since the early days of quantum mechanics that shell structure in atoms is very important. In the 1920s, quantum mechanics, including the important Pauli principle, gave an explanation of the closed shells in atoms. The derivation of the closed shell numbers (i.e., the magic numbers) in atoms and in nuclei is summarized in Box 18.1. There had been speculations concerning the possibility of shell structure in nuclei in the 1930s (which constituted the early days of modern nuclear physics) dating from the discovery of the neutron in 1932.[2] However, by 1936 it had become accepted by the majority of nuclear physicists that shell structure could not be valid in nuclei. The experimental evidence of the time, especially the discovery of closely spaced resonances and of nuclear fission, pointed in the direction of a liquid drop model where the particles are closely coupled rather than a shell model. Mayer's work on the Manhattan Project gave her insight into the properties of fission products, such as decay energies and half-lives. After 1945, she began to look at systematics of nuclear binding energies, and she discovered that they clearly indicated the presence of closed shells.[3] Closed shells for nuclei with 2, 8, 20, 28, 50, 82 or 126 neutrons (or protons) clearly emerged from her work on nuclear binding energies (more details can be found in Box 18.2). Note that all these numbers are even. It had been known before that nuclei for which both N and Z are even are slightly more stable than those where one or both of these numbers are odd. This is a manifestation of the pairing effect. Pairs of neutrons and also pairs of protons like to couple to total spin 0, with a little extra binding. (See Box 18.3 for

Box 18.1 Comparison of atomic and nuclear shell models

Atoms	Nuclei
(1) Electrons move nearly independently in a common potential due to the nucleus and the other electrons – a screened Coulomb potential.	Nucleons move nearly independently in a common potential due to the other nucleons – a well of a flat bottom and sloping sides.
(2) There are large variations in electron binding energies.	There are only small variations in nucleon binding energies, except for the lightest nuclei.
(3) The electronic spin-orbit coupling is weak.	The nuclear spin-orbit coupling is strong.
(4) Closed shell configurations that have exceptional stability occur for $Z = 2, 10, 18, 36, 54$ and 86.	Closed shell configurations of slightly greater than average stability occur for N or $Z = 2, 8, 20, 28, 50, 82$ and 126 (known as magic numbers).
(5) Interactions (long ranged, repulsive) favor states of minimum spatial symmetry.	Interactions (short ranged, attractive) favor states of maximum spatial symmetry.
(6) Ground states of atoms tend to have the maximum possible spin: $S =$ half the number of particles outside of (or missing from) closed shells.	Ground states of even-even nuclei have angular momentum $J = 0$. Each pair of identical nucleons coupling to $J = 0$ has special stability (the pairing effect). Ground states of most odd-A nuclei have $J = j$ of the odd unpaired nucleon.

Adapted from Table 2 of SAM, *Handbuch der Physik*, 1957, page 467.

more details.) The existence of closed shells goes beyond the even–odd binding energy differences. Thus, nuclei with 82 neutrons are not only more stable than those with 81 or 83, but they are also slightly more stable than those with 80 or 84. The establishment of the particular

---

Box 18.2 Magic numbers in atoms and nuclei

Atoms with 2, 10, 18, 36, 54 and 86 electrons form the inert gases, which are chemically the least active of the elements. This is because the electrons form closed shells, which leads to maximum binding. It is a simple exercise to show how we get these closed shell numbers. First, consider electrons in a pure Coulomb potential filling in accordance with the Pauli principle. We can put 2 electrons into the lowest (1s) state. The next 8 electrons go into the $N = 2$ shell, which has 2s and 2p levels at the same energy, for a total of 10. We can put $2^*(2L + 1)$ electrons into a shell of given L. Thus, the next shell, namely $N = 3$, could accommodate 18 electrons, 2 in 3s, 6 in 3p and 10 in 3d. This would give a total of 28. On the other hand, the next closed shell actually occurs instead for $Z = 18$. To account for this, we have to take into account the repulsive Coulomb interaction between the electrons, which screens off part of the potential due to the nucleus. Due to this electron screening, the potential felt by each electron is not a pure Coulomb potential. In this way the empirical closed shell numbers in atoms can be accounted for.

The case of closed shells in nuclei has similarities to those of atoms, though the potential is different. For a pure harmonic oscillator well, we find the levels grouped into subshells, similar to the case of the Coulomb potential, though with different magic numbers. A shell of given j can accommodate $2j + 1$ neutrons or protons. For a pure oscillator we have subshells with j ranging from 1/2 up to $N + 1/2$ in steps of 1. The total number of neutrons (protons) in each shell N is $(N + 1)(N + 2)$. Filling shells from 0 to N then gives the closed shell numbers 2, 8, 20, 40, 70, 112, 168. The first three numbers, 2, 8 and 20, correspond to the first three magic numbers in nuclei. However, to account for the higher magic numbers we have to take into account the deviation from the oscillator. There are two major effects here: first, the fact that the well is flat at the bottom, rather than an oscillator there, leads to the levels of large L being lowest within a shell. Second, the strong spin-orbit coupling lowers the levels with $j = L + 1/2$. For one of the levels, the one with maximum L and j, the combined effect of well flattening and spin-orbit interaction is enough to bring this level into the next lower shell, at least for $L > 3$. For example, the 1g 9/2 state, which has $N = 4$, $L = 4$, $j = 9/2$, can accommodate $2j + 1 = 10$ identical particles. This state is lowered by both well flattening and spin-orbit interaction, enough to bring it into the next lower $(N = 3)$ shell, increasing the closed shell number from 40 to 50. This leads directly to the famous nuclear magic numbers.

Box 18.3 Nuclear pairing

Mayer's next important contribution to the nuclear shell model dealt with nuclear pairing.[16] Such pairing occurs in nuclei but not in atoms. For most nuclei, all pairs of identical particles couple to angular momentum 0. (Indeed, every nucleus with even numbers of neutrons and of protons has ground state angular momentum 0.) Thus, in an odd-A nucleus, the angular momentum is just that of the last odd particle. This rule, which works well for most odd-A nuclei, was a major factor in the tremendous success of the nuclear shell model. Mayer investigated pairing with a very simple model of the interactions between nucleons, involving an attraction of zero range. The author recalls an occasion, probably in 1950, when he was present in Mayer's office. She was quite excited about her unexpectedly simple results for the particular case of three neutrons outside a closed shell. In retrospect, it appears that Mayer's very simple interaction was more realistic than might have been suspected at the time. There are indications that the zero range interaction is a consequence of subnucleonic structure. Yet, even after half a century, the problem of the interaction in nuclei is still to be resolved.

values of numbers indicating closed shells was the first major contribution of Mayer to the nuclear shell model. Even the skeptics were convinced by overwhelming evidence for the closed shells in nuclei.[4]

The next step was to explain why one got the particular values 2, 8, 20, 28, 50, 82 and 126 for the closed shell numbers. The lowest values, 2, 8 and 20, can be obtained by putting nucleons into a reasonable single particle potential, like a harmonic oscillator. However, the higher numbers were something of a challenge to explain. Several groups worked on this problem in 1948,[5] but Mayer got the answer.[6] The answer is that, in addition to the central potential, there is also a strong spin-orbit coupling. Mayer discovered that when a strong spin-orbit coupling term is added to the nuclear potential, the sequence 28, 50, etc. for closed shells emerges naturally.[7] Mayer and Jensen collaborated on a book on the nuclear shell model, which was published in 1955. Eight years later, they shared the Nobel Prize for their work.

Other important contributions

While Mayer is best known for her Nobel Prize work on the nuclear shell model, she made numerous other important contributions to physics. In fact, she was a productive and accomplished physicist from the very beginning of her career. We briefly summarize here some of her most important contributions not involving the discovery and explanation of the magic numbers.

Mayer's Ph.D. thesis was the first theoretical study of the phenomenon of double quantum emission.[8] If an electron is in an excited atomic level, it can generally decay to a lower level, for example, the ground state, by emission of a photon. However, if both initial and final states have zero angular momentum, such single photon emission cannot occur. There is one important case in nature, the hydrogen atom, where this happens. However, double quantum emission can take place. This process, which Mayer investigated in her thesis, was discovered a few years later, in planetary nebula, and it plays an important role in lasers today. Pursuing a similar approach, a few years later, Mayer was the first to propose the possibility of double beta decay.[9] Since the half-life for this process is huge

($>10^{20}$ years), it only plays a role in a few special nuclei that are otherwise stable.

While at Johns Hopkins University, Mayer collaborated with physical chemists there, and collaborated on several papers. We mention here two of them. Mayer and Herzfeld were the first to study the effect of magnetic susceptibility on the refractive index of a gas.[10] Also, the first calculations of the spectrum of an important organic molecule, benzene, were done by Mayer and Sklar.[11] Both of these contributions, and others, involved the application of the new field of quantum mechanics to problems in physical chemistry.

In 1935, Yukawa had proposed the idea of meson exchange as an explanation of the nuclear forces. Three years later Mayer and Sachs pioneered the application of this idea, the Yukawa potential between neutron and proton, to the calculation of the deuteron ground state.[12] This was Mayer's first work on nuclear physics.

At about the same time, Maria and Joe Mayer (her husband, a very distinguished physical chemist in his own right who had made important contributions to statistical mechanics of many-body systems) wrote a textbook, *Statistical Mechanics*.[13] This book turned out to be a classic, and was the standard textbook in the subject for many years.

The discovery of transuranic elements in 1940 stimulated Maria Goeppert Mayer to study them theoretically, using the atomic independent particle model of electrons.[14] Mayer was the first person to work out the atomic properties of these transuranic elements, which have interesting similarities to the rare earths. This was at the beginning of her work on the Manhattan project. Later, after the end of the war and the discovery of the nuclear shell model, she applied this shell model to various problems. For example, she co-authored an important paper with Edward Teller on the origin of elements.[15]

BIOGRAPHY AND DESCRIPTION OF SCIENTIFIC LIFE
Of all the women scientists discussed in this book, Maria Goeppert Mayer is certainly among the most well known. She was born in

Katowitz, Silesia, in 1906, and moved to Göttingen at age four. Her father was a professor of pediatrics at the University there. He himself was a sixth generation academic. Maria was studying in Göttingen during the exciting time that modern quantum mechanics was developed and applied to various problems in atomic and molecular physics. She completed her doctoral work in 1930. In that same year, she married Joe Mayer who was at the University on a fellowship. Maria and Joe Mayer came to the United States and went to Baltimore, where Joe had obtained a faculty appointment at Johns Hopkins University. However, Maria did not get a regular faculty appointment there. Nepotism rules in effect at that time did not allow the wife of a faculty member also to be appointed. The same consequences of nepotism rules later repeated themselves at Columbia University and at the University of Chicago.

While working at Johns Hopkins University as Volunteer Associate from 1931 to 1939, Maria Goeppert Mayer pioneered on a number of projects mentioned earlier. At Columbia University she was a Lecturer from 1939 to 1946 after her husband got a position there. She also worked as a part-time teacher at Sarah Lawrence College from 1941 to 1942 and again in 1945. During World War II (from 1942 to 1945), Mayer worked on the Manhattan Project, first at Columbia, and then at Los Alamos. Her work involved the analysis of fission products. This research paved the way for her later work on the nuclear shell model. Although her work on the Manhattan Project involved supervision of a group of people, again she did not have a regular paid position. That came only in 1946, when she joined the newly founded Argonne National Laboratory as Senior Physicist (1946 to 1960) and the Institute for Nuclear Studies at the University of Chicago. During the time of her Nobel Prize winning research work at the University of Chicago, her title (1946–59) was Volunteer Professor. In this period, she was elected as a member of the Heidelberg Academy of Sciences (1950) and the National Academy of Sciences (1956).

Throughout the years, in the various positions Mayer held, whatever her title may have been, she had fruitful contacts and collaborations with other physicists, which resulted in her many important joint papers.

In 1960, both Maria and Joe Mayer were offered regular Professorships at the University of California at San Diego, which they accepted. However, very shortly afterwards, Maria suffered a severe stroke. A little later, there was the memorable occasion of her receiving the Nobel Prize in Physics (with J. H. D. Jensen) in 1963 "For their discoveries concerning nuclear shell structure." Unfortunately, Maria Goeppert Mayer never fully recovered from the stroke, and she died in 1972. Yet her memory lives on. For example, a building at the University of California, San Diego, was dedicated in her name: Mayer Hall. The American Physical Society awards an annual prize in her honor, the Maria Goeppert Mayer Award, to an outstanding woman physicist. And the nuclear shell model is one of the cornerstones of modern nuclear phyics.

NOTES

1. The role of Jensen and his collaborators in their discovery of magic numbers is a fascinating story in itself. Their work was published at almost the same time as Mayer's. See O. Haxel, J. H. D. Jensen and H. E. Suess, *Phys. Rev.*, **75** (1949) 1766 and *Z. Physik*, **128** (1950) 295. Otto Haxel played a big role in this work, and some people feel that he should have shared in the Nobel Prize.

2. See, for example, a series of papers by W. Elsasser in *J. Phys. Radium*, **4** (1933) 549; **5** (1934) 389, 635; **6** (1935) 473; and by K. Guggenheimer, *J. Phys. Radium*, **5** (1934) 23, in which evidence for shell closures, for example at $N = 82$, is given.

3. Mayer was only moderately familiar with nuclear physics at the time she did her work on the Manhattan Project. She had co-authored a paper with R. G. Sachs on the neutron–proton potential, but had not worked on heavier nuclei.

4. The name *magic numbers* may have been coined by Eugene Wigner (who later shared the Nobel Prize with Mayer and Jensen for pioneering work in nuclear theory), as he was somewhat skeptical about the shell model, until the evidence for it became convincing. The name stuck, however, and, indeed, it may be quite appropriate as an indication of Mayer's great intuitive powers.

5. E. Feenberg and K. C. Hammack, *Phys. Rev.*, **75** (1949) 1877. L. Nordheim, *Phys. Rev.*, **75** (1949) 1894, trying clever ways to modify the potential so that the closed shells would appear for 28, 50, 82 and 126, but these efforts were not really successful.
6. In a footnote to her famous paper, On closed shells in nuclei II, *Phys. Rev.*, **75** (1949) 1969, Mayer credits Enrico Fermi for his remark: "Is there any indication of spin-orbit coupling?"
7. The explanation of closed shells using the spin-orbit coupling also came, independently, to Hans Jensen, then a professor at Heidelberg University. The possible role of spin-orbit coupling, by analogy to atoms, had already been considered in the 1930s. However, the nuclear spin-orbit coupling turned out to be much larger than had been expected by the analogy to atoms, and also of the opposite sign. Probably that is the reason it had not been discovered earlier.
8. Goeppert-Mayer, M., Elementary processes with two-quantum transitions. *Ann. d. Physik*, **9** (1931) 273.
9. Goeppert-Mayer, M., Double beta-disintegration, *Phys. Rev.*, **48** (1935) 512.
10. Goeppert-Mayer, M. and Herzfeld. K. F., On the theory of dispersion, *Phys. Rev.*, **49** (1936) 332.
11. Goeppert-Mayer, M. and Sklar, A. L., Calculations of the lower excited levels of benzene. *J. Chem. Phys*, **6** (1938) 645.
12. Goeppert-Mayer, M. and Sachs, R. G., Calculations on a new neutron-proton potential. *Phys. Rev.*, **53** (1938) 991.
13. Mayer, J. and Goeppert-Mayer, M., *Statistical Mechanics* (New York: John Wiley and Sons, 1940).
14. Goeppert-Mayer, M., Rare earths and transuranic elements. *Phys. Rev.*, **60** (1941) 184.
15. Goeppert-Mayer, M. and Teller, E., On the origin of the elements. *Phys. Rev.*, **76** (1949) 1226.
16. Goeppert-Mayer, M., Nuclear configurations in the spin-orbit coupling model. II Theoretical considerations. *Phys. Rev.*, **78** (1950) 22.

IMPORTANT PUBLICATIONS

Goeppert-Mayer, M., On closed shells in nuclei. *Phys. Rev.*, **74** (1948) 235.

On closed shells in nuclei II, *Phys. Rev.*, **75** (1949) 1969.

Nuclear configurations in the spin–orbit coupling model. I. Empirical evidence. *Phys. Rev.*, **78** (1950) 16.

Nuclear configurations in the spin-orbit coupling model II Theoretical considerations. *Phys. Rev.*, **78** (1950) 22.

Goeppert-Mayer, M., Moszkowski, S. A. and Nordheim, L. W., Nuclear shell structure and beta decay. *Rev. Mod. Phys.*, **23** (1951) 315.

Goeppert-Mayer, M. and Jensen, J. H. D., *Elementary Theory of Nuclear Shell Structure* (New York: Wiley and Sons, 1955).

Goeppert-Mayer, M., The shell model. In *Les Prix Nobel en 1963* (Stockholm: the Nobel Foundation, Nobel Lecture, 1964). Also in *Science*, **145** (1964) 999.

Double beta-disintegration. *Phys. Rev.*, **48** (1935) 512.

Elementary processes with two-quantum transitions. *Ann. d. Physik*, **9** (1931) 273.

FURTHER READING

Dash, J., *A Life of One's Own* (New York: Harper Row, 1973).

Sachs, R. G., Maria Goeppert Mayer – two-fold pioneer. *Physics Today*, February 1982, 2–7.

Gabor, A., *Einstein's Wife* (New York: Viking, 1995).

McGrayne, S. B., *Nobel Prize Women in Science*, 2nd edn (Secaucus, NJ: Carol Publishing Group, 1998).

Johnson, K. E., Science at the breakfast table, *Physics in Perspective*, Vol 1 (1999) pp. 22–34.

19 Helen Dick Megaw (1907–2002)

A. Michael Glazer
Department of Physics, University of Oxford

Christine Kelsey
Department of Earth Sciences, University of Cambridge

FIGURE 19.1 Photograph of Helen Dick Megaw. Supplied by author.

SOME IMPORTANT CONTRIBUTIONS

When Helen Megaw started research in 1930, the techniques used to study crystals were in the early stages of development. She designed the X-ray tube that she used in her first project, an investigation of the thermal expansion of some minerals using small single crystals. Expansion coefficients are very small, of the order of 5 parts in 10 000, and the high accuracy that she achieved reflected her experimental skills. In a citation for an honorary degree given to Megaw by Queen's

Out of the Shadows: Contributions of Twentieth-Century Women to Physics,
eds. Nina Byers and Gary Williams. Published by Cambridge University Press.
© Cambridge University Press 2006.

College, Belfast, in 2000, Professor Ruth Lynden-Bell recalled those early days of crystallography research:

> It is difficult for us to imagine the scientific environment in the nineteen thirties. It was a time of depression with little money and few jobs. Science departments were much smaller and more intimate. X-ray crystallography was a new science which attracted a number of young women such as Dr. Megaw who became distinguished scientists. In those pioneering days preparation of crystals and collection of data was more difficult and more skilled than it is today. Another big difference, which today's graduates may find hard to imagine, was that there were no computers and the tedious and detailed calculations which lead from the brightness of spots on a photographic plate to a three dimensional crystal structure were all done by hand. Dr. Megaw was one of the pioneers in this field.

Her Ph.D. work with J. D. Bernal was a study of the crystal structures of ice. Helen's accurate and demanding work on ice and heavy ice showed that the hydrogen atoms were involved in bonding between two oxygens. She and Bernal also surveyed the known structures of hydroxides and of water, and concluded that there were two types of hydrogen bond. In one type, found in ice, the hydrogen oscillates between two positions, each being closer to one of the oxygen atoms than the other. In the other type, the hydrogen is bonded more strongly to one of the two oxygen atoms. In honor of her discoveries of the nature of ice, an island off the Falklands was named Helen Megaw Island, in 1962.

In the mid 1950s in the Cavendish Laboratory, Cambridge, Helen Megaw became involved in the study of the feldspar group of minerals, the commonest minerals found in the Earth's crust. Feldspars are alumino-silicates with the general formula AT_4O_8, where A is most commonly calcium, sodium or potassium and T represents aluminium and silicon present in the correct proportions to balance the charges on the A ions and the oxygens. Thus, albite has the formula $Na(Si_3Al)O_8$, anorthite $Ca(Si_2Al_2)O_8$ and orthoclase $K(Si_3Al)O_8$. The plagioclase feldspars form a chemical series whose formulae may be written $Na_{1+y}Ca_y(Si_{3-y}Al_{1+y})O_8$ or more simply An_yAb_{1-y}, where Ab is albite and An is anorthite. Plagioclase compositions appear homogeneous when examined using an optical microscope; there is no evidence of separation into domains of albite and orthoclase. However, X-ray diffraction patterns in the composition ranges $An_{30}Ab_{70}$ to $An_{70}Ab_{30}$ show diffracted beams, called "non-Bragg" reflections, that cannot occur in perfectly ordered crystals. Megaw, in the three papers published in the Proceedings of the Royal Society in 1960, showed that the arrangement of "building blocks" in these crystals was disordered. She used diffraction theory to relate the positions and intensities of the non-Bragg reflections to the probability of the occurrence of fault planes in the structure and to the orientation of these planes within the structure. Knowledge of the detailed structural state of a plagioclase specimen is important because it allows the temperature of crystallization and the subsequent thermal history of the rock in which it occurs to be deduced.

Megaw's interest in perovskites began when she investigated the crystal structure of barium titanate, $BaTiO_3$, during her time with Philips. Barium titanate was known to have a perovskite-type ABO_3 structure. The ideal perovskite structure is very simple. The B atoms lie at the corners of a three-dimensional stack of cubes, the oxygens lie midway along the edges and the A atoms lie at the centers of the cubes. The ideal structure is uncommon because if it is to be stable the ratio of the lengths of the A–O to the B–O bonds must equal $\sqrt{2}$:1. The adjustments that occur when this condition is not held are of three kinds. If the A atoms are too small the framework "tilts," if the B atoms are too small they "off-center" (move away from the center of the octahedra). Lastly, the nature of the B–O bonds may demand a change in shape of the octahedra. The elucidation of the precise structure of a perovskite is not an easy task. Using X-ray powder diffraction data, Megaw showed that the structure was stretched by one percent along one of its cube edges while at 200 °C all the cube edges were equal. At room temperature the titanium atoms are displaced from the center of their octahedra and the compound is ferroelectric. Ferroelectricity is the electric analog of ferromagnetism and plays an important role in the electronics industry, because of the high dielectric constants of the materials, and the elongation that can be induced by an applied electric field. Barium titanate remains a most important material, as it is the archetypal ferroelectric compound, with many technological uses. Megaw was the first person to elucidate the crystal structure of this compound, and this is regarded as landmark work. She was aware of the importance of an understanding of the structural basis of the phenomenon, and her book *Ferroelectricity in Crystals* was the first to integrate this structural approach with the physical and chemical aspects of the subject. Her later detailed studies of the structures and phase transitions in $KNbO_3$, $NaNbO_3$, $LiNbO_3$ and Ca_2NbO_7, including the change in atomic vibrations near a transition temperature, have contributed much to the understanding of the structural basis of ferroelectricity.

Towards the end of her research career, Megaw worked closely with A. M. Glazer, initially appointed as her last postdoctoral assistant. Together they investigated and largely solved the structural phase transition behavior in the perovskite $NaNbO_3$, which has seven different phases. Helen realized that the clue to solving this hitherto poorly understood system was in the way the oxygen octahedra were tilted. This then led Glazer to formulate a general description of tilted octahedra in perovskites, which has now become the standard description for tilted perovskites. As Glazer looked back on his early Cambridge experiences (*Crystallography News*, June 1997) he noted:

> I soon discovered that my boss was a remarkable person:
> formidable in some ways, but also very kind and patient. She had
> a particularly interesting gift: if you wanted to know what a
> particular crystal structure looked like from a particular direction,

she could somehow turn it around in her mind and then sketch it for you. In the days before computer graphics, this was a very useful trick.

Megaw believed that "the combination of research and teaching is vital for the progress of knowledge." She regarded crystallography as a necessary working tool for those studying the properties of materials and her approach to teaching was essentially practical. The introduction of crystallography as a possible subject in the third year Natural Sciences course at Cambridge gave her the opportunity to teach the more specialist aspects of the subject to final-year students as well as to graduate students. Her book *Crystal Structures: a Working Approach* broke new ground and it has remained a valuable text for students for over 25 years.

BIOGRAPHY

Helen Megaw was born in Dublin, Ireland, where she attended the Alexandra School from 1916 until her family moved to Belfast in 1921 just before the partition of Ireland. After a brief period at the Methodist College, Belfast, she went to Roedean School, Brighton, England, from 1922 until 1925. One of her aunts was secretary to the Mistress of Girton College, Cambridge (one of the only two colleges for women), and Helen's ambition was to study there. She won an exhibition to the College in 1925 but for financial reasons decided to go to Queen's University, Belfast. The next year she won a scholarship and this time it proved possible for her take up a place. Originally she had intended to read Mathematics but she had enjoyed Chemistry at school and on her teacher's advice she opted for Natural Sciences so that she could study both Science and Mathematics. She thought that the regulations required her to study three subjects and she planned to study Chemistry, Physics and Mathematics. However, her Director of Studies, Miss M. B. Thomas, explained that she was required to study three *experimental* subjects (Mathematics being an optional extra) and she advised Megaw to choose Mineralogy as her third experimental subject. Had Megaw known that she could have chosen Geology instead of Mineralogy she would have opted for Geology and, in all probability, she would not have become a crystallographer! She achieved a Class I in Part I of the Natural Sciences Tripos in 1928. She then specialized in Physics, obtaining a Class II in Part II in 1930. When Professor Ernest Rutherford at the Cavendish Laboratory told her that there was no opportunity for her to do postgraduate work in the Physics department, Miss Thomas suggested that she approach Professor Arthur Hutchinson whose Department of Mineralogy had a strong crystallographic tradition. So it was that she became a research student under Dr. J. D. Bernal, investigating the thermal expansion of crystals, and the atomic structure of ice and the mineral hydrargillite (an hydroxide of aluminium).

Opportunities to continue research were few at the time that Megaw completed her Ph.D. in 1934, but the award by Girton College of a Hertha Ayrton Research Fellowship (see Chapter 1 of this book on Ayrton) enabled her to spend a year at the Chemisches Institut, Vienna, with Professor H. Mark followed by a year (1935–6) working with Professor Simon on the density and compressibility of solid hydrogen and deuterium at the Clarendon Laboratory, Oxford. She then taught Physics for seven years, initially at Bedford Girls' High School (1936–9) and then at Bradford Girls' Grammar School (1939–43). She was able to continue research during the school holidays, returning to Cambridge to work in a private laboratory on the crystallography of diamonds in relation to their use as wire-drawing dies. Then in 1943 she returned to full-time research as an X-ray crystallographer at Philips Mitcham Works Research Laboratory. There she studied barium titanate, $BaTiO_3$, a new ceramic capacitor from America, which had been sent to Philips by mistake. This triggered a lifelong interest in perovskites, the large group of compounds with a general formula ABO_3.

J. D. Bernal was appointed Professor of Physics at Birkbeck College in 1938. During the war his research group was dispersed and the building in which it was housed was destroyed by bombs. After the war he set up several research groups including one in the field of inorganic chemistry formed to investigate the structure of Portland cement and concrete. Megaw joined this group in the year 1945 and then was offered the opportunity of returning to Girton College as Fellow, Director of Studies and Lecturer in Physics. Initially, she felt that she could not desert Bernal so soon after her appointment but he advised her to accept the offer because it involved teaching as well as research. On her return to Cambridge she joined the Crystallography Laboratory of the Cavendish Laboratory where she worked under W. H. Taylor who was in charge of the Crystallography Laboratory. She was successively appointed Assistant in Research (1946), Assistant Director of Research (1949) and finally University Lecturer (1959). She played an active part in all aspects of life in the College and the Department and her ability to disagree strongly with her colleagues and yet to remain friends was greatly valued.

Megaw's interests were many. During her time at Philips and at Birkbeck College she was active in the Association of Scientific Workers. A keen photographer, she exhibited in the London Salon of Photography in 1943. In her younger days she was an indefatigable traveler and a keen winter sportswoman. Perhaps her most innovative project was the production of designs for textiles, carpets and other consumer goods based on projections of actual crystal structures and electron density maps. These designs were for the Pattern Group of the 1951 Festival of Britain that aimed to unite science with well-designed goods (*Architectural Review*, **94**: 236 (1951)).

She became a Life Fellow of Girton in 1970 on her retirement from her College commitments, and she retired from her University Lectureship in 1972. For some years after her retirement, she divided her time between Cambridge (she had a desk in the Department of Mineralogy and Petrology) and her retirement home in Ballycastle, Northern Ireland, where she had a garden full of interesting plants. She regarded crystallography as a necessary working tool for those studying the properties of materials and her approach to teaching was essentially practical. Her interest in teaching extended beyond Cambridge.

She was a member of the International Union of Crystallography Teaching Commission for nine years and its secretary for three of those years, during which time she edited a valuable crystallographic book list. She died in Northern Ireland in February, 2002.

HONORS

1962   Place names in Crystal Sound, Falkland Islands Dependencies, honor scientists who investigated the structure of ice crystals. An island at 66°55'S., 67°36'W was named Megaw Island in recognition of her accurate measurement of the cell dimensions of ice (*Ice*, **9**, 10–18 (1962)).

1967   Sc.D., University of Cambridge.

1989   Roebling Medal of the Mineralogical Society of America. (The first award to a woman.)

2000   Honorary Sc.D. in Crystallography from Queen's University, Belfast.

IMPORTANT PUBLICATIONS

Megaw, H. D., Cell dimensions of ordinary and 'heavy' ice. *Nature*, **134** (1934) 900.

Megaw, H. D. and Bernal, J. D., The function of hydrogen in intermolecular forces. *Proc. R. Soc. Lond. Ser-*A, **151** (1935) 384.

Megaw, H. D., Crystal structure of barium titanate. *Nature*, **155** (1945) 484.

Origin of ferroelectricity in barium titanate and other perovskite-type crystals. *Acta Crystallogr.*, **5** (1952) 739.

Ferroelectricity in Crystals (London: Methuen, 1957).

Order and disorder. I. Theory of stacking faults and diffraction maxima. *Proc. R. Soc. Lond. Ser-*A, **259** (1960) 59.

Order and disorder. II. Theory of diffraction effects in the intermediate feldspars. *Proc. R. Soc. Lon. Ser-*A, **259** (1960) 159.

Order and disorder. III. The structure of the intermediate plagioclase feldspars. *Proc. Roy. Soc. Lond. Ser-*A, **259** (1960) 184.

Crystallographic Book List (Utrecht: International Union of Crystallography Commission, 1965). (And 2 supplements, 1966; 1972.)

Crystal Structures: A Working Approach (Philadelphia: W. B. Saunders Co., 1973).

Megaw, H. D. and Glazer, A. M., Studies of the lattice parameters and domains in the phase transitions of NaNbO$_3$. *Acta Crystallogr.* A, **29** (1973) 489.

FURTHER READING

Megaw, H. D., The domain of crystallography. In *Historical Atlas of Crystallography*, ed. J. Lima de Faria (Dordrecht: Kluwer Academic Publishers, 1990). ("This is a survey explaining the relationship of Crystallography to Physics and to Chemistry – not of great interest in itself, but perhaps useful to physicists looking back at historical developments" Helen D. Megaw.)

20 Yvette Cauchois (1908–1999)

Christiane Bonnelle
Laboratoire de Chimie Physique, Matière et Rayonnement, Université Pierre et Marie Curie (Paris VI)

FIGURE 20.1 Photograph of Yvette Cauchois.

IMPORTANT CONTRIBUTIONS

Yvette Cauchois made outstanding contributions to the development of X-ray spectroscopy and X-ray optics. Her entire scientific career was dedicated to fundamental research of the interactions between matter and radiation that involved the production, absorption or reflection of X-rays. For 40 years she was at the forefront of most of the important developments that took place in this field. Her studies made possible fundamental advances in various areas of atomic physics and the physical chemistry of solids.

One of Cauchois's major contributions was the development of a methodology that made it possible to obtain a strong monochromatic convergent X-ray beam from an unfocused incident beam. Since she

Out of the Shadows: Contributions of Twentieth-Century Women to Physics,
eds. Nina Byers and Gary Williams. Published by Cambridge University Press.
© Cambridge University Press 2006.

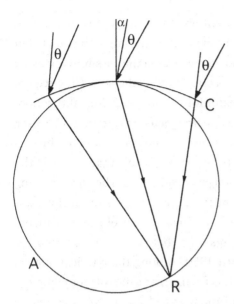

FIGURE 20.2 Principle of the Cauchois spectrometer: the figure is in a plane perpendicular to the axis of the cylinder A; θ is the Bragg angle; α is the angle between the reflecting planes and the normal to the bent crystal surface C; the radiation is reflected and focused to the point R, where it can be detected with photographic film.

wished to study the X-ray spectra emitted from heavy rare gases, still unknown in the 1930s, it was necessary for her to build an instrument capable of accurately measuring the emission of X-ray sources of very low intensity. This led her to formulate, in 1932, the principle of a new X-ray spectrometer that carries her name, described in further detail in Box 20.1. The Cauchois spectrometer combines the advantage of high intensity and high resolution in the hard X-ray region (wavelengths $\lambda < 1.5$ Ångströms) and is relatively simple to use. It met with immediate success, since Cauchois was able to demonstrate the efficacy of the instrument in atomic physics, measuring the weak X-ray emission of gases to high precision, and resolving numerous multiplets in the spectra.

The Cauchois spectrometer has been universally employed for the analysis of X-rays and γ-rays, and resulted in a major development of the physics of radiation in these spectral regions. The excitement created by her work drew the attention of spectroscopists to the problem of focusing X-rays by a bent crystal in transmission and also in reflection. It was Cauchois who demonstrated the possibility of using reflection for high-resolution spectroscopy in the soft X-ray region, and in a radiation monochromator for the study of X-ray diffraction. She used the technique to create the first focusing systems for forming X-ray images of emitting or opaque real objects. In later work, she suggested the use of wide-parameter refracting layers that extended the use of crystal spectrometers to longer wavelengths. The high resolving power and luminosity of the spectrometer were important for Cauchois in her systematic study of the X-ray spectra of heavy elements in emission and in absorption, research that she conducted from the time of her thesis in 1933 until 1950. Among the physical quantities determined from this analysis for the first time was the complete system of core level energies of the heavy elements, i.e., the ionization energies of the different subshells, which was later extended to include all atoms with atomic number ≥ 3. Particular mention should be made of the first resolution of satellite lines emitted from the inner shells of highly ionized atoms, in the vicinity of strong X-ray lines, and the determination, starting from the energies of these satellite lines, of the ionization energies of double ionization from full shells of heavy atoms.

In the period from 1938–40 systematic investigations were undertaken to search for unknown elements of the periodic system using X-ray spectroscopy. It was in collaboration with H. Hulubei that Cauchois proved the presence of the element of atomic number 93 (neptunium) in uranium minerals and that of a polonium isotope in gold and tellurium minerals; also the presence of the element of atomic number 85 (astatine) based on the radon emission spectra. Later on, she initiated studies on the X-ray spectra of transuranic elements.

As a continuation of her research on the identification of unknown elements, she started an investigation of the limits of sensitivity of qualitative and quantitative chemical analysis by X-ray emission and suggested the double spectrum method, consisting of exciting the sample and a standard made of the element to be determined. She published, in collaboration with H. Hulubei, a table of selected X-ray transitions.

In absorption work she found in the spectra $L_{II,III}$ of the transition elements unusual structures situated at the absorption limits, which she called white lines. Since her first observations, these structures have been extensively studied since they allow measurements of the density of the unoccupied d states. In the case of platinum, Cauchois demonstrated the importance of such measurements to determine whether the unoccupied states are $d_{3/2}$ or $d_{5/2}$. She contributed to the first studies of absorption structures (EXAFS: Extended X-ray Absorption Fine Structure) and showed that, based on observations in liquids, these structures are characteristic of the short-range order in the material.

With her insight into the important role played by chemical bonds in X-ray spectra, she directed in her laboratory a great deal of research into emission bands, absorption discontinuities and the structures in their close vicinity (XANES: X-ray Absorption Near Edge Structures) and their evolution as a function of the chemical bond character, as well as calculations of energy levels for various electronic configurations.

Cauchois identified and studied for the first time the soft X-ray distributions that are reflected from a crystal for X-ray energies in the vicinity of the absorption limits of the atoms in the crystal. This work constituted the first step towards the determination of the photo-absorption spectra by analyzing the reflected radiation, a long-sought goal that would open a new window for research into the electronic structure of materials. Many materials absorb X-rays too strongly to allow absorption studies, and the possibility of obtaining such information from the strong reflected beam was a very important development, as discussed further in Box 20.1.

Box 20.1   The Cauchois spectrometer

The first X-ray spectrometers in the early 1900s used Bragg reflection from crystal surfaces as a monochromator to transmit selectively only small wavelength ranges of the incident X-rays. They typically employed very narrow collimator slits for this purpose, but since these screen out most of the incident X-rays the detected signal is very small. A slitless spectrometer had been designed by H. Johann using a curved crystal in a reflection mode, but it could not be used for the short wavelengths that Cauchois wanted to study. She conceived the idea to have the radiation reflected from the inner atomic "planes" of an elastically curved crystal (also known as a "bent" crystal). The incident X-ray irradiates the convex face of the cylindrically curved crystal, and is reflected during its transmission through the crystal by planes that are inclined with respect to the lattice faces. The spectrum is focused on the concave side of a cylinder that is tangent to the crystal lattice with half its radius (see Fig. 20.2). By taking into account the usable opening of the crystal, its height and its thickness, she was able to calculate the shape and width of the observed spectral lines. The Cauchois spectrometer combines the advantages of high intensity and high resolution in the hard X-ray region (wavelengths $\lambda < 1.5$ Ångströms) and is relatively simple to use.

Anomalous dispersion

When a crystal reflects radiation whose wavelength is close to that of an absorption limit of one of the crystal elements, the reflected spectrum shows anomalous features due to dispersion of the X-rays as they travel through the crystal and interact with the atoms. These reflection anomalies were observed for the first time by Cauchois for Bragg reflection by the (1010) plane of a quartz crystal in the vicinity of the K absorption line of the silicon atoms. Her interpretation of the anomalies was confirmed by further theoretical and experimental work by her collaborators in the lab, comparing experimental spectra with quantum-mechanical calculations of the dispersion. Experiments made by specular reflection with a quartz mirror in the same spectral range demonstrated the analogy existing between the spectral distributions in both geometries. Following these first experiments made with classical X-ray sources, detailed studies of specular reflection were performed using synchrotron radiation. This was an important step toward the extraction of absorption-line information from reflection spectra. Many materials absorb X-rays so strongly that absorption spectra are impossibly difficult; only a few materials can be shaped into the extremely thin samples (a few tens or hundreds of Ångströms thickness) that would allow such measurements. If the same information could be determined from the far simpler reflection measurements, then valuable information on the electronic structure of many interesting materials could be obtained. Cauchois's research was the first to show how this might be possible.

Yvette Cauchois was the first person in Europe to grasp the potential presented by the radiation emitted by electrons rotating in a synchrotron as a source for exploring the atomic and electronic properties of matter. Starting in 1962, she initiated a research program in collaboration with the Istituto Superiore di Sanita to exploit the possibilities offered by the synchrotron at Frascati. From the first results obtained with this collaboration, the synchrotron source of strong continuous radiation with very small divergence turned out to be a very powerful instrument for obtaining absorption spectra in the wavelength region extending from X-rays to the ultraviolet, where the usual sources were simply too weak or only available at discrete energies.

Her great interest in astrophysics caused Cauchois to study extraterrestrial X-ray radiation, in particular the solar X-ray spectrum; she initiated a project for observing solar X-rays with the help of missile experiments. She participated in the first step of this research, involving the observation of images in X-ray light, where a photograph of the sun in the wavelength region around 10–20 Ångströms was obtained.

Many other investigations were started that were directly connected with her or prompted by her, among them the achievement of free electron beams in a large energy range; a cascade accelerator of 2 MeV, which she has used for studies in nuclear photoactivation and resonance; and the study of dense plasmas by X-ray diagnostics. It is impossible to cite all of the many different research interests that occupied her during her long scientific career.

BIOGRAPHY

Yvette Cauchois was born in December, 1908, in Paris, where she continued to live and carry out her research work. In July, 1928, she applied to Professor Jean Perrin to be accepted at his laboratory; she was 19 years old at that time and had just obtained her first degree in Physical Sciences at the Sorbonne, but she had been attracted to science even as a child. Under Francis Perrin's guidance she started research on fluorescence and obtained, in 1930, a Diplome d'Etudes Supérieures. Then she turned to X-ray spectroscopy for her thesis, which she defended in July, 1933, and which was entitled "An Extension of X-ray Spectroscopy: a Spectrometer Focused with a Curved Crystal; X-ray Emission Spectra from Gases." Very rapidly, her thesis work brought her international attention. She was invited to many foreign laboratories specializing in this subject and to many international conferences.

As was the custom in the 1930s, young students did not get any financial support while working on their theses. They had to support themselves under conditions that were sometimes very difficult. This was the case for Cauchois from 1928 to 1932, at which time she obtained one of the first research scholarships from the French National Center of Scientific Research (CNRS), which had just been founded. In October, 1933, as soon as she had obtained her degree, she was nominated "chargée de recherche" at the CNRS and then "maître de recherche" in 1937, which allowed her to devote herself full time to research.

As a young researcher at the Laboratoire de Chimie Physique, headed by Jean Perrin, she made many contacts with the illustrious foreign scientists that visited the laboratory. The "Monday teas" were the meeting place for important personalities of the Parisian scientific and cultural society of that period. She always spoke with enthusiasm of the prewar years that brought together Jean Perrin, Marie Curie, Irène Joliot-Curie and Frédéric Joliot, Paul Langevin, and young research workers like Francis Perrin, Pierre Auger, Leprince-Ringuet and many others.

Cauchois greatly admired and respected her mentor Jean Perrin. She spent the best part of her life at the Laboratoire de Chimie Physique, assuring its continuity during the war years 1940–5 when Perrin had to leave for the United States, and remained there when she became an assistant professor at the Sorbonne in 1945, and full professor in 1951. In 1953 she became the director of the laboratory, and she was also named the Chair of Chimie Physique. The building at the Rue Pierre et Marie Curie soon became too small for the group of Parisian physical chemists that she succeeded in attracting around her. Thus, she founded, in 1960, the Centre de Chimie Physique at Orsay that she directed for ten years. She was the President of the French Society of Chemical Physics from 1975 to 1978. She retired in 1978, and died November 19, 1999.

Cauchois's research was recognized by many prizes and honorary degrees:

Commander in orders of Palmes Académiques
Officer of the Legion of Honor
Officer in orders of the National Merit
Prix Ancel, 1933 (Société Française de Physique)
Prix Henri Becquerel, 1935 (Académie des Sciences)
Prix Girbal-Baral, 1936 (Académie des Sciences)
Prix Henry de Jouvenel (Palais de la découverte)
Prix Jérome Ponti, 1942 (Académie des Sciences)

Prix Triossi, 1946 (Académie des Sciences)
Prix Ancel, 1946 (Société Française de Photographie)
Medal of Czechoslovak Society of Spectroscopy, 1974
Gold Medal of the University of Paris, 1987
Honoris causa Doctor of University of Bucharest, 1993

IMPORTANT PUBLICATIONS

Cauchois, Y., Spectrographie des rayons X par transmission d'un faisceau non canalisé à travers un cristal courbé. *J. Phys.*, VII, **3** (1932) 320 and *J. Phys.*, VII, **4** (1933) 61.

Les niveaux extérieurs des atomes lourds révélés par leurs spectres de rayons X de grande fréquence. *J. Phys.* VIII, **3** (1942) 5.

Observations nouvelles sur les émissions X "hors-diagramme" de la série L. Application à une première détermination expérimentale, par spectrographie cristalline, des états d'ionisation profonde multiple des atomes lourds. *J. Phys.*, VIII, **5** (1944) 1.

Sur la formation d'images avec les rayons X. Possibilité de réalisation de loupes et de microscopes à l'aide de cristaux. *C. R. Acad. Sci.*, **223** (1946) 82.

Les niveaux d'énergie des atomes lourds. *J. Phys.*, **13** (1953) 113.

Les niveaux d'energie des atomes de numéro atomique inférieur à 70. *J. Phys.*, **16** (1955) 233.

Cauchois, Y., Bonnelle, C. and Missoni, G., Premiers spectres X du synchrotron de Frascati. *C. R. Acad. Sci.*, **257** (1963) 409.

21 Marguerite Catherine Perey (1909–1975)

Jean-Pierre Adloff
Centre des Recherches Nucléaires, Strasbourg, France

George B. Kauffman
Department of Chemistry, California State University;
Fresno University, Fresno

FIGURE 21.1 Photograph of Marguerite Catherine Perey. (Credit: AIP Emilio Segrè Visual Archives.)

CONTRIBUTION

In 1869, the Russian chemist Dmitrii Ivanovich Mendeleev proposed the periodic table that rationalized the known chemical elements, from hydrogen, the lightest, to uranium, the heaviest element then

Out of the Shadows: Contributions of Twentieth-Century Women to Physics,
eds. Nina Byers and Gary Williams. Published by Cambridge University Press.
© Cambridge University Press 2006.

known. The foundation of the table was the arrangement of the elements in groups and periods according to similarities in their chemical properties.

Mendeleev anticipated the existence of additional elements and predicted well-defined locations for them. The properties of 28 elements still to be discovered could be estimated from their positions in the table, which guided chemists in their search for new substances. Mendeleev coined the term "eka-element" (Sanskrit, one) for the as yet undiscovered element directly below a specified element in the same group (column).

Two years after Henri Becquerel's discovery of radioactivity in 1896, Pierre and Marie Curie found a new element solely by virtue of its emission of invisible rays: polonium (eka-tellurium), and shortly thereafter radium (eka-barium). This was followed by the discovery of three more radioactive elements: actinium, radon and protactinium. Together with uranium and thorium, also radioactive but known well before, they possessed a common property: they occupied vacant positions beyond bismuth in the terminal part of the periodic table. Hence it could be predicted with little risk that all future elements beyond bismuth should be radioactive. When the elements were later characterized by atomic number (Z), which also defines their positions in the periodic table, the radioactive elements were found to occupy the places between bismuth (Z = 83) and uranium (Z = 92).

In 1939, with three exceptions, all the elements with atomic numbers lower than uranium were known. One of these missing elements, eka-cesium (Z = 87) was the heaviest member anticipated in the column of the alkali elements, a group that consisted of lithium, sodium, potassium, rubidium and cesium. At this time this element was incorrectly supposed to be the most electropositive of all the elements.

Various workers made numerous attempts to isolate and identify element 87 from rubidium and cesium minerals by using different techniques, resulting in discredited claims of the discovery of "russium," "alcalinium," "virginium" and "moldavium." It is amazing that the search for a stable eka-cesium was pursued until the mid 1930s even though the American Nobel laureate in chemistry, Theodore William Richards, had stated as early as 1911 that eka-cesium would prove to be unstable.

However, the search for a naturally occurring radioactive element 87 was not attempted before the laws of radioactive transformations had been enunciated in 1913. It was now trivial that element 87 could be formed directly by the disintegration of a parent element with atomic number 89.

Actinium (Z = 89) was discovered in 1899 by André Debierne in pitchblende, a mineral from which the Curies had already extracted polonium and radium. For three decades knowledge of actinium was scanty because of the difficulty of handling, measuring and chemically and radiochemically purifying this elusive element.

When Marguerite Perey began working at the Institut du Radium, the first task assigned to her was the preparation of actinium-rich sources. Marie Curie's aim (see Chapter 4) was the determination of the spectrum of actinium, and she asked her young assistant to obtain a sample large enough for optical spectroscopy. Perey's preparations were used successfully for the first determination of the spectrum of actinium at Pieter Zeeman's laboratory in Amsterdam.

The detection of the element's radiation required a very careful elimination of its radioactive daughter products. When Perey measured a freshly purified sample, she observed a hitherto unsuspected radiation. By January, 1939, she was able to retrace the anomaly to the formation in actinium of a new element with chemical properties similar to those of an alkali element – undoubtedly the long-sought eka-cesium. Perey succeeded in isolating the new element and determining its chemical and radioactive properties. Her thoroughness and speed in performing the experiment enabled her to observe a phenomenon that had remained undetected for 40 years by earlier, less skillful, radiochemists.

As the discoverer of eka-cesium, Perey was entitled to name the new element: in honour of her native land she called it francium with the symbol Fr, thus echoing Marie Curie's name, polonium. Sixty years after the discovery, the knowledge of francium is still based essentially on Perey's exhaustive investigation of the physical, chemical and biological properties of the element.

Francium is the rarest and most unstable of all naturally occurring elements. Its entire content in the Earth's crust is only several hundred grams. Separated from its parent actinium, francium disintegrates completely within two hours. Today 14 of the artificial elements are more stable than francium.

Despite francium's scarcity, its discovery was a true milestone in the history of radiochemistry and of the periodic table. It was the first radioelement to be discovered since protactinium had been found in 1918. After the discovery of artificial radioactivity by Frédéric Joliot and Irène Curie (see Chapter 12) in 1934, the attention of radiochemists shifted to this new exciting field, and no major discovery was expected in the field of natural radioactivity. Thus, Perey's discovery of francium in 1939 was a totally unexpected event. Francium is the last naturally occurring radioactive element and simultaneously the very last naturally occurring element to be discovered. Thus, it completed the table of naturally occurring elements envisioned by Mendeleev 70 years earlier, because at the time no one could imagine the possibility of man-made elements.

Box 21.1   The discovery of francium

A priori, two naturally occurring α emitting actinium (Z = 89) isotopes could decay to eka-cesium: (1) ^{228}Ac (half-life, 6.15 hours), for which the emission was claimed but never confirmed, and (2) genuine actinium, ^{227}Ac (half-life, 21.8 years). An α activity had also been observed for this second isotope but was attributed to the presence of traces of protactinium. Thirty years after its discovery, knowledge of actinium was still scanty; even its half-life was uncertain. It is mixed with rare earths, especially with lanthanum, its closest homolog, from which it can be partially separated only by laborious fractional crystallizations. ^{227}Ac is a β emitter, but the very soft, readily absorbed radiation could not be detected; the element was monitored from the ß and γ-rays emitted by its long-lived daughters (^{227}Th [half-life, 18.7 days], ^{223}Ra [half-life, 11.2 days]) and the active deposit. A waiting period of three months was required to establish radioactive equilibrium.

Perey had prepared concentrated lanthaniferous actinium sources with a minimum of carrier, which might be suitable for detecting actinium's own radiation. Her idea was to search for the latter before it was smeared by the radiation of the descendants. Several minutes after carefully eliminating all daughters that are isotopes of Th, Ra, Pb, Bi and Tl by coprecipitation with carriers she began measuring the β activity.

In autumn, 1938, Perey observed in freshly purified actinium a penetrating β radiation with an intensity increasing with a half-life of about 20 minutes; it reached a plateau in the course of two hours and then increased again very slowly with the formation of long-lived decay products. The first part of the growth curve could be extrapolated to time zero (end of purification): at that moment the β activity was nil.

Perey's discovery was so unexpected and perplexing that it required a thorough verification before concluding that the new activity was due to an unidentified decay product that had separated during the purification and increased again in the purified actinium.

In January, 1939, after numerous tests, Perey concluded that part of the ^{227}Ac decay leads to a new β-emitting radioelement with a half-life of about 20 minutes. Its chemical properties were similar to those of an alkali metal, and thus it could only be the long-sought eka-cesium. Shortly afterward Perey observed α rays emitted by ^{227}Ac and the genesis of the new element was readily inferred from the radioactive displacement laws.

Perey had discovered the first isotope of element 87, with mass number 223. Her careful measurements indicated that the α-branching of ^{227}Ac is 1.2% and the half-life of eka-cesium is 21 minutes, values close to the most recent determinations, i.e., 1.38% and 21.8 minutes.

Following the tradition of the time she named it actinium K, while element 87 became francium. ^{223}Fr is the longest lived of the 31 known isotopes of the element and the only one occurring naturally.

Perey established a precise method for quantitatively determining actinium based on the separation and counting of francium. This procedure required only about two hours and was a great improvement over the previous one, which required a delay of three hours for complete equilibrium of the actinium family to be attained.

BIOGRAPHY

Marguerite Catherine Perey, the youngest child of Emile Louis Perey and Anne Jeanne Ruissel, was born on October 19, 1909, in Villemomble near Paris. A stock-market crash and the death in March, 1914, of her father, the proprietor of a flour mill, created financial difficulties for the middle-class Protestant family, which prevented the children from pursuing any higher education.

Perey attended the Ecole d'Enseignement Technique Féminine, a private but state-recognized school for technicians, from which she received her Diplôme d'Etat de chimiste in 1929. In October, 1929, she was hired by the Institut du Radium in Paris, where her intelligence, skill and eagerness to learn and understand brought her to the attention of the Director, Marie Curie. Perey soon became her personal laboratory assistant and confidante, and under her guidance she went on to become a prominent and proficient radiochemist. Perey participated in Marie Curie's last researches, which dealt with actinium, a little known and difficultly handled element. In fact, Perey was to devote her entire scientific life to studying the radioactive family of actinium. Her first years spent with Marie Curie might be regarded as a premonitory sign of destiny and the first step toward a major discovery.

After Madame Curie's death in 1934, under André Debierne and Irène Joliot-Curie's direction Perey continued research on actinium, which led to the discovery of francium in 1939. As a modest, 29-year-old technician without a university degree, she had succeeded in a search where so many experienced chemists had previously failed. No particular knowledge of fundamental science was needed, except for skill, conscientiousness and perseverance.

Perey undertook university studies at the Sorbonne during World War II. She received the licence degree and was qualified to defend her doctoral thesis in 1946. That same year she was appointed Maître de Recherches at the Centre National de la Recherche Scientifique (CNRS). In 1949, she was called to a new chair of nuclear chemistry at the University of Strasbourg, the only such chair in France outside of Paris. She established a teaching program in radiochemistry and nuclear chemistry and founded a research laboratory, inaugurated by Irène Joliot-Curie in January, 1951, which developed in 1958 into the Département de Chimie Nucléaire of the newly created Centre de Recherches Nucléaires (now Institut de Recherches Subatomiques) in Strasbourg-Cronenbourg. Perey was co-founder of this important institution, and her laboratory staff grew rapidly to about one hundred collaborators, including university faculty, research students, engineers and technicians.

Perey never married but devoted all her time to scientific and educational responsibilities in national and international committees such as the International Union of Pure and Applied Chemistry (IUPAC). Her numerous honors and awards included the Grand Prix Scientifique de la Ville de Paris (1960), the Lavoisier Prize of the Académie des Sciences (1964), the Silver Medal of the Société Chimique de France (1964), Officier of the Légion d'Honneur (1960) and Commandeur of both the Ordre National du Mérite and Ordre des Palmes Académiques. On March 12, 1962, she was elected the first woman corresponding member (*membre correspondant*) of the Académie des Sciences, which had been closed to women (including

Nobel laureates Marie Curie and her daughter Irène Joliot) for almost three centuries since it was founded in 1666 by Jean-Baptiste Colbert, Louis XIV's finance minister.

Shortly after 1946 Perey noticed a burn developing on her left hand, which she realized was radiodermatitis although her family tried to convince her that it was only an irritation caused by her work with acids. It was soon diagnosed as cancer caused by her many years of working with radioactive substances, in particular the pernicious actinium. She was forced, reluctantly, but progressively, to relinquish most of her duties. After several long stays in hospitals she moved to Nice but maintained close contact with her laboratory. Untiringly, she warned her collaborators to take precautions against exposure to radiation.

During her entire life Perey showed a great devotion to her mentor, Marie Curie. She used to recall her emotion when, after the defense of her doctoral thesis, Irène Curie said, "Today my mother would have been gladdened." In 1967, she attended her last gathering with the international community of nuclear scientists, the centenary celebration of Marie Curie's birth in Warsaw, where she encountered another famous radiochemist, Lise Meitner (see Chapter 7). In 1969, she returned briefly to Strasbourg to celebrate the 30th anniversary of the discovery of francium with her friends, colleagues and students. By July, 1973, her disease became more acute, forcing her first to stay in the Curie Hospital in Paris and finally in the Clinique du Val de Seine at Louveciennes, where she died on May 13, 1975, at the age of 65, one of the last survivors of the pre-World War II radiochemical pioneers of the laboratoire Curie. Her obituary was read at the Academy of Sciences by 1965 Nobel physics laureate and close friend Alfred Kastler.

NOTE
Reproduced in part with kind permission of the publisher from G. B. Kauffman, J. P. Adloff "Marguerite Catherine Perey (1909–1975)" in *Women in Chemistry and Physics: A Biobibliographic Sourcebook*, eds. Grinstein,

Rose and Rafailovich. © 1993, an imprint of Greenwood Publishing Group, Inc., Westport, CT.

IMPORTANT PUBLICATIONS

Perey, M., Sur un elémént 87, dérivé de l'actinium. *Compt. Rend. Acad. Sci.*, **208** (1939) 97. (Announces the discovery of a new element, eka-cesium, with atomic number 87.)

Dosage de l'actinium par l'actinium K. *Compt. Rend. Acad. Sci.*, **214** (1942) 797. (A new fast method for quantitative determination of actinium.)

L'élément 87: Actinium K. Thèse de Doctorat es sciences physiques, *Université de* Paris, 1946; also *J. Chim. Phys.*, **43** (1946) 152 and 262. (Relates the discovery and properties of actinium K, the first known isotope of element 87. The name francium is proposed for eka-cesium.)

Perey, M. and Chevallier, A., Sur la fixation de l'élément 87, francium, dans le sarcome expérimental du rat. *Compt. Rend. Soc. Biol.*, **145** (1951) 1208. (The only work on the biological properties of francium; measurement of its retention in sarcomas in rats.)

Perey, M. and Adloff, J. P., Sur la descendance de l'actinium K: (^{223}Fr). *J. Phys. Radium*, **17** (1956) 545. (Investigation of the dual decay of ^{223}Fr by α and β emission.)

FURTHER READING

Adloff, J. P., Marguerite Perey (1909–1975). *Radiochem. Radioanal. Lett.*, **23** (1975) 189. *(Also Rev. Chim. Miné.*, **12** (1975) 391.)

Kastler, A., Marguerite Perey. *Ct. R. Acad. Sci.*, **208** (1975) 124.

Kauffman, G. B. and Adloff, J. P., Marguerite Perey and the discovery of francium. *Educ. Chem.*, **26** (1989) 137.

Marguerite Catherine Perey. In *Women in Chemistry and Physics*, eds. L. S. Grinstein, R. K. Rose and M. H. Rafailovich. (Westport, Connecticut: Greenwood Press, 1993).

Adloff, J. P. and Kauffman, G. B., Marguerite Perey (1909–1975): A personal retrospective tribute on the 30th anniversary of her death. *Chem. Educator*, **10** (2005) 378–86.

Francium (atomic number 87), the last discovered natural element. *Chem. Educator*, **10** (2005) 387–94.

Triumph over prejudice: the election of radiochemist Marguerite Perey (1909–1975) to the French Académie des Sciences. *Chem. Educator*, **10** (2005) 395–9.

22 Dorothy Crowfoot Hodgkin (1910–1994)

Jenny P. Glusker

The Institute for Cancer Research, The Fox Chase Cancer Center, Philadelphia

FIGURE 22.1 Photograph of Dorothy Crowfoot Hodgkin.

SOME IMPORTANT CONTRIBUTIONS

The major scientific contributions of Dorothy Crowfoot Hodgkin were her determinations of the three-dimensional structures of several biochemically important molecules for which the chemical formula was uncertain or unknown. Among these were cholesterol, penicillin, vitamin B_{12} and insulin. She pioneered the use of X-ray diffraction of crystals to study larger and more complicated molecules than those that had been previously investigated. She concentrated her

Out of the Shadows: Contributions of Twentieth-Century Women to Physics, eds. Nina Byers and Gary Williams. Published by Cambridge University Press.
© Cambridge University Press 2006.

efforts on molecules that had high biochemical relevance, molecules that were generally viewed at that time as beyond the possibility of structure determination given the current state of the art. Her ability to interpret X-ray diffraction data and electron-density maps provided a crucial component to her success. She had the determination to insist that the problems she chose could eventually be solved, even if the answer took many years to find. For example, she obtained diffraction-quality crystals of insulin in 1935, but it was not until 1970 that the full three-dimensional structure of insulin was determined in her laboratory. During those 35 years experimental methods were greatly improved, methods for obtaining correct electron-density maps were derived, and computers were built that could handle complicated calculations for large molecules. Hodgkin pioneered many of the methods currently in use for the determination of the structures of biological macromolecules.

Mounting crystals in contact with their mother liquor
Hodgkin with J. D. Bernal showed that macromolecular crystals need to be mounted in contact with their mother liquor for good X-ray diffraction to occur. When taken out of their mother liquor, crystals tended to dry out and their X-ray diffraction patterns deteriorated extensively. Water comprises about 50% of most protein crystals and drying out may cause the regular packing from unit cell to unit cell to collapse. This problem was first noted in pepsin. Crystals of this enzyme had been grown in 1935 by John Philpot in Uppsala, Sweden. He had left a pepsin solution in a refrigerator while he went on a skiing vacation, and when he returned he was astounded to find that beautiful crystals, up to 2 mm long, had formed. He did not take them out of their mother liquor but left them in the crystallization vial and gave them to Glen Millikan who carried them back to England and presented them to J. D. Bernal. In those days crystallographers, in addition to studying the X-ray diffraction patterns of crystals, also recorded information on their refractive indices. Bernal looked at the refraction of the pepsin crystals and found that they were moderately

birefringent (had two different refractive indices in two different directions in the crystal) when in contact with mother liquor. They lost much of this property when they were taken out of the mother liquor and exposed to air. Therefore, in order not to perturb the crystal structure unnecessarily, Crowfoot Hodgkin and Bernal sealed the pepsin crystals with a drop of their mother liquor in Lindemann glass tubes. They were then able to obtain X-ray diffraction patterns and determine the unit-cell dimensions and space group of the protein crystal. This method of sealing crystals in capillaries had been developed by Helen Megaw (see Chapter 19) for her studies of ice crystals. It is still used by macromolecular crystallographers to this day.

Steroid structure

Crowfoot Hodgkin, with Harry Carlisle, was first to determine the three-dimensional structure of a steroid. This work culminated years of study of steroids in Bernal's laboratory. Steroids crystallize readily and their crystals diffract well. Measurements of unit-cell dimensions, combined with density determinations, give the weight of one unit cell. Measurements of birefringence indicated the direction of the ring system of the steroid with respect to the unit cell of the crystal. Bernal used this information to point out that the then-accepted chemical formula for a steroid had a structure that was the wrong size and shape to fit into the experimentally measured unit cell of its crystal. He suggested a different, more correct formula. His students, Crowfoot Hodgkin and Isidor Fankuchen, applied his methodology to the study of crystals of a wide variety of steroids in the days before full structure determinations were possible. They were able to show how the steroid molecules packed in each crystal form, making hydrogen bonds wherever possible. Later, Crowfoot Hodgkin and Carlisle determined the three-dimensional structure of crystalline cholesteryl iodide by X-ray diffraction. This was done under difficult conditions in wartime England in the 1940s. The resulting three-dimensional structure showed not only the overall chemical formula (see Fig. 22.2(a)), but also the geometries of the ways in which the rings are joined to each other,

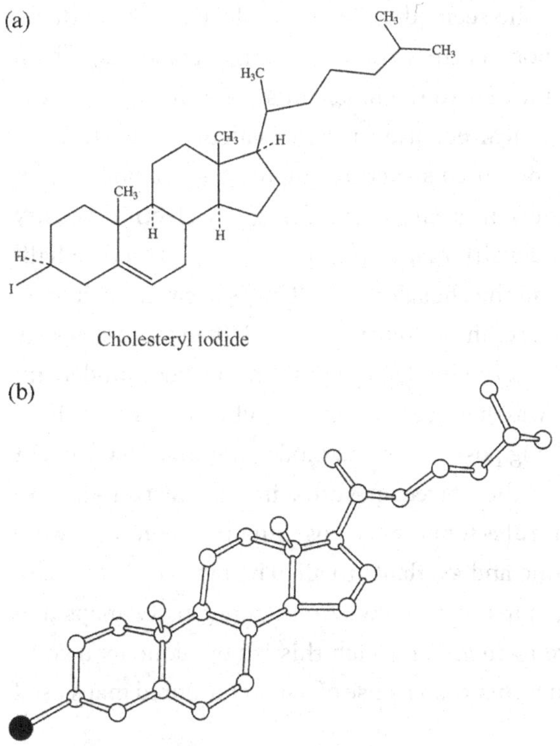

(a)

Cholesteryl iodide

(b)

FIGURE 22.2 Cholesteryl iodide. (a) Chemical formula. (b) Crystal structure.

(see Fig. 22.2(b)). The molecular structure that they obtained settled once and for all the chemical formula of this important steroid.

Interpreting electron-density maps

Dorothy Hodgkin was expert in interpreting chemical structure from the spatial distribution in electron-density maps (see Box 22.1), and this led to a three-dimensional model of the molecule under investigation. She understood the importance of the resolution of the data in the X-ray diffraction pattern, i.e., the detail that can be obtained from the analysis of the diffraction pattern, and the errors introduced by normal experimental errors in measuring these X-ray diffraction data. What she excelled at was correctly interpreting the peaks in the electron-density map; she seemed to have a sixth sense about them. When working with non-centrosymmetric crystal structures, which are the structures of interest to the biochemist, she showed that the input to the mathematical equation used to calculate an electron-density map, that is, measured structure amplitudes and phases that are calculated for atoms in a chosen model, gives electron-density peaks in the electron-density map corresponding to a structure half-way between reality and the chosen model. This is because both *measured* data and *calculated* phases have been used. If an atom is nearly in the correct position in the model, the peak in the electron-density map will appear half-way between the input and true position. If an atom is put in the wrong position in the model, a peak of half height will appear in or near the correct position in the electron-density map. She ensured that rules for non-centrosymmetric structures were worked out at that time and available to all. The present-day macromolecular crystallographer still looks at electron-density maps and fits a model structure to them, although this is now done by a computer graphics system rather than by use of hand-contoured maps used by Hodgkin.

Box 22.1   A glossary

Absolute configuration
The absolute structure of a crystal or molecule. If the structure is chiral it cannot be superimposed on its mirror image. The absolute configuration then defines which of two possibilities is correct for the object under study.

Anomalous dispersion
A discontinuity in the plot of refractive index against wavelength (that is, the dispersion) that occurs at wavelengths in the vicinity of absorption edges of the absorbing (or scattering) material. The wavelengths at which this occurs are characteristic of the absorbing or scattering atom.

Enzymes
Proteins that catalyze chemical reactions in the body.

Electron-density maps
The electron-density map is calculated by a Fourier summation. The measured intensities of the diffracted beams lead to the amplitudes of the waves, the calculation of phase angles for a derived model gives the relative phases of the waves to be summed. These relative phases are then refined to give the best possible model by inspection of features in the electron-density maps calculated with less than perfect phases. Once atomic positions are determined, the molecular architecture can be found readily by simple geometric techniques.

Heavy-atom method of structure determination
Relative phases are calculated for the heavy atom and are used to calculate an electron-density map. The heavy atom is located by use of a Patterson map.

Non-centrosymmetric structures
Structures in which the arrangement of atoms do not contain a center of symmetry. Biological molecules generally have a handedness and when they form crystals retain this handedness. This eliminates structures in which molecules with both handedness coexist.

Organometallic compounds
Compounds that have a carbon atom directly bonded to a metal atom or ion.

Patterson map
A Fourier summation using squares of the amplitudes that are used for an electron-density map. All relative phase angles are zero. The result is a map containing all interatomic vectors. Peak heights are related to the product of the atomic numbers of the two atoms involved. If there is a heavy atom in the structure then heavy atom–heavy atom vectors will dominate the map and allow the determination of the heavy atom position.

Phase problem

The problem of determining the phase angle, relative to a chosen origin, that is to be associated with each diffracted wave that is combined to give an electron-density map. These relative phases are needed for the calculation of a correct electron-density map.

Refinement of structure

The process whereby the parameters of a model structure are improved until the best possible fit is obtained between the calculated X-ray diffraction intensities and those measured experimentally. This may require many successive stages of calculation in view of the complexity of the equations involved. When completed it is assumed that the model approximates the true structure (within the estimates of error in each parameter that are also determined).

Refractive indices

Refraction is the change in direction of a beam of light when it passes from one material to another in which it has a different velocity. The refractive index is the ratio of the velocity of light in a vacuum to its velocity in the material under study. When a colorless crystal is immersed in a colorless liquid of the same refractive index, the crystal becomes invisible. If crystals have anisotropic internal structures, light will be split into two components that travel with different velocities along different paths. This is birefringence or double refraction.

Space group

The group of operations that converts one molecule or asymmetric unit into an infinitely extending three-dimensional pattern. There are 230 space groups. They are identified (with some ambiguity) by systematic absences in the X-ray diffraction pattern.

Unit-cell dimensions

Crystals that diffract X rays have a regularly repeating internal structure that can be represented by unit cells that contain the unique repeatable contents.

X-Ray diffraction

The electrons around atoms can scatter X rays. The interference between X rays scattered by the various atoms in the crystal result in intense diffracted beams in certain directions and very weak ones in others. The internal periodicity of the crystals causes these diffracted beams to be reinforced sufficiently for measurement. An analysis of the intensities of the diffracted beams (the diffraction pattern) gives the atomic arrangement within each unit cell. Unit-cell dimensions are measured from the directions of the diffracted beams. The result of the analysis is an electron density, representing the electrons that have caused the original scattering. This map is interpreted to give geometric parameters of the structure under investigation.

Structure determination using heavy atoms for phasing

An important technique used in determining macromolecular crystal structures by X-ray diffraction is the use of a heavy atom substitution to help solve the phase problem. A heavy atom is used because the amount of X-ray scattering by an atom depends on its atomic number. For a heavy atom the scattering is greater than that for carbon, nitrogen and oxygen. As a consequence, the relative phase angles of most of the diffracted X-rays will be determined mainly by the heavy atoms. Hodgkin showed in a practical way that the then-current theoretically derived lower limit of atomic number for such a heavy atom to provide useful phase information was too high. In this way, she paved the way for protein structure determination.

In her studies of vitamin B_{12} Dorothy Hodgkin and members of her laboratory worked with great care. The chemical formula was not known and it was important that there would not be any mistakes in this formula when she reported it. Therefore, when a three-dimensional electron-density map was calculated using phases of the cobalt atom (atomic number $= Z = 27$) alone, located by a Patterson map, only the peaks immediately around the cobalt atom were interpreted as atoms for phasing in the next cycle. More atoms were then added to the model until it was completed. Later it was clear that if the chemical formula had been known, many more peaks could have been chosen early in the analysis. Hodgkin had been advised by her colleagues that this method of using just the cobalt atom position to calculate relative phases for an initial electron-density map would not work because cobalt is not heavy enough to phase a structure containing around 100 atoms, as B_{12} has. However, she showed that cobalt was, indeed, sufficiently heavy for this method, and confidence in the use of heavy atoms for protein structure determination resulted.

Use of computing equipment

X-ray crystallography has some number-intensive steps that are best handled by a computer. Hodgkin realized this, and sought out people with access to the largest and most efficient computers of the time and collaborated with them.

In her work on penicillin Hodgkin used the first IBM computers in completing the X-ray calculations. This was an early application of an electronic computer to a biochemical problem. The atomic arrangement of penicillin was eventually determined through the techniques that she helped to develop.She also worked with Kenneth Trueblood, a crystallographer with the University of California at Los Angeles, because he had access to state-of-the-art computer equipment. In this work on vitamin B_{12} they used mail and telegraphs to communicate data between California and England (Dodson *et al.*, 1981). In the studies of the structure of insulin she had the insight to insist that the structure be refined, but, because the resolution was not high, she realized that the use of geometric constraints would be necessary. This was pioneering work in the early days of protein crystallography.

The use of anomalous dispersion in structure determination

Johannes Bijvoet and co-workers in 1951 used anomalous dispersion (see Box 22.1) of X rays to establish the absolute configurations of the tartrates. Hodgkin was interested in this from the start because the cobalt atom in vitamin B_{12} caused anomalous scattering of the copper radiation normally used to obtain X-ray diffraction patterns. This was used later in her laboratory to confirm the absolute configuration of the vitamin. In the work on the structure of insulin she pioneered the use of anomalous dispersion in the study of the heavy atoms in the derivatives used in the structure determination.

BIOGRAPHY

Dorothy Mary Crowfoot was born on May 12, 1910, in Cairo, Egypt. Her father, John Winter Crowfoot, had been an archaeologist in the Egyptian Ministry of Education, and later became the Principal of Gordon College at Khartoum and Director of Education and Antiquities in the Sudan. After that he worked at the British School of Archaeology in Jerusalem. Her mother, Grace May (Molly) (née Hood) was an expert on the history of ancient textiles, and wrote a book on the flowering plants of Sudan. Dorothy was the eldest of four girls. She lived in the Sudan for a while, but during World War I the girls were sent back to England to live with their grandmother. Dorothy believed that her wartime experiences gave her her independent spirit. After 1918 both parents spent each summer back in England, and the rest of the year in Sudan. Dorothy went to the Sir John Leman School in Beccles, Suffolk. Then, shortly before she went to college, she helped her parents with excavations at Jerash in Transjordan and was intrigued by the elegant mosaic patterns in the pavements there. Dorothy was encouraged in her interest in science by a family friend, Dr. A. F. Joseph; he gave her a portable chemical laboratory when she left the Sudan.

Dorothy Crowfoot obtained her university education in chemistry at Somerville College, Oxford. Her X-ray crystallographic career started with her undergraduate studies of thallium dialkyl halides with H. M. ("Tiny") Powell. On graduation she, on the advice of Dr. Joseph, went to work with the much-admired John Desmond (JD) Bernal (affectionately called "Sage") and obtained a Ph.D. at

Cambridge University in 1937. Bernal reinforced her lifelong interest in structural biochemistry.

In 1934, Dorothy, with Bernal, published the first report on the diffraction pattern of crystals of a protein, pepsin. Air-dried protein crystals gave very poor, if any, diffraction patterns, while those surrounded by the mother liquor (in which they had been crystallized) gave nice diffraction spots on photographic film. This experimental technique is now used routinely by protein crystallographers. Dorothy's interest in proteins continued and, in October, 1934, she successfully grew crystals of insulin. Although she obtained good X-ray diffraction patterns, the methods for interpreting them could not, at that time, lead to a structure determination. While in Bernal's laboratory she was also interested in steroids and their structures. Bernal had used crystallographic information to help establish the chemical structures of steroids. Dorothy, with Bernal and Isidor Fankuchen, studied crystals of over 100 steroids. They reported unit-cell dimensions, refractive indices and probable packing of molecules within a crystal. Unfortunately, it was about that time that Dorothy began to have problems with the severe rheumatoid arthritis that plagued her for the rest of her life.

When it was time for her to think about starting her own independent scientific career Dorothy Crowfoot was appointed Research Fellow of Somerville College, Oxford, in 1933. She became Tutor and Fellow of that college in 1936. After her marriage in 1937 she generally published under the name Dorothy Hodgkin. She was appointed University Lecturer and Demonstrator in 1946, but was not advanced to become a University Reader until 1955. Hodgkin was given much support by the Principal of Somerville College, Janet Vaughan, a hematologist. One of Hodgkin's students in Somerville was Margaret Roberts, later Margaret Thatcher, Prime Minister of Great Britain for many years and the only British Prime Minister with a degree in science.

Dorothy's laboratory in Oxford was in the Ruskin Science Museum, an old Gothic building. The room where she worked was the room in which Thomas Henry Huxley successfully debated Bishop Samuel Wilberforce on the subject of evolution in an historic debate on June 30, 1860, at a meeting of the British Association for the Advancement of Science. Dorothy's first graduate student was Dennis Riley. During the war she was joined by Harry Carlisle and together they determined the structure of cholesteryl iodide (Figure 22.2), confirming the chemical formula for steroids.

Her next big achievement was the determination of the chemical formula of penicillin. This antibiotic was discovered in a mold by Alexander Fleming at St. Mary's Hospital, London, in 1929. Howard Florey and Ernest Chain then isolated penicillin from this mold, in time for it to be used extensively to treat the many casualties with infected wounds in World War II. The determination of the chemical structure of this antibiotic was particularly important then. The three-dimensional chemical formula was determined by Hodgkin and her co-workers from the X-ray diffraction analyses of the sodium, potassium and rubidium derivatives of benzylpenicillin. The formula contained a four-membered ring of three carbon atoms and one nitrogen atom, that is, a β-lactam structure. This had been thought to be too unstable to be likely. The chemical formula so determined (see Figure 22.3) served as the starting point for many chemical modifications that have been successfully used as antibiotics, such as the cephalosporins and thiostreptone for which crystal structures were also determined in her laboratory.

Hodgkin continued her interest in biochemistry. The fatal disease pernicious anemia was found by Minot and Murphy in 1926 to be treatable by liver extracts, and, in 1948, the active principle from liver was isolated in crystalline form as beautiful deep-red crystals. E. Lester Smith brought them from Glaxo Laboratories to Oxford so that Mary Porter and R. C. Spiller could look at them and determine if they were the same as those isolated by Karl Folkers at Merck Laboratories in the USA. Hodgkin was excited at the appearance of the deep-red crystals and immediately measured their unit-cell dimensions and molecular weight. The chemical formula of this material,

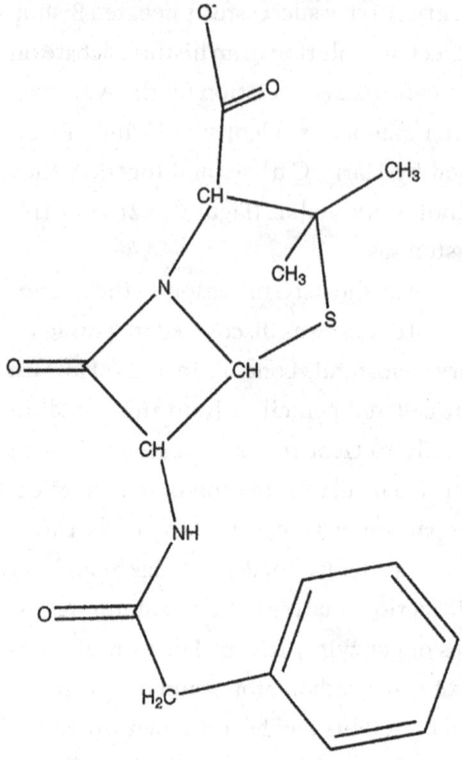

Benzylpenicillin

FIGURE 22.3 Potassium benzylpenicillin. (a) Chemical formula.

vitamin B_{12}, obtained by her and her co-workers (Figure 22.4), was that of a porphyrin-like ring with one bridging carbon atom missing so that two pyrrole rings were directly linked, and the β-positions of these rings were each fully saturated. Later, Galen Lenhert from Vanderbilt University showed in Hodgkin's laboratory that the B_{12} coenzyme (derived from the protein) contains a cobalt–carbon bond. It was the first biological organometallic compound to be identified. W. L. Bragg in *Fifty Years of X-Ray Diffraction* (1962) described the B_{12} study as "breaking the sound barrier." It led to the idea of phasing larger structures with heavy atoms, a method now used in macromolecular crystal structure determinations. Initially, Hodgkin used

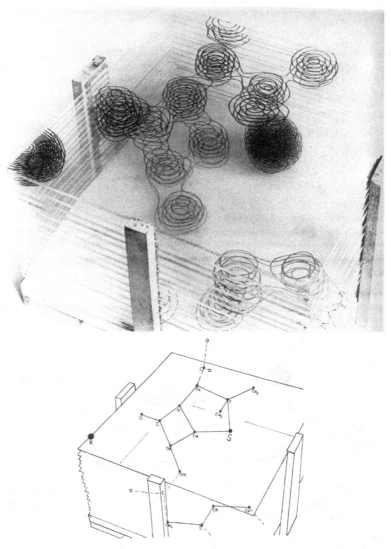

FIGURE 22.3 (b) Electron density on Perspex sheets. Photograph of a model showing the electron density calculated over the thiazolidine and β-lactam rings in potassium benzylpenicillin. The model is constructed from sheets of perspex. On each sheet contours are drawn along lines of equal electron density to record the electron density calculated at a single level in the crystal structure, parallel to the crystallographic *b* plane. The sheets are then stacked together at the correct intervals apart. Combined, they build up a representation of the electron density distribution within the unit cell. The model illustrated covers a portion only of a single molecular unit, and omits most of the carboxyl group and the amide oxygen and carbon atoms of penicillin, and also a half of the benzene ring. These atoms lie outside the volume range of the model; thus, the single closed contour marks the upper fringe of the amide oxygen. The key assists in the identification of the atoms. Closed circles indicate atoms within the limits of the model.

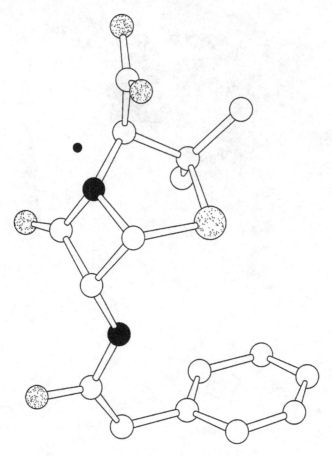

FIGURE 22.3 (c) Crystal structure (the small black circle is a potassium ion).

the cobalt atom position to phase the B_{12} work, even though the ratio of scattering power of the cobalt atom to the rest of the molecule was low. This strategy worked so well, people were encouraged to use it to tackle protein structure.

Hodgkin took the first X-ray diffraction photographs of insulin in 1935 and 34 years later she reported the complete structure. She was confident from the beginning that the molecular structure could be determined from the X-ray diffraction pattern. Chinese crystallographers, led by Tang You Chi, worked on insulin in China and,

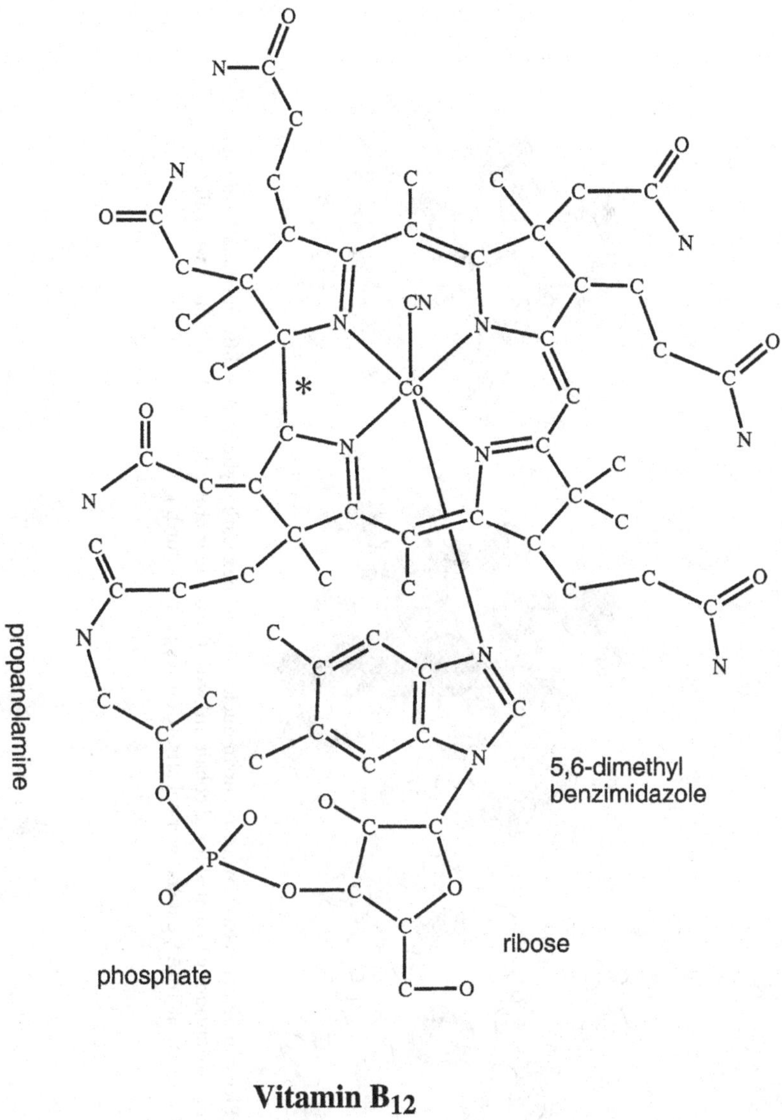

Vitamin B$_{12}$

FIGURE 22.4 (a) Vitamin B$_{12}$. Chemical formula.

when both structure determinations were completed, they compared results. They were almost identical. This collaboration led to an increased awareness in the West of scientific work in China at a time of considerable political unrest. She was honored for this in Beijing in 1993.

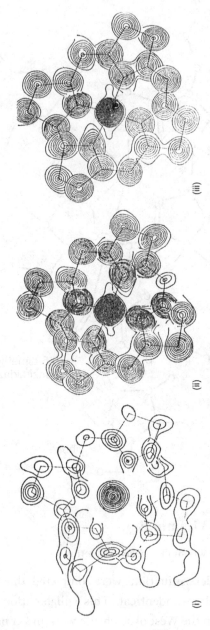

FIGURE 22.4 (b) Electron-density peaks over the corrin nucleus in the hexacarboxylic acid at different calculation stages. Terms phased on contributions calculated for (i) cobalt, (ii) with the nucleus atoms less C10, (iii) cobalt with all the nucleus atoms. Contours at 1e/A³, except over cobalts. (Courtesy J. Pickworth.)

(i)   (ii)   (iii)

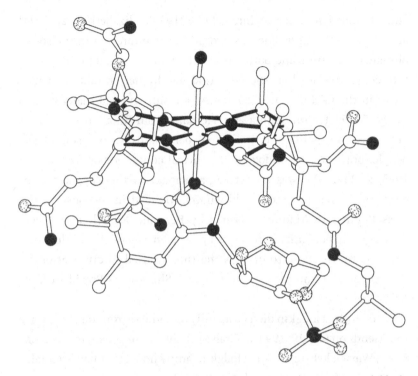

FIGURE 22.4 (c) Crystal structure of vitamin B$_{12}$ (corrin ring with black bonds). The front of this structure is marked with an asterisk in Figure 22.4(a).

Hodgkin was awarded the Nobel Prize in Chemistry in 1964 "For her determinations by X-ray techniques of the structures of important biological substances." She was the third woman to receive this Prize after Marie Curie (1911; see Chapter 4) and Irène Joliot-Curie (1935; see Chapter 12), and the only British woman to do so. She received the Order of Merit, the highest civilian honor in Great Britain, in 1965. She was the second woman to receive this honor, the first being Florence Nightingale in 1907.

In 1937, Dorothy married Thomas Lionel Hodgkin, Director of Extramural Studies in Oxford and a Fellow of Balliol College. Later, Thomas became an historian of Africa, and Dorothy was with him in Ghana when she learned that she had won the Nobel Prize. Dorothy and Thomas had three children, Luke (1938), Elizabeth (Liz) (1941) and Toby (1946). Like Thomas, Dorothy was interested in the social climate of the time and both worked hard on the social problems, an interest started for her, she once told me, by her childhood experiences in the Girl Guides (Scouts). As a result she had many friends in the Soviet Union, China and Vietnam. Tom Blundell noted that "Dorothy saw politics as individual personalities, in terms of the people, not in terms of dogma or political convictions." (McGrayne 1993, p. 252.) She was an active participant in the Pugwash Conferences on Science and World Affairs, founded in response to the Russell–Einstein Manifesto issued by Bertrand Russell and Albert Einstein. This organization called upon scientists of all political persuasions to assemble to discuss the threat posed to civilization by the existence of thermonuclear weapons. She was its President from 1975–88.

Dorothy Hodgkin died peacefully at home surrounded by family and friends on July 29, 1994, at Crab Mill, Ilmington, near Shipston-on-Stour, Warwickshire, the old Hodgkin family home that had originally been a row of cottages and which retained the charm and friendliness of the Cotswolds.

Honors
  Fellow of The Royal Society (London), 1947
  Nobel Prize for Chemistry, 1964
  Order of Merit (UK), 1965
  Chancellor, Bristol University, 1970–88
  Bakerian Lecturer, National Academy of Sciences (USA), 1972
  President, International Union of Crystallography, 1972–5
  President, Pugwash Conference on Science and World Affairs, 1975–88
  President, British Association for the Advancement of Science, 1977–8
  Fellow, Indian Academy of Sciences

Other prizes and medals
  Royal Medal, The Royal Society of London, 1956
  Copley Medal, The Royal Society of London, 1976

Longstaff Medal, British Association for the Advancement of Science, 1978
Mikhail Lomonosov Gold Medal, Soviet Academy of Sciences, 1982
Dimitrov Prize, 1984
Lenin Peace Prize, 1987

Honorary degrees and awards

Honorary D.Sc.: Universities of Bath, Brown, Cambridge, Chicago, Delhi, East Anglia, Exeter, Ghana, Harvard, Hull, Kent, Leeds, London, Manchester, Mount Sinai, Oxford, St. Andrews, Sussex, Warwick: Dr. Medicine and Surgery (Modena); D.Univ. (Open, York, Zagreb); LL.D. (Bristol); D.L. (Dalhousie).

Foreign Member: Royal Netherlands Acad. of Science and Letters, American Acad. of Arts and Sciences, Bavarian Acad., Austrian Acad., Ghana Acad. of Sciences, Puerto Rico Acad. of Sciences, Australian Acad. of Sciences, Leopoldina Acad. of Sciences, Norwegian Acad. of Sciences, Indian Acad. of Sciences, Royal Irish Acad. of Sciences, National Acad. of Sciences (USA).

Honorary Member: USSR Academy of Sciences, Royal Institution of Great Britain.

Honorary Fellow: Somerville College, Oxford; Linacre College, Oxford; Girton College, Cambridge; Newnham College, Cambridge; Bristol University (1988–1994).

Positions

1934–5 Research Fellow, Somerville College, Oxford
1935–55 Official Fellow and Tutor, Somerville College, Oxford
1946–56 Demonstrator and Lecturer, Oxford University
1956–60 University Reader, Oxford University
1956–77 Professorial Fellow, Somerville College, Oxford
1960–77 Wolfson Research Professor, Oxford University
1977–83 Fellow, Wolfson College, Oxford
1977–94 Professor Emeritus, Oxford University

IMPORTANT PUBLICATIONS

Nobel Prize speech

Hodgkin, D. C., The X-ray analysis of complicated molecules. *Science*, **150** (1965) 979.

Early work in Oxford

Bernal, J. D. and Crowfoot, D., X-ray photographs of crystalline pepsin. *Nature*, **133** (1934) 794. (The first X-ray diffraction photograph of protein crystals.)

Bernal, J. D., Crowfoot, D. and Fankuchen, I., X-ray crystallography and the chemistry of steroids I. *Trans. R. Soc.* A, **239** (1940) 135. (An extensive study of the packing of steroids in the crystalline state.)

Carlisle, C. H. and Crowfoot, D., The crystal structure of cholesteryl iodide. *Proc. R. Soc.* A, **184** (1945) 64. (The three-dimensional structure of a steroid molecule, confirming the chemical formula.)

Work on vitamins

Hodgkin, D. C., Porter, M. W. and Spiller, R. C., Crystallographic measurements of the anti-pernicious anemia factor. *Proc. R. Soc.* B, **136** (1950) 609. (The beginning of work on vitamin B_{12}.)

Brink, C., Hodgkin, D. C., Lindsey, J. *et al.*, X-ray crystallographic evidence on the structure of vitamin B_{12}. *Nature*, **174** (1954) 1169. (Further work on vitamin B_{12}.)

Hodgkin, D. C., Pickworth, J., Robertson, J. H. *et al.*, The crystal structure of the hexacarboxylic acid derived from B_{12} and the molecular structure of the vitamin. *Nature*, **176** (1955) 325. (The chemical formula and structure of vitamin B_{12}.)

Lenhert, P. G. and Hodgkin, D. C., Structure of the 5,6-dimethylbenzimidazoylcobamide coenzyme. *Nature*, **192** (1961) 937. (The structure of one of the two coenzymes from vitamin B_{12}.)

Crowfoot, D. and Dunitz, J. D., Structure of calciferol. *Nature*, **162** (1948) 608. (The structure of vitamin D_2.)

Crowfoot, D., Bunn, C. W., Rogers-Low, B. W. and Turner-Jones, A., X-ray crystallographic investigations of the structure of penicillin. In *Chemistry of Penicillin*, eds. H. T. Clarke, J. R. Johnson and R. Robinson (Princeton, NJ: Princeton University Press, 1949) p. 310. (The main report on the structure of penicillin, which clearly established its chemical formula.)

Work on antibiotics

Hodgkin, D.C. and Maslen, E.N., The X-ray analysis of the structure of cephalosporin C. *Biochem. J.*, **79** (1961) 393. (The structure of an antibiotic.)

Anderson, B., Hodgkin, D. C. and Viswamitra, M. A., Structure of thiostrepton. *Nature*, **225** (1970) 223. (Structure of an antibiotic.)

Work on insulin

Crowfoot, D., X-ray single-crystal photographs of insulin. *Nature*, **135** (1935) 591. (The beginning of her life's work on the structure of insulin.)

Crowfoot, D. and Riley, D., X-ray measurements on wet insulin crystals. *Nature*, **144** (1939) 1011. (More on insulin with her first graduate student in Oxford.)

Dodson, E., Harding, M. M., Hodgkin, D. C. and Rossmann, M. G., The crystal structure of insulin. III. Evidence for a 2-fold axis in rhombohedral zinc insulin. *J. Mol. Biol.*, **16** (1966) 227. (The structure of insulin.)

Adams, M. J., Blundell, T. L., Dodson, E. J., *et al.*, Structure of rhombohedral 2-zinc insulin crystals. *Nature*, **224** (1969) 491. (The structure of insulin.)

Bentley, G., Dodson, E., Dodson, G., Hodgkin, D. and Mercola, D., Structure of insulin in 4-zinc insulin. *Nature*, **261** (1976) 166. (The structure of insulin.)

FURTHER READING

Bernal, J. D., Carbon skeleton of the steroids. *Chem. Ind.*, **51** (1932) 466.

Bijvoet, J. M., Peerdeman, A. F. and van Bommel, A. J., Determination of the absolute configuration of optically active compounds by means of X rays. *Nature*, **168** (1951) 271.

Bragg, W. L., The growing power of X-ray analysis. In: *Fifty Years of X-Ray Diffraction*, ed. P. P. Ewald (Utrecht: Oosthoek, 1962) pp. 120–35.

Dodson, G. and Chothia, C., Fifty years of pepsin crystals. *Nature*, **309** (1984) 309.

Dodson, G., Glusker, J. P., Ramaseshan, S. and Venkatesan, K. (eds.), *The Collected Works of Dorothy Crowfoot Hodgkin* (Bangalore, India: Indian Academy of Sciences, 1994).

Dodson, G., Glusker, J. P., Sayre, D. (eds.), *Structural Studies on Molecules of Biological Interest. A Volume in Honour of Dorothy Hodgkin* (Oxford, UK: Clarendon Press, 1981).

Ferry, G., *Dorothy Hodgkin: A Life* (London: Granta Books, 1998, also New York: Cold Spring Harbor Laboratory Press, 2000).

McGrayne, S. B., *Nobel Prize Women in Science* (New York: Carol Publishing Group, 1993).

Trueblood, K. N., Structure analysis by post and cable. In *Structural Studies on Molecules of Biological Interest. A Volume in Honour of Professor Dorothy Hodgkin*, eds. G. Dodson, J. P. Glusker and D. Sayre (Oxford, UK: Clarendon Press, 1981) pp. 87–105.

23 Gertrude Scharff Goldhaber (1911–1998)

Alfred Scharff Goldhaber

C. N. Yang Institute for Theoretical Physics, State University of New York

FIGURE 23.1 Photograph of Gertrude Scharff Goldhaber.

IMPORTANT CONTRIBUTIONS

Gertrude Scharff Goldhaber made her principal contributions in the field of nuclear physics, focusing on phenomena in the range of energies 1 MeV and below (1 MeV is the energy gained by one electron upon passing through an electric potential difference of one million volts). These are energies close to the threshold energy for creation of an

Out of the Shadows: Contributions of Twentieth-Century Women to Physics,
eds. Nina Byers and Gary Williams. Published by Cambridge University Press.
© Cambridge University Press 2006.

electron–positron pair, but substantially lower than the typical energy required to liberate a proton or neutron from a nucleus smaller than that of an iron atom. In this arena her principal contributions were detecting the emission of neutrons during the spontaneous fission of a uranium nucleus; demonstrating the identity of electrons with the beta particles emitted in weak decay processes; and her extensive, systematic studies on the collective behavior of nuclei associated with rotation and deformation from spherical shape.

Physics in the MeV range

During her doctoral studies in Munich culminating in 1935, Gertrude Scharff studied the effects of mechanical stress on ferromagnetism, and in London up to 1939 she worked on electron diffraction. With her marriage to Maurice Goldhaber in that year she moved to the University of Illinois, and turned to nuclear physics. Each subject she studied involved phenomena at a somewhat higher energy scale than the previous one, but once she got to nuclear physics she continued to focus on what today would be called low-energy aspects of the subject. That is, a nucleus may be viewed as a droplet in which the individual "molecules" are the neutrons and protons. Because quantum effects are important, the nucleus also may be viewed as a molecule in which the individual "atoms" are neutrons and protons. The best picture often seems a hybrid between these two analogs. In 1942 Trude (as she was known to friends and colleagues the world over) showed that spontaneous nuclear fission is accompanied by the emission of one or more neutrons. While such neutron emission was widely suspected to occur (and well established for induced fission) she appears to have been the first to make a direct observation of the phenomenon. Neutron emission is critical to the fission chain reaction because a neutron (especially a slow neutron) striking another uranium nucleus can stimulate fission. For that reason, publication of her result was postponed until after the close of World War II.

Shortly after the war, she and Maurice Goldhaber collaborated on a fundamental experiment about the nature of elementary particles. The electron was the first elementary particle identified, accepted as a constituent of atoms and therefore of all familiar matter. In certain rare processes (known as weak decays) so-called "beta" particles are produced, looking very much like electrons. Clearly, it is worth knowing whether betas and electrons are identical. In this experiment Trude and Maurice let betas impinge on lead, whose atoms have a large atomic number. If the betas were different from electrons, they could occupy the same locations as deeply bound atomic electrons, and therefore could fall in towards the nucleus of a lead atom, releasing a large amount of energy in the process. However, if electrons and betas were the same, then the exclusion principle (which says that no two electrons can occupy the same quantum state at one time) implies that there could not be any descent of betas towards the lead nucleus. The lack of high energy X-rays in the experiment showed that the beta particles must be indistinguishable from electrons.

A little later Trude started to focus in on her lifework, concerning the properties of nuclei when they are only "tickled" or excited a little bit. In many cases, this excitation may be pictured as a slow rotational motion of the nuclear medium as a whole, rather than just one nucleon jumping into a higher-energy state like an excited electron in an atom. Her early studies of such systems were an important component of the background for the collective theory of nuclear motion, which earned a Nobel Prize for Bohr and Mottelson in 1975. As time went by, she developed a more and more detailed understanding of the most prominent band of collective states, eventually codified in her Variable Moment of Inertia model. This also led to what may have been the first mother–son collaboration in physics. Famous precedents existed for all the other parent–child combinations, but apparently not for this one. Her joint work with Alfred Goldhaber was based on the idea that if the nucleus has a distorted shape, that fact should be seen both in its rotational behavior and in the electromagnetic radiation due to a related distortion in its distribution of electric charge. Two papers about this made a case that there is a simple and instructive relation between the electrical and the mechanical deformation of the nucleus.

The observations that she made in her systematic studies required creating a nucleus in the appropriate excited state, and then detecting the radiation, which signaled its drop to the ground state. Among the techniques for producing the initial state was hitting a nucleus with a neutron (often from a reactor), thus producing a radioactive nucleus, which after beta emission would become the desired state, quickly de-exciting by radiating a photon. Such photons (in the high-energy X-ray or gamma-ray range) then could be detected in various ways including by Geiger counters.

Trude contributed to science in a number of ways beyond her own research. She served with a National Research Council panel on women in science. She organized programs for high school science teachers and for students at Brookhaven National Laboratory, initiated a monthly series of lectures at the laboratory in which the speakers explained their research in terms suitable for a general scientific audience, and was a co-founder of Brookhaven Women in Science. She was a national Phi Beta Kappa lecturer, making college visits where she spoke both about her own research and about life in science.

BIOGRAPHY

Gertrude Scharff was born in Mannheim, Germany, July 14 (Bastille Day), 1911. Perhaps because of that date, she always had an affinity for things French. Only a short time after her birth, the relative serenity of the previous century was shattered by the outbreak of World War I. By the end of the war she was attending public school, in what must have been quite challenging conditions. She soon developed an interest in science with the support of her parents. This may have been partly because her father had been forced to abandon a plan to study chemistry, as his own father died just when he would have been ready for university, making it necessary for him to go into the family business, food wholesaling. Even with the end of the war, life in Germany was difficult, as hyperinflation in the mid twenties was followed by worldwide depression. Nevertheless, Trude went on with her studies, entering the University of Munich at the age of 18, and fairly soon coming to focus on physics.

A frequent pattern then was for students to spend semesters at various universities, and she did that three times, visiting Freiburg, Zürich and Berlin. At Freiburg, she shared a room and much of her extracurricular activity with a cousin, herself quite a strong personality, who said that she was in awe of Trude's dedication to her studies as well as her mathematical powers. On skiing holidays, people provided their own ski lifts, i.e., before you skied down you had to climb up! This may have been good training for climbing over other kinds of obstacles later on. During her semester in Berlin, she joined a "proseminar", where students gave talks on recent journal articles under the supervision of Max von Laue. One of the other students in the seminar was Maurice Goldhaber, and so they first became acquainted by hearing each other's physics talks.

After these sojourns, Trude returned to Munich and began thesis research with Walther Gerlach, studying the effects of mechanical stress on ferromagnetic properties of certain materials. Apparently, at that time in Munich, there were no local female mentors or role models, but she had good relations with her professors. The really big problem was the accession of the Nazis to power in 1933. Life became increasingly awkward for Jews who had any visibility, such as working at a university or running a business. Her parents went to Switzerland but rather soon returned. They never succeeded in leaving Germany again, and both were to perish in the Holocaust.

Meanwhile, Trude attained her Ph.D. in 1935, and immediately sought to leave Germany. Inquiries with a number of contacts in England were almost uniformly discouraging, the exception being a hopeful and supportive note from Maurice Goldhaber in Cambridge. She headed for London where she found that having a degree actually was a disadvantage, as there was more room for refugee students than refugee professionals. For six months she lived on the proceeds of selling a Leica camera and translating German documents into English, eventually getting a place in the lab of G. P. Thomson. In 1939 she and Maurice Goldhaber married, and she moved to join him at the University of Illinois, in Urbana. Their original encounter in Berlin had evolved into what would be a long and fruitful life together.

Now an issue of women in science came to the fore. Maurice already was on the faculty at Illinois, and so, according to strict interpretations of state anti-nepotism laws, she could not be hired as well. Beyond that she was not even allowed lab space, so her only options were to quit research, or to accept her husband's invitation to join him in his lab, of course with no salary. Not only was she unpaid, which meant that Maurice's salary had to cover all costs, soon including child care for sons Alfred and Michael, but also she had to switch fields to make use of the resources in the lab, which was devoted to nuclear physics. As indicated earlier, she made the transition with considerable flair.

In 1950 the family moved from Urbana to Long Island, where Maurice and Trude both joined the scientific staff of Brookhaven National Laboratory. It was then, fifteen years after obtaining her doctorate, that Trude for the first time received a regular, long-term, paid position. During almost all of the preceding interval she had been engaged in experimental physics research, but without the recognition one might otherwise think would go with her accomplishments. The only time she had not done physics research was during part of the war, when she had joined an engineer at Illinois in applied work. Finally, some time after the war, she had been placed on a soft-money position in the physics department at Illinois.

After having started in a regular employment position so late, she was compelled to retire (in a way that couldn't happen now) at the age of 66, meaning her regular employment lasted only 27 years. She continued her research for some years in collaboration with colleagues holding grants at other institutions, and later as a consultant at Brookhaven.

Men as well as women have described how she reached out to help and encourage them when they were facing difficulties, whether professional or personal. She frequently did scientific mentoring. Her connections with women tended to be more personal, listening and counseling, and when necessary intervening if she felt another woman was facing remnants of the discrimination that had been a fact of life in her own career.

She received a number of honors in recognition of her work. When she was elected to the National Academy of Sciences, Maria Goeppert Mayer (see Chapter 18) and Chien-Shiung Wu (see Chapter 24) were the only female physicists who had been elected earlier. She gave numerous invited talks, was a Fellow of the American Physical Society and served on the Council of the Society. She was a member of the board of University Research Association, which oversees Fermilab. Besides these official appointments, she maintained extensive correspondence with leaders in a wide range of sciences who had learned to value her reactions and insights. As one example, she was invited several times to participate in discussions at the Kinsey Institute on issues relating to the role of hormones in human sexuality. Her influence also is clear for succeeding generations in her own family. Both Alfred and Michael hold doctorates in theoretical physics, while David and Sara (the children of Alfred and his wife Suzan) are engaged in physics and medicine, respectively.

Box 23.1 Rotation of quantum systems

The physics in Trude's work is important for understanding nuclear structure, but it also is a rich source of insight into other physical systems. To get at the essence, consider two familiar objects: a soccer ball and an American football. The ideal version of the former is completely spherical, while the latter is stretched along one axis. If one imagines trying to rotate either of these two objects, there is very little difference between them. However, indulge in a bit of fantasy and imagine that both could be made perfectly smooth. In that case, one could not get any grip on the round ball, and so it could not be made to rotate, while a simple push would set the long axis of the football spinning. For familiar systems, such perfect smoothness is unattainable. Even the surface of a ball of ice is rough at the microscopic level, and so the ball can be spun. However, quantum physics, with its discreteness in place of classical continuity, permits ideal smoothness in a system such as an atomic nucleus.

To get a smooth spherical ball to rotate now requires two things at once: at the same time the ball must stretch in some direction, and the resulting long axis must start rotating. Once accomplished, this rotating configuration maintains itself, because the forces trying to restore sphericity are exactly those needed to keep the rotating matter moving in curved paths rather than straight lines. The upshot is that the moment of inertia, which gives the ratio of angular momentum to angular velocity, behaves similarly to that of a stone whirled around at the end of an elastic string. The faster the rotation the more the string is stretched, and the bigger is the moment of inertia. The main qualitative difference from a familiar classical system is that the quantum smoothness can allow the moment of inertia to be zero for the lowest rotational energy state, meaning that there is a minimum angular velocity required before there can be any rotation at all. The resulting description of the rotational behavior in terms of just two parameters for each nucleus gives a wonderfully simple and revealing systematic interpretation of rotational properties for pretty much the entire span of nuclides.

A tool that Trude developed to explain her findings to others has evolved into something so commonplace in this computer age that its origins may have been forgotten – three-dimensional plots, often called "lego" plots. In her case, a quantity such as the energy of the lowest nuclear excitation would be plotted versus the number of protons and the number of neutrons in the nucleus. These latter two quantities formed axes for the chart of the nuclides widely distributed by the

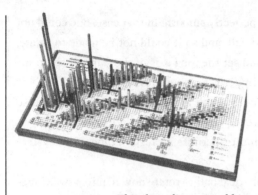

FIGURE 23.2 This three-dimensional histogram is made with square wooden dowels whose heights indicate the energy of the lowest rotational excitation of each nuclide with even neutron and proton number. Note the peaks, most dramatic for the magic nuclei, which according to the shell model (see Chapter 18 on Maria Goeppert Mayer, discoverer of the shell model) are exactly spherical in the ground state. Secondary peaks led to identification of 'pseudomagic' nuclei, which because of pairing effects not included in the shell model also should be considered spherical. The uppermost diagonal row corresponds to the lowest mass nuclei, with neutron number increasing to the right parallel to the base of chart, and proton number increasing upwards parallel to the left side of the chart.

research laboratory of the General Electric Corporation for many years in the postwar period. To describe the nuclear excitation energies, she erected above the squares corresponding to individual nuclides solid vertical towers (with height proportional to energy) using materials such as wood, so that one could see peaks and valleys, which gave an immediate intuition about the changes in behavior from one set of nuclides to another. Today this kind of visualization is done with computer graphics, and as a result is much easier and even more useful.

IMPORTANT PUBLICATIONS

Scharff Goldhaber, G. and Klaiber, G. S., Spontaneous emission of neutrons from uranium. *Phys. Rev.,* **70** (1946) 229. (Voluntarily withheld in May 1942.)

Goldhaber, M. and Scharff Goldhaber, G., Identification of beta rays with atomic electrons. *Phys. Rev.,* **73** (1948) 1472.

Scharff Goldhaber, G., Dover, C. B. and Goodman, A. L.,The variable moment of inertia model and theories of nuclear collective motion *Ann. Rev. Nucl. Sci.*, **26** (1976) 239.

Goldhaber, A. S. and Scharff Goldhaber, G., Electric and dynamic quadrupole moments of even-even nuclei. *Phys. Rev. C*, **17** (1978) 1171.

Scharff-Goldhaber, G., Pseudomagic nuclei. *J. Phys. G: Nucl. Phys.*, **5** (1979) L207.

Gertrude Scharff Goldhaber, obituary, *The New York Times*, February 6 (1998) p. D18.

Bond, P. D. and Henley, E. M., Gertrude Scharff Goldhaber, *Biographical Memoirs of the National Academy of Sciences*, **77** (1999).

24 Chien-Shiung Wu (1912–1997)

Noémie Benczer-Koller
Rutgers University

FIGURE 24.1 Photograph of
Chien-Shiung Wu.

IMPORTANT CONTRIBUTIONS
The early days: the years at Berkeley

Chien-Shiung Wu's first scientific publications resulted from her
Ph.D. thesis research at Berkeley in 1940. She identified two
radioactive isotopes of Xe produced in decay chains following uranium fission. Shortly after she finished her work, during World War
II, scientists who were building a nuclear fission reactor discovered
that the self-sustaining reaction, necessary for continuous operation
of the reactor, was quenched. They remembered Chien-Shiung Wu's
work and suggested she might have a solution to that vexing problem. Indeed, she had determined the decay characteristics of the
two isotopes, information that proved helpful to understanding the

Out of the Shadows: Contributions of Twentieth-Century Women to Physics,
eds. Nina Byers and Gary Williams. Published by Cambridge University Press.
© Cambridge University Press 2006.

FIGURE 24.2 The electron magnetic spectrometer used to determine the shapes of β spectra.

mechanism causing the poisoning of the reactor and hence led toward a solution.

The shapes of beta-decay spectra

The beta decay of nuclei takes place by emission of electrons that exhibit a continuous energy spectrum. The effect had been observed by J. Chadwick and C. D. Ellis and confirmed by L. Meitner (see Chapter 7) and W. Orthman. Such a spectrum contradicted the expected conservation of energy in the decay process. W. Pauli postulated the simultaneous emission of a massless, penetrating, neutral particle, the neutrino, which restored conservation of energy and momentum. E. Fermi proposed in 1934 a comprehensive theory of beta decay, which included the neutrino and disagreed with the theory. The experimental data were inconsistent. Chien-Shiung Wu recognized the importance of reconciling these discrepancies in order to test Fermi's fundamental theory. She had carried out significant work on fission in California and had worked on perfecting radiation counters during the war. She knew that she needed more sophisticated instrumentation. An iron-free magnetic spectrometer had been constructed before the war at Columbia University. Realizing the potential of the instrument, she retrieved it from storage, rebuilt the coils that had frozen because of having been stored with water in them, and started her experiments on the shapes of the spectra of electrons emitted in beta decay. Through her exquisite sense of experimental physics, utter care for precision and reproducibility, she realized that what was required to obtain the correct spectrum of low energy electrons was to prevent electron scattering in the source. Her systematic approach to obtaining ever thinner and thinner sources enabled her to complete a series of superb experiments, which confirmed the Fermi theory of beta decay for allowed transitions. Not satisfied with having developed the method to determine whether there was an adequate theory of beta decay and actually to prove it, she pursued more stringent tests, the study of the shapes of so-called unique forbidden transitions for which the theory is very specific. She showed that indeed the theoretical predictions were confirmed by experiment.[1] As T. D. Lee described her and her work, "C. S. Wu was one of the giants of physics. In the field of beta decay, she had no equal."

The overthrow of space symmetry in weak interactions
After these experiments, which placed her at the top of the profession and brought her worldwide recognition, she was poised to handle the next challenge. It came in the form of a question posed by her colleague at Columbia, T. D. Lee. The field of high energy physics carried out at accelerators was burgeoning. Many new particles were being discovered. Two of these, named at the time the θ and τ, presented a particularly nagging problem for physics. These particles had the same mass, spin and lifetime, and yet one decayed into two pions while the other decayed into three pions. These two decay modes are of opposite parity; reflection symmetry was not obeyed. This symmetry, a property of physical systems seen in all observed interactions

studied up to that time, requires that a mirror image of a process be identical to the original process. Associated with this symmetry is a quantity called parity. The $\tau-\theta$ puzzle led T. D. Lee and C. N. Yang to question the accumulated evidence for conservation of parity in various decay processes. After extensive discussions with Chien-Shiung Wu, the undisputed experimentalist in beta decay and weak interaction physics, they found that there was no evidence for either parity conservation or non conservation in weak interactions. Chien-Shiung Wu immediately decided that she should perform the experiment that would, for the first time, test this symmetry in weak interactions. She was remarkably agile in choosing the experiment that could settle this question. It required low-temperature measurements, a technology with which she was not experienced. She decided to study the asymmetry of electron emission in directions forwards or backwards with respect to the spin direction of polarized ^{60}Co nuclei. She formed a collaboration with the experts in low-temperature spin polarization at the National Bureau of Standards in Washington D.C., E. Ambler, R. W. Hayward, D. D. Hoppes and R. P. Hudson. She adapted a detector for low-energy electrons to fit in the cryostat, and prepared the largest crystals of paramagnetic salts ever produced, crystals that were essential for polarizing the ^{60}Co nuclei. They demonstrated unambiguously that more electrons were emitted backwards with respect to the ^{60}Co spin than forwards.[2] This observation was just the expected result if mirror symmetry was violated. They found that this symmetry was maximally violated in the weak interaction responsible for beta decay.

On the basis of her early measurements of the magnitude of this parity violating effect, colleagues at Columbia University recognized that a pion-muon-electron decay chain would also exhibit the effect. They carried out their experiment in a very short time with an elegant and much simpler apparatus. Nevertheless, they agreed to publish their results together with C.-S. Wu and her collaborators, whose first results had stimulated the latter work. Remarkably, in spite of the intense pressure to publish, she refused to do so for a whole week until she was absolutely sure, through further tests, of the validity of the data. She never waivered from her commitment to understand every facet of an experiment, and to eliminate all artifacts that could display effects mimicking the elusive signal.

The beautiful and definitive work on beta decay, which confirmed the Fermi theory of weak interactions, culminated in the most original and momentous experiments, which proved, without ambiguity or doubt, that parity was not a conserved quantity in weak decays. Lee and Yang received the Nobel Prize in physics in 1957 for their theoretical work. The bold experimental work of Chien-Shiung Wu and her collaborators helped propel the daring experiments of the next generation, as clearly acknowledged by J. Cronin, one of the solvers of the $\tau-\theta$ puzzle: "Chien-Shiung Wu's great discovery inaugurated the golden age of particle physics."[3]

After this great discovery, she pursued investigations in the same style, searching for rare double-beta decays; conserved vector current contribution to weak decays; conservation of light particles, the leptons, exotic atoms in which muons or antiprotons or K mesons play the role of electrons in normal species; as well as a search for the breakdown of time reversal invariance. In a departure from standard particle and nuclear physics, but in a direction that had potential societal benefits, she carried out microscopic studies of hemoglobin in order to understand the structure of changes in the molecules that are responsible for the onset of sickle cell anemia.

In a profession where women are rare and not always fully recognized, Chien-Shiung Wu, along with Marie Curie (see Chapter 4), Lise Meitner (see Chapter 7), Irène Joliot-Curie (see Chapter 12) and Maria Goeppert Mayer (see Chapter 18), was a major architect of the foundation of nuclear and particle physics in the twentieth century.

BIOGRAPHY
Chien-Shiung Wu's life spanned much of the twentieth century. Her accomplishments revolutionized long-held scientific beliefs and

opened the gateway to major experimental tests of the new perspectives. Her personal growth as a scientist serves as a model for women scientists striving to reach parity with their male colleagues.

Like many of the women who achieved their potential in that century, she was fortunate to have been nurtured in a supportive family. Chien-Shiung Wu was born in LiuHe, Jiangsu Province, China, on May 31, 1912, to Wu Zhongyi and Fan Funhua, in an environment where education was paramount. Her father was trained as an engineer, but at that time he was the Director of a School for Girls. He wanted women to be educated and "that every girl have a school to go to."[4]

Wu graduated from this school in 1922. In order to continue her education she had to leave home and travel to Nanjing to attend the Soochow School for Girls. She had enrolled in the Normal School program, which was favored by women pursuing teaching careers. However, by then she already knew that she was interested in science, especially physics and mathematics. She studied most of these subjects by herself, with books borrowed from other students. Her father's advice, "ignore the obstacles and keep walking ahead," was the motto she followed then, as well as later in life, to overcome the many hurdles in her path throughout her professional life.

In 1930, determined to study physics, she registered at the National Central University in Nanjing. Upon graduation, she taught for two years and carried out research in X-ray crystallography at the National Academy of Sciences, in Shanghai. One of her instructors recommended that she consider continuing graduate work at the University of Michigan. With financial support from an uncle she sailed for America in 1936. Her first stop in the US was at the University of California, in Berkeley. Within a few days, she had visited a number of laboratories, met her future husband who was also a student recently arrived from China, was offered the possibility of continuing her studies and found out, in addition, that women were not allowed in the University of Michigan Student Union building!

She promptly enrolled at Berkeley where she earned a Ph.D. in 1940 under the guidance of Ernest Lawrence and the direct supervision of Emilio Segré.

Two years later, the nation was at war. Dr. Wu joined the teaching staff at Smith College. But she was interested in research. While it was very difficult at the time for a woman to obtain a position in a research institution, instructors were needed because most men were serving in the war effort. Thus, she moved to Princeton in 1943. Finally, in the last move of her career, she was hired by Columbia University in 1944 to work on radiation detectors for the Manhattan Project and she found herself again in the laboratory.

The end of the war allowed her to take the direction of her career in her own hands. She needed a problem that would be hers and, most important, that would make significant advances in our understanding of nature. She felt very strongly that, as C. N. Yang mentioned in a statement describing her approach to science, "if you choose the right problem you get important results that transform our perception of the underlying structure of the Universe."

With characteristic flair, she chose to study the mechanism and the underlying structure leading nuclei to decay by emission of electrons or beta decay. After completion of this work, she reached further to the frontier of knowledge and explored mirror symmetry in beta decay. In spite of the international recognition of her beta-decay work, she was still an associate professor without tenure, but, nevertheless, a key member of the large nuclear physics laboratory that had been supported during the war as part of the Manhattan Project. It wasn't until after the parity work, in 1958, that she was named Professor of Physics at Columbia. In the same year, she was elected to the National Academy of Sciences and was awarded the National Medal of Science and the Research Corporation Award. She received the John Price Wetherill Medal of the Franklin Institute in 1962. But it took many more years before she was able to enjoy the recognition, honors, awards, honorary degrees and appreciation she so amply deserved, most notably the Cyrus B. Comstock Award (1964), the Tom Bonner Prize of the American Physical Society (1975) and the Wolf Prize from the State of Israel (1978). She was the first woman to receive an honorary doctorate from Princeton University (1958).

Chien-Shiung Wu's insatiable thirst for understanding fueled her scientific drive. She was a quintessential experimentalist and had an uncanny flair for singling out the crux of a problem in a field that was fogged with erroneous evidence. She had an exquisite evaluative sense, and could build sophisticated experimental apparatus. She had unusual charm and conquered small groups as well as filled lecture halls with her enthusiasm. At the same time she delivered down-to-earth lectures, which ranged from serious professional presentation of the work to a display of her obvious delight in having penetrated a secret of nature. With a twinkle in her eyes and a light touch of humor she would summarize her observations in a human form and marvel at the subtleties of nature, which is "much more mischievous than we imagine." She became a public figure. She was the first woman to serve as President of the American Physical Society (1975). In that role, she had the opportunity to reflect privately and publicly. In later years, she devoted much of her time to develop the scientific infrastructure and educational programs in both the People's Republic of China and Taiwan. Surprisingly, for a person who was so close to her experiments, she, nevertheless, appreciated the importance of "big physics" and supported the development of large accelerators for particle physics and of synchrotron radiation facilities that were to be used in large part for biological and condensed matter studies.

Beauty and aesthetics were major ingredients of her work, of her demeanor, of her relationship with friends and of her home. Besides her scientific contributions she left behind an unparalleled human legacy. She was deeply concerned with the education of her son Vincent Yuan and she derived immense pride from the fact that not only did he obtain a Ph.D. in physics, but that his career also focused on studies of parity nonconservation.

She was extremely close to her husband Luke C. L. Yuan who supported her unflinchingly, emotionally as well as in her efforts at using and developing state-of-the-art electronics. And she beamed when she described the academic achievements of her granddaughter. She nurtured 33 graduate students with whom she worked very closely. She appeared in the laboratory early in the morning and stayed very late, and expected everyone to be as totally devoted to physics as she was. Her students played the role of an extended family. She cared about their lives and relationships, even wrote to parents to relate good news on the progress and achievements of her students. Most of her graduate students would recognize that she opened for them the doors of the physics community, inspired them with the highest professional standards, endowed the advisor–student relationship with grace, love and affection and showered them with friendship, encouragement, nurturing and continuous support long, long after they had left her laboratory.

Chien-Shiung Wu's contributions have been critical to many of the discoveries in nuclear physics in the second half of the twentieth century. Her pioneering work in weak interactions and, in particular, in beta decay, have established nuclei as a unique laboratory for the study of fundamental symmetries. She combined the human aspects of research with the technological infrastructure, features that were exhibited in her deep appreciation of history and that render her review articles such wonderful "stories" to read. She has been a teacher and mentor to many a generation of nuclear and particle physicists. She supported the improvement of teaching of physics, mathematics and science. She supported efforts to uphold the freedom of scientists elsewhere in the world. There has been much progress since Chien-Shiung Wu first landed in California in 1936, both in physics and in the recognition of professional women, much of it due to the perseverance, enlightenment and accomplishment of women like her. The spirit of dedication to science and to her people that characterized Chien-Shiung Wu will have a lasting effect on future generations.

NOTES

1. "Recent investigation of the shapes of beta-ray spectra," Chien-Shiung Wu, *Rev. Mod. Phys.*, **22** (1950) 386.
2. "Experimental test of parity conservation in beta decay," C.-S. Wu, E. Ambler, R. W. Hayward, D. D. Hoppes and R. P. Hudson, *Phys. Rev.*, **105** (1957) 1413.
3. Cronin, J. J., *International Conference on Physics since Parity Symmetry Breaking in Memory of Professor C. S. Wu* (People's Republic of China, Nanjing: World Scientific, August 16–18, 1997).
4. McGrayne, S. B., *Nobel Prize Women in Science* (Birch Lane Press, 1992).

FURTHER READING

Wu, C.-S. and Moszkowski, S. A., *Beta Decay* (New York: Interscience Publishers, 1966).

25 Eleanor Margaret Burbidge (1919–)

Virginia Trimble

Department of Physics and Astronomy, University of California, Irvine and Las Cumbres Observatory

FIGURE 25.1 Photograph of E. Margaret Burbidge taken from her autobiographical article "Watcher of the Skies", *Ann. Rev. Astron. Astrophys.*, 1994, **32**:1–36.

SIGNIFICANT CONTRIBUTIONS

Eleanor Margaret Burbidge (née Peachey and always called Margaret) is an optical spectroscopist. That is, most of her career has been devoted to the analysis of the visible light coming to us from stars, galaxies and quasars. She has contributed a great deal to our understanding of the chemical compositions of the stars and the structure of galaxies and quasi-stellar objects. An astronomer like Burbidge uses a large telescope to gather light from a source and a spectrograph to spread

Out of the Shadows: Contributions of Twentieth-Century Women to Physics,
eds. Nina Byers and Gary Williams. Published by Cambridge University Press.
© Cambridge University Press 2006.

the light out into its component colors (wavelengths) and records the resulting spectrogram (spectrum for short) on a photographic plate or, since about 1970, more often on a Charge Coupled Device (CCD) or other electronic detector.

Margaret Burbidge began her work in astronomy in the shadow of World War II, receiving her Batchelor's degree in Summer 1939 and her Ph.D. in 1943 at University College London, with a thesis on the spectrum of the Be star Gamma Cassiopeiae, under the nominal direction of C. C. L. Gregory. She was largely in charge of the care, maintenance and use of the University College Observatory 24" telescope, since Gregory and other male members of the staff were assigned to war-related work for the duration.

What is a Be star? The "B" part indicates a surface temperature near 20 000 K (our sun, a G star, is at about 5600 K) and the "e" indicates emission lines, as well as the usual stellar absorption lines, that must come from a shell or ring of hot gas orbiting the star. The emission lines vary in strength in months and sometimes disappear completely for a while, returning after a few years. The Be stars typically rotate rapidly, and many of them are paired with other stars in binary systems. Some are very young, and others (including Gamma Cassiopeiae) are somewhat evolved. The thesis concluded that all of these factors were relevant to explaining her data, and the modern view (e.g., Kaler 1989, p. 191 ff) is much the same. Her thesis and first published paper (also 1943) appeared under the name E. M. Peachey, apparently a name of French Hugenot derivation, remarks Burbidge (1997) in her autobiography, noting that her ancestors were perhaps fishermen, expecting the reader to catch the similarity of the name to the French word pecheur without further explanation. All her later work appears under the name Burbidge.

It is usually said that Margaret Burbidge is the observer and Geoffrey Burbidge the theorist, but even a cursory reading of their papers, a large fraction of which are joint and most of the rest of which thank each other for various kinds of input, reveals that the actual division of labor was rarely that sharp. Both struggled over the trials of photographic emulsions and the tribulations of non-equilibrium atomic processes. G. R. Burbidge also received his Ph.D. from University College London at about the time of their marriage in 1948, and they began observing additional Be stars from the much better site of the Observatoire de Haute Provence (France) in 1949. First results appear in Burbidge and Burbidge (1951), and a review of the topic in Burbidge and Burbidge (1956).

Postdoctoral fellowships in the United States, beginning in 1951, gave them access to the 40″ refractor at Yerkes and the 82″ reflector at McDonald Observatory (in Texas, but then owned by University of Chicago). Both also spent time at Harvard, where Margaret Burbidge absorbed the lore of stellar abundances and variability from Cecilia Payne-Gaposchkin (see Chapter 14). She continued to collect spectra of Be stars but also to investigate some of the chemically peculiar stars (see Kaler 1989, p. 172 ff for a modern view). These are a heterogeneous collection of stars with surface temperatures near 10 000 K, and spectra revealing very large excesses and deficiencies of chemical elements seemingly almost chosen at random. Some also have magnetic fields over their surfaces of thousands to tens of thousands of gauss. (The solar average is only about 1 gauss, though stronger fields are found in sunspots.) Babcock (1947) had only then recently recognized the first of these strong fields, and magnetism was very much in the air at the University of Chicago (fanned also by Enrico Fermi's work on the acceleration of cosmic rays by magnetic fields, and the discovery of interstellar polarization by Hall and Hiltner).

The Burbidges concentrated on α^2 Canum Venaticorum, not at 10 000 gauss the strongest of the magnetic fields, but surely the strangest of the spectra, with lines due to the rare element europium dominating those of much commoner elements. Their analysis (Burbidge and Burbidge 1955) and that of other astronomers gradually established that these peculiar spectra result from concentrations (and sometimes deficits) of particular elements that are confined to the stellar surface, that is, the anomalies do not reflect the operation of nuclear processes inside the stars. But, in the same 1951–3 time frame, they attended two workshops concerned with stellar abundances and evolution and met both Fred Hoyle (who had been working on nuclear reactions in stars since before 1939) and William A. Fowler of the California Institute of Technology. Fowler was part of a group of nuclear physicists who were beginning to measure, at Kellogg Laboratory, the properties of atomic nuclei that would be needed to understand the full range of nuclear reactions in stars.

The most conspicuous fruit of these new interests and friendships was the classic 1957 paper, "Synthesis of the Elements in Stars" (Burbidge, Burbidge, Fowler and Hoyle, 1957, very generally called B^2FH within the astronomical community). It lays out the full range of nuclear reactions (and astronomical sites for them) needed to account for the observed abundances of the elements in the sun and other stars. An important aspect of their achievement was the recognition that three separate processes, involving the capture of neutrons and perhaps protons, were needed to match the full range of the isotopes of the heaviest elements, from germanium to uranium. A. G. W. Cameron reached many of the same conclusions at about the same time, but published less extensively, so that his work was less influential.

The Burbidges moved back to England in 1953 and then from Cambridge to Caltech and Mt. Wilson Observatory in the middle of this work, in 1955. There they were able to obtain spectra of a wide variety of kinds of stars whose surface abundances reflected the operation of nuclear reactions in the stars themselves or in earlier generations, e.g., their work on HD 46407, a so-called Barium II star, whose anomalous spectrum revealed excesses of barium and other products of one of the neutron-capture reactions they had been thinking about.

Back at Yerkes from 1957 to 1962, they again had access to the McDonald 82" reflector but by now they were at least as interested in whole galaxies as in individual stars. This was partly because of the implications of their work on element synthesis for the evolution of stellar populations, but also because of Geoffrey Burbidge's interaction with astronomers studying radio emission from galaxies during their stay in Cambridge. And there was at McDonald, though not much used, a prime focus spectrograph that would permit collecting light from the whole length of a galaxy at once. The spectra showed emission lines (from hydrogen, the most abundant of all elements, and from nitrogen) whose wavelengths were shifted by the rotation of the galaxies, so that one side is coming toward us and the other going away from us at the same time. The balance between this rotation and the force of gravity required to keep the galaxies stable, in turn, reveals how much mass is at various positions in the galaxy.

The outcome was a series of papers with titles frequently of the form, "The Rotation and Mass of . . ." followed by the name of a galaxy. Many of these were collaborations with theorist Kevin H. Prendergast (now at Columbia University), who devised a mathematical framework for calculating the distribution of mass with distance from the center of the galaxy required to account for the measured rotation speeds at various positions.

Most of the galaxies they studied had just about the mass distribution that they had expected from the stars that contribute to the continuum part of the spectrum (Burbidge, Burbidge and Prendergast, 1965). The method is introduced in Burbidge (1962) and the results summarized in Burbidge and Burbidge (1975; see Freeman, 1975, for another, contemporary view of the problems). The Burbidges also remarked early on that nothing could then be said about the outer parts of the galaxies, whose light was too faint to record with photographic plates in a rather old-fashioned spectrograph attached to a medium-sized telescope. But the "hand over" to a slightly younger observer who would push galactic rotation curves out to where they begin to reveal the presence of dark matter can be spotted in the paper Burbidge, Burbidge and Rubin (1964) on the velocity field of M82. (See Rubin, 1987, for a brief discussion of what happened next in this field, and see also Chapter 31).

In 1962, the Burbidges settled into permanant positions at the University of California, San Diego, where they remain to this day (though with intermediate excursions back to England). Here, Margaret Burbidge was entitled to apply for time on the 120″ telescope at Lick Observatory, owned by the University of California. The next year, astronomers at Mt. Wilson and Palomar Observatories announced the discovery of a new class of astronomical object, the quasars. This was originally an attempt to pronounce the acronym QSRS (quasi-stellar radio source). The term QSO (quasi-stellar object) is used more or less interchangeably, though most are not radio sources.

Quasars are defined by (a) quasi-stellar (i.e., very compact) images on the sky, and (b) emission lines that display redshifts much larger than are found for any star in our own galaxy. The conventional interpretation is that they are nuclei of distant galaxies, with distances proportional to redshift as part of the overall expansion of the universe, in which a central black hole accretes material from surrounding stars and interstellar gas. Much of the energy released during the accretion or infall is reprocessed into fast-moving particles, strong magnetic fields and radiation over the whole range from radio to optical to X-rays.

Many of the observations obtained by Margaret Burbidge over the years since 1963 have been interpreted by the rest of the community as part of this picture (e.g., Burbidge, Lynds and Burbidge, 1966, the co-discovery of absorption lines in a QSO, and Wampler, et al., 1973, reporting a red shift of 3.53 for OQ 172, which remained the largest known for nearly a decade). She wrote one of the early reviews of the subject (Burbidge, 1967). It is clear, however, that even quite early in the quasar story (e.g., Burbidge, et al., 1966, on the significance of the strength of ionized magnesium emission lines in QSO spectra) that they both had doubts about the conventional wisdom. See also the book, Burbidge and Burbidge (1967).

Through the intervening years, both Burbidges have become more firmly convinced that many or most QSOs are much closer to us than would be implied by their redshifts, and that a variety of other aspects of the conventional interpretations of the observations in cosmology need rethinking (e.g., E. M. Burbidge, 1997). Most of the astronomical community continues to believe that evidence strongly favors QSO distances proportional to redshift and a universe with a hot, dense big bang in its past. The community would, however, also agree that E. Margaret Burbidge has earned the right to her scientific opinions and conclusions.

For updates on nucleosynthesis, see Trimble (1991) and Wallerstein, *et al.* (1997). The conventional view of cosmology and quasars appears, for instance, in Bahcall and Ostriker (1997).

BIOGRAPHICAL MATERIAL

Eleanor Margaret Peachey was born on August 12, 1919, in Davenport, near Manchester, England, the daughter of Stanley John Peachey, an instructor in chemistry at Manchester School of Technology, and the considerably younger Marjorie Stott, who had been his student there. Her parents encouraged an interest in science through gifts of suitable books, a microscope and (naturally!) a chemistry set.

She attended, as did most middle-class Englishwomen of her generation, an all-girls school, developing interests in poetry, languages and mathematics, as well as the sciences, and continued on to University College London (UCL) in 1936, where the genders were much more thoroughly integrated than at Cambridge or Oxford in the same period. She received the degrees B.Sc. in 1939 and Ph.D. in 1943. Though her official thesis advisor was C. L. Gregory, she also worked with Elizabeth Williamson at UCL. The war-time reassignment of many UCL scientists to tasks in Army, Airforce and Admiralty gave Miss/Dr. Peachey earlier and wider range of experience of scientific independence and responsibility than would otherwise probably have been the case.

E. M. Peachey married Geoffrey R. Burbidge, a slightly younger fellow student at UCL, in April, 1948. Their daughter Sarah, born in 1956, made a firm, early decision not to be a scientist, though family anecdotes show her to be a keen observer of her surroundings. She eventually took a degree in law.

E. M. Burbidge's life in astronomy has strongly exemplified her stated principle of finding a way around obstacles rather than complaining. At the end of the war, she sought a Carnegie Fellowship at

Mt. Wilson Observatory, only to discover that these were available only to men. When the Burbidges finally arrived in Southern California in 1955, Geoffrey (nominally the theorist) held a Carnegie Fellowship at the Observatory, and Margaret (nominally the observer) a postdoctoral position in the Kellogg Radiation Lab at the California Institute of Technology. Not surprisingly, they shared both the observing and the theory tasks. During stays before and after this at the University of Chicago, where nepotism rules prohibited simultaneous faculty appointments for husband and wife, the Burbidges worked around that too, as Maria Goeppert Mayer and Joseph Mayer had done a little earlier (see Chapter 18).

The Burbidges shared the Helen B. Warner Prize of the American Astronomical Society in 1959, at the beginning of their Chicago stay. Margaret Burbidge remains (in 2005) the only woman recipient of this prize, directed at astronomers in the early stages of their careers. In 1971, she declined the Annie J. Cannon Prize of the American Astronomical Society, restricted to women astronomers, on the grounds that it was no longer either appropriate (with which most of her women colleagues agreed) or needed (less obvious). This forced a reassessment of the Prize, which has since been administered by the American Association of University Women and given, upon application, to women astronomers not far past the Ph.D.

Burbidge was the first woman elected president of the American Astronomical Society. During her term, she made the first presentation of the Society's Russell Lectureship to a woman, Cecilia Payne-Gaposchkin (who had been the first winner of the Cannon Prize in 1934) (see Chapter 14). Margaret herself was the Russell Lecturer in 1984, and the third (and so far last) woman Russellist was Vera Rubin, discussed earlier, who learned from Margaret Burbidge how to do long-slit spectroscopy.

The Burbidges (also like the Mayers) have spent much of their latter careers at University of California, San Diego. They tried the return to England in 1972–3, where Margaret Burbidge had been asked to take up the directorship of the Royal Greenwich Observatory, but not the honorary position of Astronomer Royal, which had always previously been associated with it. The splitting of the roles was perceived by most of the community as a gender-based slight, though this is probably not the whole story. In any case, some combination of science politics, a leg broken in an automobile accident and the difficulties of living and working in England, where shops were open such restricted hours that every family just about had to have one stay-at-home member, sent her back to San Diego after 18 months. She became a US citizen shortly after being elected into the American Astronomical Society presidential sequence in 1975.

E. M. Burbidge is a Fellow of the Royal Society, the American Academy of Arts and Sciences and the US National Academy of Sciences. She has received the National Medal of Science, the Albert Einstein Award from the World Cultural Council and recognition from a number of other organizations, including honorary doctorates from roughly a dozen colleges and universities in England and the United States, beginning with Smith College in 1963.

Box 25.1   Astronomical spectroscopy

An astronomical spectrum normally includes a continuum (rainbow) and absorption and/or emission features (i.e., missing light and/or extra light at specific wavelengths caused by changes in the atoms and molecules of specific elements and compounds). Spectra can be analyzed fairly directly to tell us:

(1) the temperature and gas density of the source,

(2) its chemical composition (which elements are present and how much of each),

(3) whether the star, galaxy, or whatever is moving toward us or away from us and how fast,

(4) how fast it is rotating,

(5) the strength of its magnetic field, if any, and

(6) what sort of gas or dust is between us and the source. Somewhat less directly, the analysis can also often provide information about the size and mass of the object and its brightness and distance.

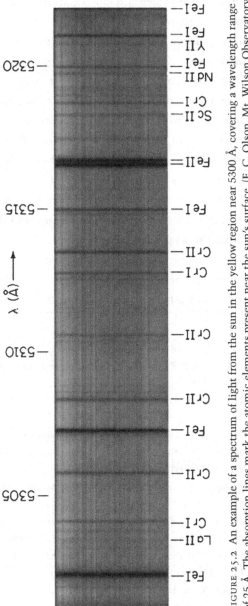

FIGURE 25.2 An example of a spectrum of light from the sun in the yellow region near 5300 Å, covering a wavelength range of 25 Å. The absorption lines mark the atomic elements present near the sun's surface. (E. C. Olson, Mt. Wilson Observatory.)

Once the light from the focus of the telescope is in the spectrometer it is reflected from a precisely ruled diffraction grating, which spreads out the light according to its wavelength, and the amount at each wavelength is recorded either on photographic emulsions or with electronic CCD detectors. Figure 25.2 shows an example of an emulsion spectrum recorded from the sun over a rather narrow range of wavelengths in the yellow part of the visible spectrum. The light background is part of the continuous "rainbow" emission characteristic of the 5600 K surface temperature of the sun. The sharp black lines occur at wavelengths where the thermal radiation is strongly absorbed by atomic transitions of elements near the sun's surface, blocking the transmission. Since the positions of the atomic lines are well known from laboratory studies, the presence and abundance of the various elements in the star can be identified from the analysis of the spectrum. The Burbidges and others used this technique to obtain measurements of the amounts of the various elements in the stars, and to show that the most abundant elements are the ones produced by the commonest nuclear reactions.

IMPORTANT PUBLICATIONS

Burbidge, E. M. and Burbidge, G. R., Hydrogen and helium line intensities in some Be stars. *Astrophys. J.*, **113** (1951) 84.

Burbidge, G. R. and Burbidge, E. M., *Astrophys. J.*, **122** (1955) 396. (And *Astrophys. J.* suppl. 1, 431).

Burbidge, E. M. and Burbidge, G. R., *Vistas in Astronomy* 2, (1956) 1446.

Burbidge, E. M., Burbidge, G. R., Fowler, W. A. and Hoyle, F., The synthesis of elements in stars. *Rev. Mod. Phys.*, **29** (1957) 547.

Burbidge, E. M., In *Problems of Extragalactic Research*, ed. G. C. McVittie (New York: Macmillan, 1962) p. 85.

Burbidge, E. M., Burbidge, G. R. and Rubin, V. C., A study of the velocity field of M82 and its bearing on explosive phenomena in that galaxy. *Astrophys. J.*, **140** (1964) 942.

Burbidge, E. M., Burbidge, G. R. and Prendergast, K. H., The rotation and mass of NGC 681. *Astrophys. J.*, **142** (1965) 154.

Burbidge, E. M., Lynds, C. R. and Burbidge, G. R., *Astrophys. J.*, **144** (1966) 447.

Burbidge, E. M. and Burbidge, G. R., *Quasi-stellar Objects*. (W. H. Freeman, 1967).

The masses of galaxies. In *Galaxies and the Universe*, eds. A. Sandage *et al.* (Chicago: Univ. Chicago Press, 1975) p. 81.

Burbidge, E. M., Observations of QSOs which are critical for cosmology. In *The Universe at Large*, eds. G. Munch *et al.* (Cambridge, UK: Cambridge Univ. Press, 1997) p. 137.

FURTHER READING

Babcock, H. W., *Astrophys. J.*, **105** (1947) 105.

Bahcall, J. N. and Ostriker, J. P., eds., *Unsolved Problems in Astrophysics*, (Princeton: Princeton Univ. Press, 1997).

Freeman, K. C., In *Galaxies and the Universe*, eds. A. Sandage *et al.* (Chicago: Univ. Chicago Press, 1975) p. 409.

Kaler, J. B., *Stars and their Spectra* (Cambridge, UK: Cambridge Univ. Press, 1989). (Corrected paperback edition, 1997.)

Trimble, V., Origin and abundances of the elements, revisited. *Astron. Astrophys. Rev.*, **3** (1991) 1.

Rubin, V. C., Constraints on dark matter from optical rotation curves. In *Dark Matter in the Universe*, J. Kormendy & G. R. Knapp (IAU Symp. 117) (Kluwer, 1987) p. 51.

Wallerstein, G. *et al.*, Synthesis of the elements in stars: forty years of progress. *Rev. Mod. Phys.*, **69** (1997) 995.

Wampler, E. J., Robinson L. B., Baldwin, J. A. and Burbidge, E. M., *Nature*, **243** (1973) 336.

26 Phyllis StCyr Freier (1921–1992)

Cecil J. Waddington

School of Physics and Astronomy, University of Minnesota

FIGURE 26.1 Phyllis StCyr Freier.

SOME IMPORTANT CONTRIBUTIONS

Phyllis Freier made a number of very significant contributions to physics, some of which are described here. She also was an educator, with students ranging from Girl Scouts to graduates. In addition, she raised a family with two children, who have both had successful careers.

Out of the Shadows: Contributions of Twentieth-Century Women to Physics, eds. Nina Byers and Gary Williams. Published by Cambridge University Press. © Cambridge University Press 2006.

Professionally she established an internationally respected nuclear emulsion laboratory. She trained and supervised technicians and graduate students. With her colleagues, she published nearly 70 refereed papers and a large number of conference proceedings. The majority of the topics that she studied were concerned with the nature and properties of the particles that make up the cosmic radiation, but in later years there was also a significant emphasis on the nuclear physics of the interactions of relativistic nuclei. Her formal contributions can be illustrated by discussing a few of the more significant topics.

The discovery of heavy nuclei in the cosmic radiation

Her first major discovery occurred while she was still a graduate student at the University of Minnesota studying for her Ph.D. She worked with a group who were developing the capability of placing instruments on balloons able to reach altitudes of >90 000 ft and stay there for several hours. With this technology it was possible to study and identify the particles that were present in the cosmic radiation above the atmosphere of the earth, known as the primary cosmic radiation. At that time it was generally assumed that these particles were protons, with a possible admixture of helium nuclei.

In April, 1948, on the first balloon flight to reach a high altitude and remain there for several hours, two different detectors of particles were included. These consisted of a cloud chamber and a small package of nuclear emulsions. Analysis of both suggested that there were energetic nuclei in the primary cosmic radiation with charges as great as that of iron originating from some extraterrestrial source.

Freier had the responsibility of processing and analyzing the nuclear emulsions. Although these emulsions basically resembled ordinary photographic emulsions, they were modified so that they could record the passage of individual energetic charged particles. The processing and examination of these emulsions was difficult and poorly understood at this time. Procedures for processing had to be developed by trial and error. The analysis depended upon examination of the developed emulsion under high powered microscopes looking for the trails of silver grains produced by the fast particles. The analysis was time consuming, requiring dedicated patience and the establishment of systematic procedures. The energy deposited in the emulsions, which formed the particle trails, gave these trails a thickness that increased as the square of the charge on the particle. As a result, highly charged nuclei could be readily distinguished from those of lesser charge. Freier was the first person to see such a trail produced by a highly charged particle and to recognize it for what it was.

She developed procedures to estimate charges of these nuclei, and was able to show that the abundances of nuclei in the primary cosmic radiation roughly resembled those of the matter in the solar system. This observation proved that the cosmic radiation consisted of nuclei that had been accelerated to relativistic energies, but that had a composition rather similar to that found in normal main sequence stars. The exceptions from being precisely similar have occupied cosmic-ray studies ever since. The proof of the existence of heavy nuclei in the primary cosmic radiation represented a major step forward in trying to understand the baffling problem of the origin of this radiation – a problem that is still not completely solved.

The results of these pioneering studies were first described in a paper published in 1948[1] and in her 1950 Ph.D. thesis. A general survey of the influence of this discovery, with many of the original references, can be found in the proceedings of a symposium held in 1988 on the 40th anniversary of these discoveries.[2]

The abundance of secondary elements in the primary cosmic radiation

A decade later, general features of the relative abundances in the cosmic radiation of the elements from lithium (^{3}Li) to nickel (^{28}Ni) had been largely determined. However, a serious controversy had developed regarding the abundances of the light elements, lithium, beryllium (^{4}Be) and boron (^{5}B). These elements are very rare in the universe. They are easily destroyed in stellar interiors where nucleosynthesis creates essentially all the heavy nuclei. Some observers reported that these elements were also essentially absent from the primary cosmic radiation, while others claimed that they were as abundant as the heavier elements carbon (^6C), nitrogen (^7N) and oxygen (^8O). It was assumed that any light nuclei seen in the cosmic radiation would be secondary fragments resulting from the breakup of heavier nuclei due to nuclear interactions occurring during passage through interstellar matter. Thus, the relative abundances of these light and heavier elements were a measure of the amount of matter traversed by these nuclei moving at close to the speed of light. From the assumed matter density in interstellar space it was possible to estimate the time since the nuclei were accelerated, and hence the "age" of the cosmic radiation. The experimental situation was complicated by the fact that all these measurements were made on high-altitude balloons. These typically floated under an amount of residual atmosphere that was similar to or greater than the amount of interstellar material needed to generate a significant abundance of these light elements. As a result, the corrections for the fragments produced in the overlying atmospheric matter were critical, and controversial. In 1958, Freier and co-workers were able to make a series of very careful measurements on nuclear emulsions exposed at a higher altitude than had been done previously, thus minimizing the effects of the residual atmosphere.[3] These measurements proved that there was a finite abundance of light nuclei in the primary radiation and that the age of the nuclei was a few million years.

The modulation of the cosmic radiation by solar activity
Ground measurements had shown that the cosmic-ray intensity was modulated during the solar cycle, with the intensity being greatest when the sun showed minimum activity. By 1965, it was possible to study the variations of the intensity of cosmic-ray helium nuclei over a complete 11-year cycle of solar activity. The helium nuclei are the most abundant multiply charged particles in the cosmic radiation and thus can be used to trace the effects of solar modulation on all the heavy nuclei. In a 1965 review of the available data, Freier and Waddington showed that the variations in intensity

could be represented as being the results of an energy loss as the particles entered into the region affected by the solar wind known as the heliosphere.[4] The intensity is reduced in the heliosphere. Low-energy particles cannot get as far into the heliosphere as the earth. Higher-energy particles arrive but with lower energies, hence the whole energy spectrum is moved down and the differential and integral intensities reduced. In this paper, the loss was modeled as being due to a retarding electric potential, since this was mathematically convenient, although acknowledged as being physically implausible. Later work has shown that an energy-loss model does describe the modulation, although it is not due to an electric field but to the effects of the outflowing solar wind.

Cosmic-ray isotopes

By 1980, it was clear that in order to define the subtle differences between the cosmic-ray abundances and those of the matter in main sequence stars such as the sun, it was necessary to determine the isotopic abundances in cosmic rays of some of the more abundant elements. Such measurements represented a serious technical challenge when applied to an isotropic flux of relativistic nuclei using detectors capable of being flown on balloons or satellites. One of the first attempts to measure these abundances for the nuclei heavier than oxygen was reported in a paper[5] that discussed the isotopes of neon (^{10}Ne) and magnesium (^{12}Mg). The abundances of these isotopes had a particular significance in that they demonstrated that there was an excess of neutron-rich nuclei in the cosmic radiation as compared with the solar abundances. These observations helped to define the difference between the source of the cosmic rays and main sequence stars such as the sun.

Fragmentation of very heavy nuclei

With the discovery in the 1960s of the presence of nuclei in the cosmic radiation with charges significantly heavier than iron or nickel, it became important to consider the effects of the interstellar medium on such heavy nuclei. These nuclei would be more rapidly fragmented than the lighter nuclei, and hence provided a better probe of the effects of the interstellar medium on the cosmic radiation. At the same time, artificial accelerators such as the LBL Bevalac and the Brookhaven AGS began to produce beams of energetic nuclei as heavy as gold and uranium. Exposures of nuclear emulsions to these beams resulted in a long series of experiments to study the effects of interactions. These studies were not only concerned with the implications for cosmic-ray studies, but also with the nuclear physics of high-energy interactions of heavy nuclei. Most of these studies involved relatively large international collaborations, such as the KLM (Krackow, Lousiana, Minnesota) collaboration, with much of the analysis of the emulsions being performed in the University of Minnesota lab directed by Freier. An initial study of the interactions of 1.0 AGeV gold nuclei (A is the atomic mass number so the total kinetic energy of a ^{197}Au nucleus with 1 AGeV is 197 GeV)[6] produced at the Bevalac,[7] provided a starting point for many of the later studies.[8]

Other studies
Freier was also involved in a number of other related studies. These included topics such as the energetic particles produced in solar flares,[9] gamma-ray astronomy[10] and the propagation of nuclei through the interstellar medium.[11]

BIOGRAPHY
Phyllis Freier was born Phyllis Mary StCyr on January 19, 1921, in Robbinsdale, Minnesota and, apart from a brief period during World War II, remained in Minnesota for all her life. She was educated at local high schools and at the University of Minnesota, being awarded a B.S. in 1942. As a graduate student she was a teaching assistant at the University of Minnesota from 1942–4. She met and married another teaching assistant, George Freier, in 1942, who was then drafted to the Naval Ordinance Laboratory, Washington DC. In 1944, she was awarded her M.A. and then joined George in Washington, from 1944–7,

working in a different section of the same Laboratory. On their return to Minnesota, in 1948, George was awarded his Ph.D. and was appointed an Assistant Professor in 1949. Phyllis was awarded her Ph.D. in 1950, but was prevented by the nepotism rules in place at that time from being appointed to a faculty position until 1970. For these 20 years her national and international reputation grew, but she was not allowed to teach undergraduate or graduate students, and could not be awarded her own research grants. While this allowed her to devote more of her energies to her research, it was clearly an unsatisfactory situation. Fortunately, her colleagues were supportive, and she not only received the funding needed for her programs, but was also able to act as an unofficial adviser and mentor to the graduate students involved in the emulsion group as well as those in associated fields.

In 1970, she was finally appointed to a tenure position as an Associate Professor and then as a Full Professor, in 1975. She rapidly proved her talents as an excellent teacher and over the next few years was awarded all the available honors for teaching in the University. In 1979, she was the CLA University College Distinguished Teacher, in 1986, she was given the I. T. Student Board Outstanding Professor Award and, in 1988, the Horace T. Morse-Minnesota Alumni Association Award for Outstanding Contributions to Undergraduate Education. Typically, she would receive standing ovations from her students at the end of each quarter and the highest possible ratings. As one of the very few women Professors on the Physics faculty she acted as a role model and mentor for many of the women graduate students in the school. Many of these students have gone on to take their own distinguished positions in science and society.

Freier was elected a Fellow of the American Physical Society (APS) and of the American Association for the Advancement of Science. She played an important role on the national scene, being elected an early Chairman of the Cosmic Physics Division of the APS and later serving as a Divisional Councilor of the APS. In addition, she served on the APS Committee on the Status of Women in Physics and on the APS Nominating Committee. At the University level, she served on the Senate Judicial Committee from 1975–6 and was chair 1978–9. She also served on the Senate Consultative Committee 1982–5. She chaired the University Sexual Harassment Committee 1980–2 and was on the Internal Review Committee for Women's Intercollegiate Athletics 1980–1. At the College level she served on the IT Consultative and Appeals Committee, The Grievance Committee and the Planning Committee, as well as on the CLA Consultative Committee.

In spite of all her research, teaching and administrative duties, she still found time to raise two successful children, look after a husband, work with a Girl Scout Troop of minority girls, teach Sunday School and play an active role in her community. Her energy was legendary, inspiring her colleagues and graduate students to greater efforts than they had originally intended. Even at the end, when her physical activities had been cruelly limited by Parkinson's disease, she retained her intellectual curiosity and was still playing an active role in the analysis of data.

NOTES

Only a few of Freier's many publications in refereed journals are listed. She also made numerous contributions to conference proceedings, most of which can be found in the proceedings of the biannual IUPAP international Cosmic Ray Conferences. These contributions were typically in the form of short, four-page papers and are frequently quoted as primary references.

1. Freier, P., Lofgren, E. J., Ney, E. P. and Oppenheimer, F., Evidence for heavy nuclei in the primary cosmic radiation. *Phys. Rev.*, **74** (1948) 213–17.
2. Freier, P. S., The cosmic abundances of matter. *AIP Conf. Proc.*, **183**, ed. C. Jake Waddington (1988).
3. Freier, P. S., Ney, E. P. and Waddington, C. J., Lithium, beryllium, and boron in the primary cosmic radiation. *Phys. Rev.*, **113** (1959) 921–7.
4. Freier, P. S. and Waddington, C. J., The helium nuclei of the primary cosmic radiation as studied over a solar cycle of activity, interpreted in terms of electric field modulation. *Space Sci. Rev.*, **4** (1965) 313–71.

5. Freier, P. S., Young, J. S. and Waddington C. J., The neutron-rich isotopes of cosmic ray neon and magnesium. *Astrophys J.*, **240** (1980) L53–8.

6. In order to compare energies of different nuclei, energies are always expressed in AGeV or GeV per nucleon (same thing). Two nuclei with the same AGeV have the same velocity.

7. Waddington, C. J. and Freier, P. S., The interactions of energetic gold nuclei in nuclear emulsions. *Phys. Rev.*, **C 31** (1985) 888.

8. Csernai, L. P., Freier, P., Mevissen, J., Nguyen, H. and Waters, L., Identification of collective flow by transverse-momentum analysis of emulsion data for Au + AgBr and Xe + AgBr. *Phys. Rev.*, **C34** (1986) 1270.

9. Biswas, S., Stein, P. S. and Stein, W., Solar protons and alpha particles from the September 3, 1960, flares. *J. Geophys. Res.*, **67** (1962) 13.

10. Dahlbacka, G. H., Freier, P. S. and Waddington, C. J., Gamma ray emission from the region of the galactic center. *Astrophys. J.*, **180** (1973) 371.

11. Brewster, N. R., Freier, P. S. and Waddington, C. J., The propagation of ultraheavy cosmic ray nuclei. *Astrophys. J.*, **264** (1983) 324–36.

27 Rosalyn Sussman Yalow (1921–)

M. S. Dresselhaus
Department of Physics, Massachusetts Institute of Technology

F. A. Stahl
Department of Physics and Astronomy, California State University, Los Angeles

FIGURE 27.1 Photograph of Rosalyn Sussman Yalow. (Credit: AIP Emilio Segrè Visual Archives, W. F. Meggers Gallery of Nobel Laureates.)

IMPORTANT CONTRIBUTIONS

Few physicists start new fields of physics. Rosalyn Yalow is one of the pioneers who were instrumental in developing the field of modern biomedical physics. She brought her knowledge of nuclear physics to bear on problems in medicine, and expanded biomedical physics to become a contemporary subfield of physics. Before her

Out of the Shadows: Contributions of Twentieth-Century Women to Physics,
eds. Nina Byers and Gary Williams. Published by Cambridge University Press.
© Cambridge University Press 2006.

entry, there were scientists with biomedical backgrounds working in this field who could apply physics, including the use of radioisotopes as probes; however, they did not use the rigorous quantitative approach of physics in biomedical research that she brought to the field. Her Ph.D. thesis research was as nuclear physicist at the University of Illinois under the direction of Professor Maurice Goldhaber (see Chapter 23 on Gertrude Goldhaber). She became familiar with medical applications of radioisotopes while teaching at Hunter College. Yalow worked as a volunteer for Edith Hinkley Quimby in order to gain experience working on medical application of radioisotopes. She left teaching to work in the Radiotherapy Service at the Bronx Veterans Administration Hospital, where she made her great discoveries in collaboration with Solomon Berson over a 22-year period, and was to remain for her entire career.

Her research career can be divided into three phases. Phase I led up to her discovery with Berson of the basic physics and clinical applications of radioimmunoassay (RIA). During Phase II she worked closely with Berson, until his death in 1972, to develop and extend the methodology of RIA. In Phase III she worked in new directions, alone or with other collaborators.

The first period of her research was characterized by a sequence of experiments, with Berson. Their first biomedical research paper described the development of in vivo blood volume determination. A number of investigations of the human blood volume, important for assessing the well-being of patients, had been attempted previously, but there was disagreement in the literature about the methodology and the interpretation of the results. The Yalow–Berson method was based on the introduction of a radioactive tracer into the bloodstream and monitoring the decay rate of the radioactive signal. They used a radioactive isotope of potassium (^{42}K) tagged to red blood cells, taking advantage of two specific properties. First, ^{42}K is a beta-emitter with energies of 3.6 and 2.0 MeV, providing easy detection in liquid samples so that only extremely low doses were needed to obtain accurate measurements. Second, its short half-life of 2.44 hours allowed blood

volume and other determinations to be made at short intervals, leading to a scientifically reliable determination of human blood volume.

Building on the methodology of this early work, their further
studies were on clinical diagnosis of thyroid diseases and the kinet-
ics of iodine metabolism. The first of these involved an analysis of
how the thyroid gland metabolizes iodine and how the kidneys eliminate the metabolites from the blood. Using the same basic radioactive tracer approach, she designed instrumentation and developed a
method for determining the clearance rate of ^{131}I by the thyroid gland
following its injection into the blood stream. Based on a measurement
of the dilution of the ^{131}I isotope, the removal rate of iodine provided a
simple and quick determination of the activity of the thyroid gland. To
this day, it remains the best method for measuring the clearance rate
of iodine by the thyroid gland. From a practical medical standpoint,
the procedure they developed could be used to provide an assessment
of thyroid activity in a 35-minute laboratory test. This method could
also be extended to assay small peptides, such as hormones and other
substances of medical significance. Yalow's entry into the study of
insulin by radioisotope techniques, again using ^{131}I, was a turning
point in her career and the start of Phase II.

The interest of Yalow and Berson was particularly drawn to
the malfunction of insulin in Type II diabetics, who have plenty of
insulin. However, their insulin does not properly remove sugar from
their blood, and consequently these patients suffer from a high level
of blood sugar. The Yalow–Berson measurement involved a determination of the fraction of ^{131}I administered to patients that remained
after a period of time as free ^{131}I and the fraction that became attached
to proteins. The work carried out by the Yalow–Berson team of the
protein-binding site of the insulin-^{131}I complex showed it to be a large
gamma globulin protein.

From their subsequent experiments Berson and Yalow concluded that the gamma globulins were antibodies produced by the
immune system in response to repeated injection of foreign insulin
proteins. Later work confirmed their full claims that the binding

protein was indeed an antibody, and that the human immune system can recognize small foreign molecules by producing a protective antibody. This 1955 paper was the first to demonstrate the use of radioisotopes to study the primary reaction between an antigen and an antibody. It opened up a new and important research field in medicine and biomedical physics.

This work soon led to their discovery of the scientific basis for RIA, namely that the binding of insulin-^{131}I to a fixed concentration of antibody is a quantitative function of the amount of insulin present in the system (whether insulin-^{131}I or unlabeled insulin). RIA very soon became an important clinical method. For the remainder of their years together, Yalow and Berson worked intensively at perfecting the assay for insulin and at understanding the scientific details of the antigen–antibody reaction of insulin. Concurrently, Yalow and Berson worked on extending the principle of the RIA assay to many other antigens, such as hepatitis B, and they applied RIA to a variety of other proteins located at competitive binding sites, such as vitamin B_{12}.

With the sudden death of Berson in 1972, Yalow entered Phase III of her professional career. She developed new directions for RIA, and advanced her leadership in the field. Her publication record continued at the same active pace, and became part of the record that was considered in awarding her the Nobel Prize in Medicine, in 1977. On that occasion she emphasized that the RIA discovery was the work of the Yalow–Berson team, and she graciously acknowledged the contributions of younger colleagues who had contributed to the work after Berson's death. These younger colleagues continued to work with her actively during her Phase III career until she finally closed the laboratory in 1992 and retired from active research.

BIOGRAPHY

Rosalyn Sussman was born in New York City on July 19, 1921, the second child of Clara (Zipper) and Simon Sussman. Both precocious and very determined, she prospered in one of the strong public education tracks that New York offered to girls at that time, first in high school and then at Hunter College. During that progression, her interests grew from mathematics to chemistry and then to physics. She excelled at Hunter College, majoring in chemistry until her senior year, when the physics major program was initiated. As a result, the physics portion of her curriculum was minimal, so she took additional courses after completion of her baccalaureate degree in 1941 in order to prepare for graduate study. She was elected to Phi Beta Kappa and received her A.B. degree magna cum laude. Even as an undergraduate, she was distinguished for her leadership, mentoring other aspiring physics majors including one of the authors (F. A. Stahl), a student in the class of 1942.

In 1941, women were not readily admitted to doctoral programs in physics. Indeed, Rosalyn had been advised to take a course in shorthand so that she might – perhaps! – become a secretary to a physicist. Sensing that a back-door entry was better than no entry, she accepted a part-time secretarial position at Columbia University's College of Physicians and Surgeons (P & S), hoping for eventual admission to graduate study in physics at Columbia University through that connection. She enrolled concurrently in a stenography class. But she also applied to several universities for graduate study in physics, and in February 1941 received an offer of a teaching assistantship at the University of Illinois, beginning the following September. She stayed on the job until June, spending the intervening summer taking two physics courses at New York University.

At the University of Illinois, Rosalyn Sussman found that she was the only woman graduate science student in physics, the first since 1917, which suggests at least an historical correlation between the military draft for men and opportunities for women. In that group of students entering in the fall of 1941, she met Aaron Yalow, and their friendship eventually became courtship. They were married in 1943, following an arrangement of their respective positions at the university in which Aaron received a research assistantship and was no longer technically on the state institution's payroll. This arrangement allowed them to avoid the anti-nepotism regulations then in effect.

She added wartime household management to her load of undergraduate teaching, graduate courses and experimental research. Officially, her thesis adviser was Maurice Goldhaber. His wife, Gertrude Goldhaber (see Chapter 23), was a distinguished physicist but was barred from an official university appointment because of anti-nepotism regulations. The Goldhabers provided steady encouragement to the young Yalows. In 1945, Rosalyn completed her Ph.D. requirements in nuclear physics and returned to New York seeking employment. Though trained for a career as a research physicist, her first professional position was as an engineer in the telecommunications industry. In 1946, when the company left New York, she resigned and was then appointed to a temporary teaching position at Hunter College. Although most of her classes were for male veterans returning to higher education with support from the G. I. Bill of Rights, she did teach some courses for regularly enrolled women. In February, 1949, in a sophomore class in modern physics, one of those undergraduate women students was one of the authors (M. S. Dresselhaus, see Chapter 32), and they established a strong mentoring relationship that lasted through her graduation from Hunter College and continued into her professional years. Rosalyn Yalow had an unusually strong dedication and commitment to people and causes that she felt were important to her.

Aaron at that time was working in medical physics at Montefiore Hospital, and Rosalyn was seeking a research field where her training in nuclear physics could be utilized. Through him, Rosalyn met Edith Quimby, a medical physicist at P & S, and she volunteered to work in Quimby's laboratory in order to gain research-oriented experience in medical physics. Soon after beginning that unofficial association, she met Giacchino Failla, the chief medical physicist at P & S. Following his interview with her, he telephoned Dr. Bernard Roswit, the director of radiotherapy at the Bronx Veterans Administration Hospital (VA), and told him that he had the right person, Rosalyn Yalow, available for the task of setting up the radioisotope service that the VA was planning. She was hired on the strength of that recommendation. This decision turned out to be great for Rosalyn, for the VA and for science.

Yalow started at the VA in 1947 as a part-time consultant, while still teaching at Hunter. The space assigned to her was a janitorial closet, which she equipped using funds from Dr. Roswit's small grant. She initiated several research projects in collaboration with him and other physicians, serving several clinical specialties. In January, 1950, she left Hunter for full-time research at the VA, and she remained with the VA until her retirement 42 years later.

Once engaged fully with radioisotope projects, she realized that in order to build a comprehensive research program she would need a collaborator with complementary strengths in medicine. As a physicist with expertise in radiological physics, she sought an equivalent expert in internal medicine who was also drawn to research. The VA Chief of Medicine consequently introduced her to a physician with these characteristics, then completing a residency in internal medicine at the VA Hospital, Dr. Solomon Berson. He was comprehensively knowledgeable and strongly interested, and as dedicated to quantitative rigor as Yalow. In July, 1950, he joined the radioisotope service, and they began the program of quantitative biomedical research that eventually led to the development of radioimmunoassay (RIA). By 1959, their techniques had evolved into clinical diagnostic practice, and leading physicians from many hospitals visited Yalow and Berson at the VA to learn the techniques. To serve humanitarian needs, they chose not to patent RIA.

In the 1960s Yalow and Berson expanded their research program, extending the RIA principles to many new research directions, while they also ran the VA clinical services in what had become known as nuclear medicine, involving diagnostic scanning techniques for all the organs that could utilize radioisotope scans. The medical practice provided most of the support for their research in that era. Yalow designed much of the innovative instrumentation; the Radioisotope

Service initially was supported by the VA without external grants, and that continued throughout her entire research career in biomedical physics.

During the 1950s the Yalows became parents of two children, son Benjamin (1952) and daughter Elanna (1954), and Aaron had left Montefiore Hospital in 1951 for a faculty position in physics at Cooper Union. They lived a mile from the VA facility, which was essential to their family life because Rosalyn Yalow was as compulsive as a mother and homemaker as she was as a scientist, completely dedicated to both. In their home life, the family observed the rituals of Jewish orthodoxy, which had characterized Aaron's background though not Rosalyn's. Both domestically and professionally, Rosalyn dismissed obstacles to every task she set for herself, and she expended the energy that was necessary to accomplish all her goals, working nights and on weekends when necessary. Her concept of family included the younger research associates and fellows who worked professionally with her and Berson over the years.

In the course of Yalow's association with Berson, they were equals in the research partnership, but in public he was dominant and she was deferential, in part by choice, and in part because in medical circles the Ph.D. is denied the cachet accorded the M.D. In contrast, her husband and fellow physicist provided the emotional support she needed, and took a keen interest in all of her research projects, frequently critiquing her research and encouraging her as one physicist to another.

In 1968, Berson left the VA for an appointment as Professor and Chairman of the Department of Medicine at the Mount Sinai School of Medicine. Yalow was emphatically opposed to his decision, predicting that the burdens of administration would end his research career. She also turned down the idea of moving her lab to Mount Sinai, certain that as a Ph.D. at Mount Sinai, she would have neither the research support nor the professional freedom that she had become accustomed to at the VA Hospital. Yalow's predictions about the frustrations he would experience were quickly realized by Berson, but he did not return to the VA. However, he did encourage those of his residents who showed an interest in academic medicine to get into research. Early in 1972, Berson recommended Dr. Eugene Straus, who had trained at Mount Sinai, to Yalow, who accepted him in the VA research group at that time. In time he advanced to the position of Yalow's medical partner, and later became her confidant and biographer. In April, 1972, Berson, an inveterate smoker, suffered a severe heart attack in Atlantic City, New Jersey, where he was attending a conference. He had made the trip immediately after an aggravating altercation with the Medical School president, newly appointed as the successor to the man who had lured Berson to Mount Sinai. Alone in his hotel room, Berson was unable to summon help and died. He was only 54. Yalow was devastated. In addition to her personal grief, she was professionally shaken to learn that both she and Berson, for whom a Nobel Prize had long been predicted, had been removed from candidacy because the Prize is not awarded posthumously. The gossip that arose around this setback depicted Berson as the "brain" and Yalow as the "brawn," whose contributions were merely the instrumentation and lab work that established ideas conceived by Berson.

At that juncture Yalow knew that she would have to continue in the pursuit of cutting-edge research, in new directions of nuclear medicine, but now she would be working on her own. She reorganized the group and expanded its membership with new associates, several from foreign countries as well as Americans. These young associates in turn expanded the research program under her direction, particularly in studying many new hormones, going far beyond her early emphasis on insulin. She participated very actively in all the activities in her group.

On April 4, 1974, two years after Berson's death, Yalow held the dedication ceremony in which her lab was renamed the Solomon A. Berson Research Laboratory of the Bronx Veterans Administration Hospital. Henceforth, all the publications covering the work accomplished by members of Yalow's research teams would bear Berson's name as part of their affiliation. By this time, Yalow had emerged from Berson's professional shadow and had become the sole intellectual leader of the Laboratory, and she began receiving the invitations for public appearances and for invited papers that previously had gone to Berson.

Yalow was finally recognized for her own work and stature in the late 1970s. In 1975, she was elected to the National Academy of Science. In 1976, she was selected to receive the Albert Lasker Award for basic medical science, the first woman to receive that award. In 1977, she received the Nobel Prize for Physiology or Medicine, shared with Roger Guillemin and Andrew Schally, who had worked on other aspects of endocrinology. She was only the second woman, after Gerty Cori in 1947, to receive the prize in medical science, and only the sixth woman in the sciences overall, after Marie Curie, Irène Joliot-Curie, Maria Mayer and Dorothy Hodgkin, as well as Cori. In 1979, she was elected to the American Academy of Arts and Sciences. In 1986, the Berson Laboratory was designated a Nuclear Historic landmark by the American Nuclear Society. She received the National Medal of Science in 1988. The list of organizational awards bestowed on Yalow has grown to many pages. She has also received nearly 60 honorary doctorates, and has been elected to honorary memberships in societies regularly open only to M.D. degree holders. She has held professorships at several medical schools in New York, including Montefiore and (ironically) Mount Sinai. Her last formal positions were as the Director of the Solomon A. Berson Research Laboratory and the Solomon A. Berson Distinguished Professor-at-Large at the Mount Sinai School of Medicine.

Aaron Yalow died in 1992, after nearly 50 years of happy life together. Rosalyn Yalow remained in her home and in her lab. But in 1993 she suffered a stroke, and then a second stroke in 1995. At that time she was in a coma for five days and was not expected to live. But the clot in her brain dissolved and she rallied. After two more weeks of recuperation, she returned to her home and began a severely restricted routine, in which she did not use the upper story of her house and was driven to and from her office. She struggled to walk when it was unavoidable. Then, in 1997, she fell at home and sustained a hip fracture and other injuries. Several weeks after hip surgery, she was moved to a convalescent facility, and there she slowly regained some mobility. After about two months she was able to go home, and a few weeks later, in May, she was driven to Monmouth University in New Jersey to receive her 57th honorary doctorate. In September 1997, she returned to her office at the VA lab, laboriously but in good spirits. After breaking her right hip a second time on April 29, 2002, she lost the mobility to go to work at the VA.

Rosalyn Yalow had never developed avocational interests, in contrast with Berson's broad interests and his devotion to music and chess. When her homemaking activities decreased as her children grew to adulthood, she gave that much more time to her scientific work. Her daughter lives on the west coast and her son, who lives nearby, has assumed the responsibilities for her care. She maintains some correspondence, but no longer pursues research. Rosalyn Yalow's legacy continues through the tremendous impact of her findings on medical practice, of her methodology on how biomedical physics is carried out today, and of her mentorship and training of younger researchers.

NOTE
The authors thank Eugene Straus for advice and encouragement.

IMPORTANT PUBLICATIONS

Yalow, R. S. and Berson, S. A., The use of K^{42}-tagged erythrocytes in blood volume determinations, *Science*, **114** (1951) 14–15.

Berson, S. A., Yalow, R. S., Bauman, A., Rothschild, M. A. and Newerly, K., Insulin-I^{131} metabolism in human subjects: demonstration of insulin binding globulin in the circulation of insulin-treated subjects, *J. Clin. Invest.*, **35** (1956) 170–90.

Yalow, R. S. and Berson, S. A., Assay of plasma insulin in human subjects by immunological methods, *Nature*, **184** (1959) 1648–9.

Yalow, R. S., Radioimmunoassay: a probe for the fine structure of biological systems. *Science*, **200** (1978) 1236–45. (Nobel Prize lecture.) Also same title in *Les Prix Nobel En 1977* (Nobel Foundation, 1978) pp. 237–41.

A physicist in biomedical investigation, *Physics Today*, **32** (1979) 25–9.

FURTHER READING

McGrayne, S. B., *Nobel Prize Women in Science: Their Lives, Struggles, and Momentous Discoveries* (New York: Birch Lane Press, 1993).

Straus, E., *Rosalyn Yalow, Nobel Laureate: Her Life and Work in Medicine* (New York: Plenum Press, 1998).

Yalow, R. S., Autobiography in *Les Prix Nobel En 1977* (Nobel Foundation, 1978) pp. 237–41.

28 Esther Conwell (1922–)

Lewis Rothberg

Department of Chemistry, University of Rochester

FIGURE 28.1 Photograph of Esther Conwell.

IMPORTANT CONTRIBUTIONS

During her long and productive career, Esther Conwell has made major contributions to scientific theory in three main areas. Her early work, around the time of the invention and maturation of the transistor, was in the technologically important area of charge motion in

Out of the Shadows: Contributions of Twentieth-Century Women to Physics,
eds. Nina Byers and Gary Williams. Published by Cambridge University Press.
© Cambridge University Press 2006.

semiconductors, which now forms the basis for the microelectronics industry. Her attention then turned to light-propagation physics and the understanding of new phenomena made possible by the invention of optical fiber and lasers that now form the core of modern long-distance communications networks. Most recently, Conwell has concentrated on charge motion and optical properties of quasi-one-dimensional organics, a new class of materials that promises to be important in applications as diverse as photocopying, solar cells and light-emitting diodes.

In the late 1940s, scientists understood that solid-state materials, particularly the semiconductors silicon and germanium, had promise for making miniature electronic devices to replace bulky vacuum-tube technology. Conwell's theoretical work on electron motion in silicon and germanium was aimed at addressing experimental observations that the deliberate introduction of charge impurities into these materials ("doping") had the effect of reducing the ability of electrons to move in response to an electric field. In particular, measurements of the resistivity at low temperatures in heavily doped materials could not be understood in terms of the existing theories of the scattering of electrons from crystal lattice vibrations. Working as a graduate student with Victor Weisskopf, Conwell derived the behavior expected for electrons traveling in a crystal containing charged impurities and showed that it was able to account for the experimentally observed resistivity of doped semiconductors (Conwell and Weisskopf, 1950). This theory continues to be of practical importance today, as nearly every semiconductor device now in use is based upon heavily doped silicon (solid-state transistors) or doped compound semiconductors like gallium arsenide (semiconductor lasers).

The invention of the transistor in 1948 at Bell Laboratories followed some of the work described earlier, and Conwell was invited to Bell as a member of the staff in 1951 to continue studying charge motion in semiconductors with Shockley and co-workers. The new silicon and germanium solid-state devices were found to display unusual phenomena when large electric fields were applied to the conduction electrons. If the electric field is pulsed on only for a very short time it is possible to raise the average energy or temperature of the electrons without raising the temperature of the crystal lattice. Conwell (1964) was able to calculate the electron temperature and explain many of the observed effects. Her book *High Field Transport in Semiconductors*, published in 1967, continues to be the authoritative text on charge motion in high fields, a circumstance relevant to nearly every modern semiconductor device.

Another set of observations that attracted Conwell's interest in the mid 1950s involved the large variations in low-temperature behavior of doped semiconductor samples; heavily-doped samples maintained high conductivity while lightly-doped samples became almost insulating. Conwell was able to show that this could be explained in terms of a band of allowed electron energies due to the impurities. With low doping, the impurities are separated enough so that conduction requires hopping of electrons from one impurity site to another nearby that is unoccupied. At low temperatures, this process becomes exceedingly improbable. With increasing concentration, however, the band associated with the impurities widens until it overlaps the conduction band of the pure semiconductor and the resistivity drops. Her theory of this phenomenon is an early example of an insulator-to-metal transition as impurity concentration is increased, still a phenomenon of great interest and study in disordered materials.

The early 1960s saw the invention of the laser and of pure glass optical fibers that could guide laser light. The implication for information transmission and the communications industry were immediately clear and Conwell, then working at General Telephone and Electronics Labs, applied herself to the study of light propagation in waveguides. One method for waveguide fabrication at the time involved diffusing impurities into glass fibers to create a spatial profile of the index of refraction capable of confining light. Conwell calculated the spatial mode patterns to be expected for the characteristic index profiles that can be obtained through impurity diffusion.

The high intensities obtainable with lasers also led to the discovery of so-called nonlinear optical effects such as generation of the second harmonic frequency of the light. These new phenomena seemed to hold promise for making integrated optical devices such as switches, where the light intensity could be modulated by an applied electric field or other light beams. Again, a practical geometry for these devices would be in waveguides where the propagation of the optical beam can be confined and controlled. Conwell worked on predicting the behavior of several of these types of devices, for example the theory of harmonic generation in waveguides (Conwell, 1973), and she holds several patents for integrated optical devices based on her theoretical work.

With her move to Xerox Research Labs, Conwell turned her attention to charge propagation in organic metals and semiconductors. Molecular and polymeric systems are now the active materials in electrophotography, and charge mobility in photoreceptors is a critical part of photocopying technology. A fascinating class of organic charge conductors is quasi-one-dimensional in nature; long-chain molecules can act as "wires," with electrons preferring to remain on a single chain rather than hop to a neighboring chain. The one-dimensionality allows the chain to deform locally when an extra charge is introduced. That deformation travels with the electron, forming an entity called a polaron. In some polymers a different type of deformation may form around a charge, known as a soliton. Jeyadev and Conwell (1987) did calculations of the mobilities of charged solitons and polarons in trans-polyacetylene, which is prototypical of a large class of organic semiconducting polymers that have come to be called "conducting polymers." These materials have exhibited conductivities higher than that of metals when doped, and Conwell's theoretical work is useful in estimating the maximum conductivity that might be achievable in such a polymer. It is currently believed that if these materials could be made in an ordered state and doped stably, they might exhibit conductivities three orders of magnitude higher than that of copper.

Gartstein and Conwell (1994) modeled the hopping of polarons in disordered materials. Disordered molecularly doped polymers are the essential charge transport medium used in photocopying and have been studied extensively. In a wide range of disordered materials the charge mobility has been found to increase as the square root of the electric field intensity over many orders of magnitude in field. Despite years of work the reason for this dependence was not understood. Gartstein and Conwell were able to show that the dependence results from partial correlation in the energies of nearby hopping sites, rather than the complete randomness that was usually assumed. This important and illuminating work has inspired a good deal of further study of the effects of correlation, reinvigorating theoretical work on charge transport in disordered materials.

The conducting polymers described earlier also exhibit promise for making light-emitting diodes. In these devices, as in the well-known light-emitting diodes made from the usual inorganic semiconductors, electrons are extracted at one contact, leaving holes that travel as positive charges, and electrons are injected at the other contact. An electric field pushes the electrons and holes toward each other. In the conducting polymers, when an electron and hole meet they attract each other to form a short-lived bound state, an exciton. The excitons are a source of luminescence, many of them decaying with the emission of visible light. Conwell has made significant theoretical contributions concerning the binding energy and luminescent properties of the excitons.

Although the conducting polymers are usually thought of as one-dimensional, separate chains do have effects on each other. There is evidence that a nearby chain can cause the exciton to break up into an electron on one chain and a hole on another. The resultant entity is usually significantly less luminescent than the exciton, making this a detrimental effect in a light-emitting diode. Wu and Conwell (1997) pioneered the theoretical investigation of such interchain pairs in these polymers, predicting the absorption spectra by which they could be identified. Most recently she is chiefly concerned with calculating the effects of the very high fields that must be applied to the diodes for maximum light output, and possibly to make them function as injection lasers.

BIOGRAPHY

A native New Yorker, Esther Conwell is the eldest of three sisters, all of whom are professionals, and all of whom were encouraged in professional education and careers by a singularly intelligent and far-seeing father who, as early as the 1930s, saw that the needs and status of women in American society were due for a radical change. Because she advanced so rapidly in school, she was already studying physics and mathematics at Brooklyn College in her middle teens, conspicuous not only because of her youth and because she was the only woman in her science classes, but also because she showed such intellectual gifts and promise. As a consequence, Brooklyn College professors and physicists Bernhard Kurrelmeyer and Walter Mais took her under their wing to groom her for graduate study in physics. They helped her to apply for graduate assistantships, and, in 1941, she was awarded an assistantship in the doctoral program in physics at the University of Rochester. World War II interrupted her studies, but already she had, with V. F. Weisskopf, produced the Conwell–Weisskopf formula. For a time she worked on the Manhattan Project and for Western Electric Corporation before returning to continue her work toward the doctorate at the University of Chicago. Her thesis work was on wave functions in hydrogen and oxygen negative ions, work done under the tutelage of Nobel Prizewinner, Subramanyan Chandrasekhar.

When the war was over, Conwell returned to her native city to take up a career in teaching physics at Brooklyn College. She completed her thesis work in absentia while undertaking what was then the normal 16-hour-a-week course load. After she was granted tenure and a leave of absence, Conwell took a job in 1951 at Bell Laboratories in Murray Hill, New Jersey, working with Nobel laureate William Shockley's group and did important work in condensed-matter theory during the exciting time immediately following invention of the junction transistor. In 1952, when Sylvania Electric (later to become General Telephone and Electronics) Laboratories wanted to enter the solid-state electronics business, they brought her to their Bayside Laboratories to continue her research in semiconductors.

In 1956, pregnant and in her ninth month, working right up to her due date against corporate policy, she was sitting in the company lunchroom when the head of the labs asked her about her due date. She replied, "Yesterday," and, in a panic, he sent her home then and there. That weekend, she gave birth to a son. She took a three-month leave of absence, then worked half-time at the laboratories for another three months, before she was able to return to full-time research. Promoted to head a department, she later turned down further offers of promotion because she preferred to continue her scientific and technical research rather than move into administration.

In 1972, she was honored with a term as the Abby Rockefeller Mauzé Chair of Physics at MIT, where she taught integrated optics before moving on to the Xerox Wilson Research Center in Rochester, New York. At Xerox, Conwell devoted herself to research in semiconducting and metallic organic materials, and for her scientific and technical achievements was made a corporate Fellow. In appreciation of her contributions to the corporation's research, she was awarded the Xerox Presidential Prize.

In recognition of the scope and originality of her work, Conwell was elected to the National Academy of Science, the National Academy of Engineering, the American Academy of Arts and Sciences, the New York Academy of Science and was made a Fellow of the Institute of Electrical and Electronic Engineers (IEEE). She was the first woman in the 93-year history of the Thomas Alva Edison Medal of the IEEE to receive the prize awarded to such distinguished scientists as Alexander Graham Bell, Nikolai Tesla, George Westinghouse and Vannevar Bush. She was only the second woman to be elected to both the National Academy of Science and the National Academy of Engineering. Conwell also received the Distinguished Alumni Award and an honorary Doctor of Science from her alma mater, Brooklyn College.

She is a member of Phi Beta Kappa, Sigma Xi and is listed in *Who's Who in America.*

Conwell has taught and lectured in many parts of the world, including Russia, Lithuania, Italy, China and Japan. She was a guest professor at Stanford University in Palo Alto, the Hans Christian Anderson Professor at the Technical Institute outside of Copenhagen and a Professor of Physics at the graduate school of the Sorbonne, the Ecole Normale Superieure. She has also served as a scientific and technical consultant to the government in the National Science Foundation, the National Research Council, the Commission on Human Resources, the Defense Science Board, the National Materials Advisory Board, the Presidential Young Investigators Panel and many other government agencies and commissions.

In 1988, she was appointed Adjunct Professor of Chemistry and an Associate Director of the National Science Foundation Center for Photoinduced Charge Transfer at the University of Rochester, positions she held for a decade until her retirement from the Xerox Wilson Research Center in 1998. She continues to retain her University positions and remains productive in research on charge transport and the optical properties of organic polymers.

Conwell is best known for her outstanding scientific work, including over 200 published papers, but throughout her professional career she has been a consistent advocate of important causes for the scientific community, most particularly for women in science and engineering. A founding member of the American Physical Society Committee on Women in Physics, she chaired that Committee in 1973. She has also served on the National Research Council's Committee on Women in Science and Engineering as vice-chair from 1991 to 1993. In that capacity she chaired a conference devoted to the particular problems of women scientists and engineers in industry. Out of that conference came the book *Women Scientists and Engineers Employed in Industry: Why So Few?* published by National Academies Press.

Recognizing the many difficulties women encounter in pursuing a career in the sciences, she remains committed to attracting more women to such careers, most particularly to physics and engineering, and to improving their opportunities and status in those fields. Although she has served and continues to serve as a mentor and an inspiration to many of her younger women colleagues in physics and engineering, she declines to be considered a "role model," pointing out that there are many different ways to forward a career and many different personalities that choose to do so. When asked to what she attributes her considerable accomplishments, she replies with characteristic modesty and understated humor: "A wise and encouraging father, good teachers and colleagues, a helpful husband and son, and just plain tenacity and perseverance, what used to be called keeping the seat of one's pants glued to the seat of the chair."

IMPORTANT PUBLICATIONS

Conwell, E. and Weisskopf, V. F., Theory of impurity scattering. *Phys. Rev.*, **77** (1950) 388.

Conwell, E., Impurity-band conduction in germanium and silicon. *Phys. Rev.*, **103** (1956) 51.

Phonon and electron distribution in high electric fields. *J. Phys. Chem. Solids*, **25** (1964) 593.

Esther C., *High Field Transport in Semiconductors* (New York: Academic Press, 1967).

Conwell, E., Theory of second-harmonic generation in optical waveguides. *IEEE J. Quantum Electron*, **QE-9** (1973) 867.

Conwell, E. and Burnham, R., Materials for integrated optics: GaAs. *Ann. Rev. Mater. Sci.*, **8** (1978), 135.

Epstein, A. J., Kaufer, J. W., Rommelmann, H., *et al.*, Evidence for solitons in conducting organic charge transfer crystals. *Phys. Rev. Lett.*, **49** (1982) 1037.

Jeyadev, S. and Conwell, E., Soliton diffusion in trans-polyacetylene. *Phys. Rev. Lett.*, **58** (1987) 258.

Gartstein, Y. N. and Conwell, E., High-field hopping mobility of polarons in disordered molecular solids. *Chem. Phys. Lett.*, **217** (1994) 41.

Wu, M.-W. and Conwell, E., Effect of interchain coupling on conducting polymer luminescence: excimers in derivatives of poly (phenylene vinylene). *Phys. Rev. B: Condens. Matter*, **56** (1997) R10060.

29 Cécile DeWitt-Morette (1922–)

Bryce DeWitt (deceased)
Department of Physics, University of Texas at Austin

FIGURE 29.1 Photograph of
Cécile DeWitt-Morette.

SOME IMPORTANT CONTRIBUTIONS

DeWitt-Morette's professional expertise is in the interplay between
physics and mathematics, where mathematics offers new concepts
in the formulation of physical problems and where physics presents
unexplored territories to mathematicians. A prime example of this
interplay is her work on path integration. Introduced by Richard P.
Feynman in the 1940s, path integration, after slow general acceptance in the early years, has become the dominant algorithm for defining the quantum theory of all systems: strings, fields, nuclei, atoms,
molecules and condensed matter systems.

Out of the Shadows: Contributions of Twentieth-Century Women to Physics,
eds. Nina Byers and Gary Williams. Published by Cambridge University Press.
© Cambridge University Press 2006.

Her initial paper on the subject, published in 1951 and written while she was a postdoctoral fellow at the Institute for Advanced Study in Princeton, was the first paper to use the "background method" as well as the first to establish path integration on a rigorous basis in non-relativistic quantum mechanics.[1] The background method allows one to break free of the constraint of having to compute transition probabilities in powers of coupling constants and permits a global perspective on the path integral. In a more general functional-integral setting it leads to the "quantum effective action," which in later years came to play a key role in gauge theories, renormalization theory, the theory of spontaneous symmetry breaking and non-perturbative studies such as lattice quantum field theory.

In 1971, DeWitt-Morette published, with her student, Michael G. G. Laidlaw, a work of major importance that investigated path integrals for paths taking their values in topologically non-trivial spaces.[2] They derived the basic theorem for path integration on multiply connected spaces and applied their result to the theory of indistinguishable particles. In the course of their work they found that indistinguishable particles in two dimensions need be neither fermions nor bosons but can have intermediate statistics. Frank Wilczek later called such particles "anyons."[3] This paper was the first step in a field of research on the interplay of topology and path integration that has become of intense interest to both physicists and mathematicians.

In 1978, DeWitt-Morette noticed that functional integrals over paths emanating from a fixed point in a space of arbitrary topology can be mapped into functional integrals over paths in Euclidean n-space, leading to powerful methods of evaluation. In the 1980s, she began to adapt the methods of stochastic calculus to the computation of path integrals. These methods led her to recognize that the domain of the path integral is most usefully taken to be from a fixed point.

This development plays an important role in her effort, which began in 1972 and continued, since 1993 with Pierre Cartier, to provide path integration with a firm axiomatic foundation so that even mathematicians may accept it as a legitimate tool rather than an heuristic guide. Already, in 1972, she showed that the path integral may be defined rigorously without limiting procedures, and without analytic continuation from a Wiener integral.[4] This is a fact of which the majority of physicists seem unaware to this day. The extension of such definitions to quantum field theory is much more difficult, and is the focus of her present efforts with Cartier.

The techniques developed for the axiomatization effort have also been put to practical use. They often suggest new ways of dealing with cases in which the commonly used "semiclassical" expansion breaks down, for example when the critical point of the action is degenerate, as in the presence of caustics, or when there are constraints imposed by conservation laws. DeWitt-Morette has applied her methods to dealing not only with these cases, but also with the case in which there is a critical point of the action outside the domain of integration. She and her students have applied her methods to the calculation of rainbow scattering and glory scattering. They have studied the scattering of waves of arbitrary helicity by black holes in the glory regime, and have obtained simple explicit expressions for the glory cross-section for scattering by black holes.

The collaboration with Cartier has brought to the heuristic approach of physicists the perspective of a mathematician, resulting in robust methods applicable to a greater variety of problems (e.g., Poisson processes) than are usually considered to be describable in path-integral terms.

Spin and Pin groups
Since 1988, DeWitt-Morette has studied the physical implications of the existence of multiple Spin or Pin structures, first in string theory, then in space-times with non-trivial topology, and finally in Minkowski space-times. To describe fermion fields one must endow space-time with a "Spin" or "Pin" structure, i.e., a principal fibre bundle of which the typical fibre is a double cover of the Lorentz group. Not all potential space-times (complete globally Lorentzian manifolds) admit Spin or Pin structures; certain so-called Stiefel-Whitney

restrictions must be met. But even when such structures exist they are often not unique. Of importance in all these investigations is the existence of two Pin groups, i.e., two distinct double coverings of the extended Lorentz group that includes space or time inversion. The two groups have been found to yield distinct physical results in the topological context (which has applications in Kaluza–Klein theory). In the Minkowski context the only process known so far that can be used for distinguishing the two Pin groups is neutrinoless double beta decay; with the likely observation of neutrinoless double beta decay, one knows that the neutrino emitted and reabsorbed during the interaction must be described by the group Pin (3,1) not by the group Pin (1,3).

Pedagogy

DeWitt-Morette works closely with her students. Twenty-six of her research papers have been written in collaboration with them. She has written, in collaboration with Yvonne Choquet-Bruhat (Chapter 30) and Margaret Dillard-Bleick, an important and widely used reference book of two volumes,[5] the first of which has gone through five reprints. The books are written in a style that has allowed many selected portions to be successfully used as substitutes for more narrowly focused textbooks. They included many new applications of recent developments in mathematics.

BIOGRAPHY

Cécile Andrée Paule Morette was born on December 21, 1922, at l'Ecole des Mines, Paris, to Marie-Lousie Ravaudet and André Morette. In those days the School of Mines was strictly a male preserve. Her father had been asked to lecture there for the year and had been given family quarters within the school grounds.

Cécile did not always intend to study physics. As a young woman, she originally wanted to study medicine but her mother dissuaded her because it was "too physically demanding." Consequently, she began to study physics, first at the University of Caen and later at the University of Paris. By the end of World War II she was "house theorist" in a laboratory directed by Frédéric Joliot, and was occasionally supervised by Irène Joliot-Curie.

Her family suffered many losses during the war. Her mother, grandmother and a sister died in Caen under allied aerial bombardment on June 6, 1944, while Morette was taking an examination in Paris. Her brother died in combat in 1940. Two other brothers, as well as a sister, survived. French physics, which was already ailing before the war, also suffered heavy losses. Apart from the work of a few experimentalists like Frédéric and Irène Joliot-Curie (see Chapter 12), French physics was a disaster by the end of the war. This was particularly true of theoretical physics.

In 1946, Cécile was able to widen her theoretical horizons by traveling to England, where she met Paul Dirac. She then went to Ireland, where she became a member of the Institute for Advanced Studies, in Dublin. In Ireland she progressed rapidly under the tutelage of Walter Heitler and H. W. Peng, who later became the director of the Institute for Theoretical Physics of the Academia Sinica in Beijing. She computed meson production cross-sections using Heitler's radiation damping theory and wrote three papers.

After receiving her doctorate from the University of Paris in 1947, she was a member of the Institutet for Teoretisk Fysik, in Copenhagen, for a year. This was followed by an appointment in the Institute for Advanced Study, in Princeton, from 1948 to 1950. Her time in Copenhagen and Princeton brought her into contact with most of the top theoretical physicists in the world. At Princeton, Morette learned of path integration directly from Richard Feynman.

In the fall of 1949, her future husband, Bryce DeWitt, arrived at the Institute. With youthful irrationality she accepted DeWitt's proposal of marriage, causing a profound change in her career trajectory. She refused several offers of jobs in Europe because her new life was based in the United States.

Nevertheless, she was determined to do something for French physics and conceived the idea of founding a summer school of theoretical physics, with a distinguished and international team of lecturers. Through the help of a friend, she secured a place to house the school near Les Houches, a village in Haute Savoie, whose township boundaries include the summit of Mont Blanc. But the crucial ingredient was money. Needing the support of her many physicist friends, she encouraged them to appropriate her ideas about the school, even though she alone paced the hallways of the ministries in Paris. Because nothing of the kind had ever been done before, she received mainly scorn from officials: why should the French government pay foreign professors to come and have a vacation in the Alps? But one man, Pierre Donzelot, in the Ministry of Education believed in her. It was his decision that made the school possible.

The first session of l'Ecole d'Eté de Physique Théorique was held in 1951, under primitive conditions. Madame la Directrice, whose first pregnancy had come with a rush, was suffering from morning sickness, the hotelier who provided the meals was grumbling and the participants had to wash their own dishes, their own laundry and themselves in cold water. But the enthusiasm was intense. There was no problem in keeping the professors (who included Wolfgang Pauli, Leon Rosenfeld, Res Jost and Leon Van Hove) from wandering away through the magnificent surroundings. The problem was to keep them from talking too much. There were 30 students, half of them French and the rest international, a formula that was maintained during the first few years of the school's existence. Another formula, which lasted considerably longer, was the insistence upon basic courses on advanced topics, each of 10 to 20 lectures, during an eight-week period.

The school's high quality is demonstrated by the fact that, at the latest count, 25 of the lecturers have become Noble laureates. Among the students whose first exposure to advanced topics in theoretical physics occurred at Les Houches are included three subsequent Nobel laureates, two Fields medalists and a dozen members of the French Academy of Sciences, as well as innumerable distinguished foreign physicists. A year after the school was founded, the Italian Physical

Society asked DeWitt-Morette for help in starting something similar in Italy, and the well-known school at Varenna was born.

In 1958, NATO formed a Science Committee under the direction of Norman F. Ramsey. The Committee established a program for the support of summer study institutes in Western Europe, citing Les Houches as a model and gave the school one of its first unrestricted institutional grants.It also provided grants to cover the travel expenses of participants from NATO countries. Later on, well over 200 study institutes were supported by NATO each year.

In September, 1951, DeWitt-Morette accompanied her husband to the Tata Institute of Fundamental Research in Bombay. While there, they wrote a joint paper and their first daughter, Nicolette, was born. In 1952, after her husband accepted a position at the nuclear weapons laboratory in Livermore, California, DeWitt-Morette secured a position as lecturer at the University of California at Berkeley where she guided doctoral students and gave courses on quantum field theory. A second daughter, Jan, was born in 1955.

In January, 1956, DeWitt-Morette's husband accepted the directorship of the Institute of Field Physics, a new center at the University of North Carolina at Chapel Hill created by the industrialist Agnew Bahnson. At the beginning she and her husband each bore the title of visiting research professor. Within a few years her husband was given a regular professorship, and later, upon the death of Bahnson, he held a chair named after the latter. DeWitt-Morette was actually demoted in 1967 from visiting research professor to lecturer, despite the fact that she had played a crucial role in attracting money to the university, in organizing conferences and in serving as the director of the university's Institute of Natural Science from 1957 to 1966. The University cited nepotism rules as the impetus for the demotion despite the fact that no such published regulations existed.

During this time DeWitt-Morette not only taught courses and guided the work of Ph.D. candidates but also continued her active supervision of the Les Houches School, a task not made easier by her distance from Europe. From 1966 to 1968, she was a member of the organizing committee of the European Physical Society, and in 1967 she initiated, together with John Wheeler, the first of the Battelle "Rencontres" in Seattle, remaining on the organizing committee for these summer institutes until 1972. Her primary summer commitment, of course, remained Les Houches, to which she traveled each year with her children, sometimes accompanied by her husband. Two more daughters, Christiane and Abigail, had been born in 1957 and 1959.

Although she was not totally separated from research during her years at the University of North Carolina, DeWitt-Morette became more widely known for her administrative abilities than for her research. However, her situation at North Carolina became so intolerable that she and her husband left at the end of 1971 to accept tenured positions at the University of Texas. (This was her first tenured post!) Even at Texas, she was given only a half-time position and shunted into the astronomy rather than the physics department, perhaps because of last-minute fears of nepotism. Fortunately, the two departments were not only physically close but on the best of terms, and the astronomy department (particularly its chairman, Harlan J. Smith) was very kind to both her and her husband. The DeWitts were co-leaders of an eclipse expedition to Mauritania mounted by the astronomy department and the NSF in 1973. By 1983, the last vestiges of anti-nepotism had disappeared from the university, and she moved to the physics department. In 1987, her position was changed from half-time to full-time. Since her move to Texas she has published 47 papers and written two major books.

With the change in her professional status in 1972 (at age 49!) she resolved to devote full-time to research. At the end of that year she resigned the directorship of l'Ecole d'Eté de Physique Théorique, after 22 years of service to France and to the world. She remains on the Board of Trustees. She left her successors a significantly larger school and the intellectual legacy of the school is immeasurable. In 1981, the French government named her Chevalier de l'Ordre National du Mérite, and in 1999 she became Chevalier de la Légion d'Honneur.

In 1992, she was awarded the Prix du Rayonnement Français, and in 1996 was appointed to the Board of Trustees of the Institut des Hautes Etudes Scientifique in Bures. In July, 2002, DeWitt-Morette and her husband were awarded a Marcel Grossmann Prize for promoting General Relativity and Mathematics research and inventing the "summer school" concept.

DeWitt-Morette's other activities during the past 50 years are legion. They have included skydiving, judo (brown-belt level), trekking in remote areas, skiing, windsurfing, sewing and darning, editing 29 books, serving on countless committees, lecturing at many research centers and working with mental health organizations.

NOTES

Portions of this chapter have previously appeared in *Women in Chemistry and Physics: A Bibliographic Source Book*, edited by Louise S. Grinstein, Rose K. Rose and Miriam H. Rafailovich, copyright 1993 by Grinstein, Rose and Rafailovich, reproduced with permission of Greenwood Publishing Group, Inc., Westport, CT.

1. DeWitt-Morette, C., On the definition and approximation of Feynman's path integral, *Phys. Rev.*, **81** (1951) 848.
2. Laidlaw, M. G. G. and DeWitt, C. M., Feynman functional integrals for systems of indistinguishable particles. *Phys. Rev.*, **D3** (1971) 1375.
3. Wilczek, F., *Fractional Statistics and Anyon Superconductivity* (Singapore: World Scientific, 1990).
4. DeWitt-Morette, C., Feynman's path integral; definition without limiting procedure. *Commun. Math. Phys.*, **28** (1972) 47.
5. Choquet-Bruhat Y., DeWitt-Morette, C. and Dillard-Bleick, M., *Analysis, Manifolds, and Physics* 1977; Revised Edition 1982. Part 11. 92 Applications, 1989 (North Holland, Amsterdam).

IMPORTANT PUBLICATIONS

Choquet-Bruhat, Y., DeWitt-Morette, C. and Dillard-Bleick, M., *Analysis, Manifolds, and Physics*, 1977; Revised Edition 1982. Part 11. 92 Applications, 1989 (North Holland, Amsterdam).

DeWitt-Morette, C., Maheshwari, A. and Nelson, B., Path integration in non-relativistic quantum mechanics. *Phys. Rep.*, **50** (1979) 266.

DeWitt-Morette, C., Feynman path integrals. From the prodistribution definition to the calculation of glory scattering. In *Stochastic Methods and Computer Techniques in Quantum Dynamics*, eds. H. Mitter and L. Pittner, *Acta Phys. Austriaca Suppl.*, **26** (1984) 101.

Cartier, P. and DeWitt-Morette, C., A new perspective on functional integration. *J. Math. Phys.*, **36** (1995) 2237.

Berg, M., DeWitt-Morette, C., Gwo, S. and Kramers, E., The Pin groups. In Physics, C, P and T. *Rev. Math. Phys.* **13** (2001) 953–1034.

FURTHER READING

Dyson, F. J., Feynman at Cornell. *Physics Today*, **42** (2) (1989) 32.

30 Yvonne Choquet-Bruhat (1923–)

James W. York, Jr.
Physics Department, Cornell University

FIGURE 30.1 Photograph of Yvonne Choquet-Bruhat. In this photograph, Mme Choquet-Bruhat is conversing with colleagues at Les Houches, France, in June, 1982. The theoretical physics institute at Les Houches was founded by her friend and collaborator Cécile DeWitt-Morette (see Chapter 29), who was co-author with Choquet-Bruhat of the three-volume mathematical physics books *Analysis, Manifolds, and Physics* 1 & 2, plus a revised supplement.

SOME IMPORTANT CONTRIBUTIONS

Prelude to a masterpiece

General relativity (GR) is a remarkable theory of gravity that replaces Newton's theory because it more accurately describes observations and experiments. It has required new mathematics and great prowess to decipher its implications for physics. Several of Mme.

Out of the Shadows: Contributions of Twentieth-Century Women to Physics, eds. Nina Byers and Gary Williams. Published by Cambridge University Press.
© Cambridge University Press 2006.

Choquet-Bruhat's most important contributions concern a precise mathematical treatment of certain elements of GR: causality and restricted initial conditions, as well as the presence of superfluous variables that permit the use of arbitrary coordinate systems. One has to face difficult nonlinear partial differential equations in Einstein's theory of gravity. She carried out this task at a new level of mathematical rigor that she herself created. This led to a far-reaching enrichment of both mathematics and physics. To get a glimmer of what she has accomplished, we sketch in this "prelude" how such questions arose.

In the nineteenth century, largely in the hands of Faraday and Maxwell, the concept of an electric or magnetic force acting instantaneously at a distance was replaced by that of a space-filling electromagnetic field acting by "direct contact" of the field on electric charges and the currents resulting from motions of the charges. Maxwell was able to infer from this field theory the existence of waves that travel at the speed of light "c," about 300 million meters per second (or about 186 000 miles per second). These waves were found experimentally to make up the spectrum of microwave, infrared, visible, X-ray and gamma-ray radiation. A search for a medium, tentatively named the "luminiferous æther," to carry these waves found nothing. The keenest search was the famous experiment of Michelson and Morley carried out in Cleveland, Ohio, in the USA. The conclusion is that such an æther does not exist. This was unaccountable without the ideas introduced by Einstein and Minkowski.

Einstein was able to embrace the established experimental and theoretical facts of the Maxwell–Faraday–Lorentz field theory of electrodynamics with his Special Theory of Relativity of 1905 – as well as to make a number of startling predictions of new phenomena, such as the equivalence of mass and energy. Einstein found, in accord with the Michelson–Morley and other experiments that all observers, regardless of *their* relative velocity and also of the velocity of the source of the waves, will obtain the same value "c" (above) for the speed of light. Furthermore, the speed of light is the upper limit on the speed determined by any observer of any physical object or signal.

These rather counter-intuitive attributes discovered about nature are not evident in our everyday observations, but they make possible the existence of a concept of *causality* that is absolute, that is, causality independent of the observer or apparatus by which it is determined that, say, an event "A" causes another event "B."

A few themes from the masterpiece
Minkowski showed that Einstein's results in Special Relativity can be understood geometrically by supposing the existence of a flat four-dimensional space-time. ("Flat" in analogy to the fact that a rubber patch is flat while a rubber balloon has a surface that is not flat.) This geometrical idea was the perfect one to describe the phenomenon of Special Relativity, but flat geometry did not include the phenomenon of gravitation just as Special Relativity did not treat any effects of gravity. Thus, flat space-time was given up by Einstein in favor of curved four-dimensional space-time. Curvature is present in order to take into account the *gravitational* effects of the masses present in space. Einstein's use of curvature was his way of incorporating the Principle of Equivalence of gravitational and inertial mass. An object has only one kind of mass. This is an experimental fact that is not obvious. One concomitant of this principle is that a test particle falling freely follows a geodesic of space-time (the "straightest possible path" in space-time, but not in space). An observer falling with it would have no clue about gravity unless, for example, a nearby similar particle was undergoing a relative change in position or velocity with respect to the first one. But this *geodesic deviation* is in mathematics a measure of *curvature*. The curvature tensor is the local measure of the gravitational field. This "tensor" can be determined in "empty" space by observing the relative motions of non-coplanar freely falling particles. A more technical discussion of the work is discussed in Box 30.1 on the Ricci tensor.

Of very great significance is that Yvonne Bruhat's analysis enabled her to prove rigorously *for the first time* local-in-time existence and uniqueness of solutions of the Einstein equations. The work

Box 30.1   Ricci tensor

The field equation found to be suitable for a region devoid of matter is the vanishing of the two-index symmetric Ricci tensor, denoted **Ricc** (ten components). This tensor is a linear combination of components of the curvature tensor, or Riemann tensor **Riem**, whose vanishing would lead us back to gravity-free flat Minkowski space-time. The Ricci tensor in turn is formed from the components of the symmetric two-index metric tensor **g**, whose coefficients form a quadratic form built from the sum of the squares of (dt, dx, dy, dz) with a negative sign for the square of dt ("t" is time) and a positive sign for the squares of the others. (Here, units have been chosen such that the speed of light c is equal to 1.) Also involved in the structure of **Ricc** are the first and second derivatives of g and the inverse of g, denoted here **inv(g)**. The equation **Ricc(g) = 0** is nonlinear and contains superfluous variables because it allows the "symmetry" or freedom of permitting any coordinate system. This freedom is analogous to a similar but simpler one in the theory of electromagnetism that allows an unobservable freedom in the relation between the fields and their corresponding potentials.

On physical grounds, and from approximations to the theory, one expects that the equations should imply the existence of gravity waves that travel, remarkably, at the same speed as those of light. Such phenomena are described mathematically by hyperbolic second-order equations with null characteristic directions. But this is not what one finds upon a first look. Therefore, Yvonne Bruhat invoked what are called the four "harmonic" coordinate conditions:

$$\mathbf{div}\,[(\mathrm{sqrt}(-\det g))\,\mathbf{inv}(g)] = 0$$

to fix the superfluous coordinate-related variables. She found a "reduced" quasi-linear second-order hyperbolic ("wave") equation (from the "reduced" Ricci tensor) with null characteristic directions. This result implies that the changes of the metric and its derivatives propagate as a wave at the speed of light. The experimentally observable gravity wave is found by detecting the oscillating curvature

> **Riem**, which is computed from the changing quantities mentioned earlier.
>
> As further steps she showed that, if the correct initial conditions are imposed, then the reduced equations preserve the harmonic gauge conditions. Secondly, if the initial value constraints hold at some initial time, then they will hold for "all" times by virtue of an *identity* satisfied by the full (ungauged) Ricci tensor. This last step has recently been put on a rigorous footing by the author.

employed very general function spaces, such as Sobolev spaces and their embedded function spaces, to establish the results. This work inspired far-reaching developments in mathematics, quite apart from her result in GR. Principally, perhaps, there is the (Jean) Leray theory of hyperbolic systems. Choquet-Bruhat was herself instrumental in the elaboration of the work inspired by *her* finding.

In another, later, work Mme. Choquet-Bruhat extended early work of Lichnerowicz in giving an account of the four equations restricting the allowed values of the initial data, which can be recognized as the Gauss–Codazzi conditions for embedding a spacelike hypersurface into a space-time with vanishing Ricci tensor. A method employing conformal transformations was used. Her analysis yielded the constraints as an elliptic system, analogous to the elliptic conditions satisfied by the scalar and vector potentials in electrostatics and magnetostatics. Her analysis was essential in pointing the way to other elliptic systems for the constraints, the definitive one using the conformal method having been given by this author in 1999.

Mme. Choquet-Bruhat also continues to remain at the forefront of research on new hyperbolic systems to represent time evolution in General Relativity. As this biographical sketch is being written, such work is of great interest not only in pure mathematics, but also in providing a sound basis for numerical efforts using computers to explore the Einstein theory and its astrophysical consequences.

I have not been able here to touch on Mme. Choquet-Bruhat's extensive and profound work in other areas of mathematical physics, including fluids and supergravity. For example, she carried out the first study of causality in the theory of supergravity. I have been able only to scratch the surface of a brilliant continuing life in science.

BIOGRAPHY

Yvonne Choquet-Bruhat was born in Lille, France, on December 29, 1923, with the given name Yvonne Suzanne Marie-Louise. Her mother, Berthe Hubert, a Professor of Philosophy Agrégée, strongly encouraged the intellectual growth of her daughters, at a time when this was not commonly done, as well as that of her son, who is a retired mathematician of distinction. Choquet-Bruhat's father was Georges Bruhat, Professor of Physics at the Sorbonne and author of a number of influential physics textbooks. This author's distinguished friend, Eugen Merzbacher, among other things a former president of the American Physical Society and the author of a famous text on quantum mechanics, first studied physics using Professor Georges Bruhat's texts.

Following her secondary school diploma, received in 1941, Yvonne won the silver medal in physics in the Concours General, a nationwide competition among all the best students in France. She continued her studies during the war. The course of daily life and lack of international contacts did not cause real difficulties. However, a truly hard blow fell when the Nazis deported her father, a participant in the French resistance. This cruel act struck a bitter blow to the young woman and foreshadowed a hard life. The record shows, however, that her fortitude and genius won out. She would soon offer brilliant and original contributions of far-reaching significance to her chosen fields of mathematics and mathematical physics.

Choquet-Bruhat studied at the Ecole Normale Supérieure and received her Ph.D. in 1951 from the Université de Paris. She was a researcher at the Princeton Institute for Advanced Study for two years before joining the faculty of the Université de Marseille, in 1953. She has been a professor at the Université de Paris since 1968, where she currently holds emeritus status. She has also held numerous visiting professorships, for example at Cornell, Berkeley, Princeton, MIT, Cambridge University, Bonn, Rome, Florence, Turin and Fudan in Shanghai, China.

Her doctoral research advisor was André Lichnerowicz, but she worked more closely with the mathematician Jean Leray, with whom she remained in communication until his recent death. In this early period, she established in her thesis crucial results reaching between mathematics and classical physics when the physics is governed by certain types of systems of nonlinear partial differential equations. The mathematics of such systems, except in limiting approximations, was unfamiliar. Her results enabled her to give the first proof of the local existence and uniqueness of the full Einstein equations on differentiable manifolds of a quite general and useful type. Her paper summarizing these findings appeared in 1952 (q.v.). A conference celebrating the 50th anniversary of the publication of this masterly work was held in Cargèse, Corsica, in 2002.

The French school of mathematicians has elaborated her findings in many works over the years. Of special significance was the development by Jean Leray of a general theory of hyperbolic systems based on Yvonne Bruhat's insights. Hyperbolic systems have remained of wide interest in mathematics and physics. Some of her many other fundamental studies are listed later under Important publications.

Choquet-Bruhat is now professor emeritus. Her formal retirement was followed by an international symposium "Analyse, Variétés et Physique" held in her honor in June,1992, at the Collège de France, in Paris. (A book of invited lectures given at the symposium, *Physics on Manifolds*, edited by M. Flato, R. Kerner and A. Lichnerowicz, was published in 1994 by Kluwer Academic Publishers in Dordrecht, The Netherlands.)

Mme. Choquet-Bruhat has remained very active following her formal retirement. She has continued to publish papers of the highest quality and to travel extensively, visiting her collaborators and friends, old and new, at meetings, universities and research institutes around the world. She finds great pleasure visiting her children and grandchildren at their homes or at her beach house. She enjoys bicycling, hiking and camping. All the while she is gracious and pleasant, proud of her work but not in the least arrogant or aggressive. She is a very good friend indeed.

She has never even considered exploiting her status as a woman during the recent heightened awareness of the status of women in the historically male-dominated intellectual culture of physics. She obtained her status the old-fashioned way – by doing work of such superior quality that it could not be ignored. However, we cannot but be aware that she is one of the greatest female mathematical physicists in history. Women (and men!) cannot be expected to match Yvonne to achieve deserved recognition and status in the community of scientific researchers and scholars.

At present she is working on a monograph in collaboration with this author: *The Dynamics of General Relativity*. Her encyclopedic knowledge, judgment, cleverness and quickness radiate all the qualities of a genius, an appellation that is rightly applied to her. Mme. Choquet-Bruhat has, not surprisingly, won numerous prizes and awards over the years. In 2003, she and the author received the Dannie Heineman Prize in Mathematical Physics given by the Dannie Heineman Foundation, the American Physical Society and the American Institute of Physics. She will also receive the 2003 Marcel Grossmann prize given in recognition of contributions to the understanding of relativity and gravitation. The prize is awarded every three years at the Marcel Grossmann meeting, by a distinguished international committee chaired by Professor Remo Ruffini, of Rome.

IMPORTANT PUBLICATIONS

Choquet-Bruhat, Y., Theorem d'existence pour certains systems des équations aux derivees partielles non lineaires. *Acta Math.*, **88** (1952) 451.

Theorems d'existence en mécanique des fluids relativistes. *Bulletin de la Soc. Math. de France*, **86** (1955) 155.

Sur l'integration des equations de la relativité general. *J. Rational Mech. Anal.*, **5** (1956) 951.

Ondes asymptotics et approachees pour des systems d'equations aux derivees partielles non lineaires. *J. Maths. Pures et App.*, **48** (1969) 117.

Choquet-Bruhat, Y. and York, J. W., The cauchy problem. In *General Relativity and Gravitation: An Einstein Centenary Survey*, ed. A. Held (New York: Plenum Press, 1980) pp. 99–172.

Choquet-Bruhat, Y. and Christodoulou, D., Existence of global solutions of the Yang-Mills, Higgs, and spinor field equations in $3 + 1$ dimensions. *Ann. E. N. S*, 4th Series, **14** (1981) 481.

Choquet-Bruhat, Y., Causalité des theories de supergravite. *Soc. Math. France, Asterisque*, **79** (1984).

Choquet-Bruhat, Y. and York, J. W., Geometrical well-posed systems for the Einstein equations. *Comptes Rendus de L'Academie des Sciences, Serie 1, Paris*, **321** (1995) 1089.

Choquet-Bruhat, Y., Isenberg, J. A. and York, J. W., Asymptotically euclidean solutions of the Einstein constraints. *Phys. Rev. D*, **61** (2000) 1.

FURTHER READING

Choquet-Bruhat, Yvonne. CWP website (http://www.physics.ucla.edu/-cwp).

Noether Lecture, 1986 (http://www.awm-math.org/noetherbrochure/Choquet86.html).

31 Vera Cooper Rubin (1928–)

Robert J. Rubin

Mathematical Research Branch, National Institutes of Health, Bethesda, MD

FIGURE 31.1 Photograph of Vera Cooper Rubin.

SOME IMPORTANT CONTRIBUTIONS

Vera Rubin is an observational astronomer who has made fundamental contributions to the fields of galactic and extra-galactic astronomy. Her contributions have had a major impact on our understanding of the internal dynamics of spiral galaxies, of their evolution in a cluster

Out of the Shadows: Contributions of Twentieth-Century Women to Physics,
eds. Nina Byers and Gary Williams. Published by Cambridge University Press.
© Cambridge University Press 2006.

environment and of their large-scale streaming motions. Her studies of the rotation of spiral galaxies suggest that each galaxy contains an additional unseen component, a dark matter component, whose mass is ten times the mass of the visible galaxy and whose presence is inferred by its gravitational effect in the visible galaxy. These observations have had a profound effect on contemporary cosmology and particle physics.

In Rubin's student days, the then current view of the universe of galaxies was expressed by the Perfect Cosmological Principle: "The universe is uniform and expands isotropically according to the law of red shifts." (The term "red shift" refers to the observed shift toward longer wavelengths of spectral lines emitted from galaxies. A red [or blue] shift arises from a velocity of recession [or approach].) It was, of course, understood that there could be small local deviations in density and velocity from this majestic view of the universe. In 1965, Penzias and Wilson detected a cosmic background radiation,[1] which provided a fixed frame of reference (a sea of photons) against which motions could be measured; shortly thereafter, Partridge and Wilkinson reported a negligible motion of the sun with respect to this rest frame. In 1976, Rubin and her colleagues Kent Ford, Norbert Thonnard, Morton Roberts and John Graham decided that the combination of new observing techniques and theoretical understanding warranted readdressing a question that Rubin had asked earlier in her 1951 M.A. thesis. Are the recession velocities of galaxies of similar type and brightness in different directions on the sky the same, or are there systematic deviations from the smooth, isotropic Hubble expansion? Rubin and collaborators discovered that there is a velocity anisotropy across the sky, which they interpreted as a large velocity of our local group of galaxies of $450\,\mathrm{km\,s^{-1}}$, directed at right angles to the direction to the Virgo Cluster of galaxies. The initial reaction of astronomers was one of skepticism. Since the then current view was that the random motion of individual galaxies was small, being at most 75–$150\ \mathrm{km\,s^{-1}}$, how could the motion of the local group of galaxies be so large?

Despite the skepticism, the investigation of large-scale motions was subsequently undertaken by many groups. Within a decade of the report of the Rubin–Ford effect, a similar velocity anisotropy of the microwave background radiation was reported by two independent groups corresponding to an even larger velocity of our local group, 540 km s^{-1} directed at 45° to the direction to the Virgo Cluster.[2] In addition, similar investigations of large-scale motions of more distant galaxy samples, spiral and non-spiral, were undertaken by many groups. In summary remarks at the International Astronomical Union symposium on observational cosmology held in Beijng, China, in 1986,[3] Malcolm Longair wrote: "Another hot topic was the observation of large-scale streaming velocities of galaxies. The history of these studies dates back to the much-discussed Rubin and Ford effect and is a good example of an observation, which many people wished would go away, suddenly becoming eminently respectable."

Most of the observations for the large-scale motion study were made at Kitt Peak with the 84-inch telescope. Starting in 1973 after the 4-meter telescope was completed, Rubin began to map velocity fields within distant galaxies. By 1980, she had assembled velocity fields for 21 Sc spiral galaxies with a large range of luminosity and radius. The velocity field of each galaxy was analyzed in terms of a rotation curve, the local circular velocity within the spiral disk vs. distance from the center of rotation (see Box 31.1). The set of rotation curves provided convincing evidence for the existence of an invisible component of these galaxies, whose mass is up to ten times the mass of the visible component. At present this invisible, or dark matter, component has been detected only through its gravitational effect on the motion of the visible component. Since 1980, the gravitational effect of dark matter has been observed on larger and larger scales, in X-ray emission of dense clusters of galaxies and in the gravitational lensing of distant quasars and galaxies by foreground clusters of galaxies.

The significance of Rubin's contributions has been recognized in citations accompanying major awards. She was awarded the National

Medal of Science by President Clinton in 1993, "For her pioneering research programs in observational cosmology which demonstrated that much of the matter in the universe is dark and for significant contributions to the realization that the universe is more complex and more mysterious than had been imagined." The American Astronomical Society awarded Rubin its highest honor as the Henry Norris Russell Lecturer of 1994 citing, "For her several major contributions to extragalactic astronomy. Tirelessly amassing accurate galactic rotation curves, she has over the last quarter century helped to establish unanticipated and very important new results, advancing our knowledge of the dynamics of galaxy rotation and the prevalence of non-luminous matter."

Box 31.1   The rotation curve of a spiral galaxy

A spiral galaxy consists of stars, gas and dust distributed in the form of a flattened circular disk. With few exceptions, all disk components move in circular orbits around the center. For a component of mass m to move with velocity $V(r)$ in a stable circular orbit of radius r, there must be a balance between the forces acting on it: i.e., the gravitational force of attraction from all the matter inside its orbit and the opposing centrifugal force, i.e.,

$$\frac{GM(r)m}{r^2} = \frac{mV(r)^2}{r}$$

where G is the constant of gravitation and $M(r)$ is the total mass within the orbital radius r. $V(r)$, the rotational velocity as a function of r, is called the rotation curve of the galaxy. From the above balance-of-forces equation, measurement of $V(r)$ gives a direct determination of the mass distribution within the galaxy.

   If a rotating galaxy is viewed edge-on, the line-of-sight velocity relative to the velocity of the center is directed toward the observer on one side of the center and away on the other. If a rotating galaxy is viewed face-on, the component of its rotational motion in the direction of the observer is zero and no information about its rotation can be

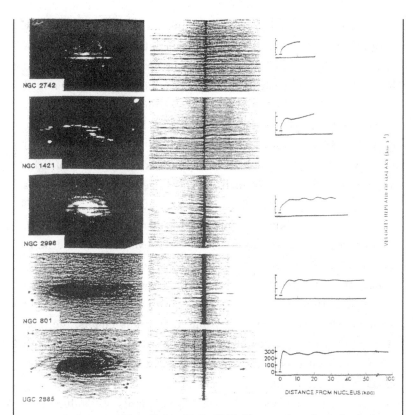

FIGURE 31.2 From Rubin, V. C., 1983. The rotation of spiral galaxies. *Science*, **220** (1983) 1339.

obtained. By aligning the slit of a spectrograph along the major axis of a suitably inclined spiral galaxy and measuring the hydrogen line Doppler shift from point to point along the slit, Rubin determined its rotation curve. Figure 31.2 displays rotation curves for five galaxies.

Alongside each rotation curve, an image of the galaxy and its spectrum are shown. The top three galaxy images are taken from the television screen, which displays the image of the galaxy reflected from the slit jaws of the spectrograph. The dark line crossing these galaxy images marks the slit through which light enters the spectrograph. Wavelength increases from the bottom to the top in each spectrum. The strongest step-shaped line in each spectrum is from hydrogen in the galaxy and is flanked by weaker lines of ionized nitrogen. The strong vertical line in each spectrum is continuum emission from stars in the nucleus. The undistorted horizontal lines are produced by molecules in the earth's atmosphere, principally OH. Rotational velocities are expressed in kilometers/second and distance from the nucleus in kiloparsecs (one kiloparsec = 3260 light years). In the case of UGC 2885, the largest galaxy in Fig. 31.2, the measured rotational velocity over most of the disk is 0.1 percent of the velocity of light, and the rotation period of matter at the edge of the visible disk is 1.5 billion years, roughly one-tenth the age of the universe.

The most striking feature of these rotation curves is that the velocity, after rising, remains essentially flat over the visible disk. In earlier studies of rotation of spiral galaxies,[7] generally only the inner rising part of the rotation curve of bright nearby galaxies had been observed. Astronomers had expected that in the faint outer part of a galaxy where the density of visible matter falls rapidly, the rotation curve would correspondingly turn over and fall. However, Rubin's observations carried out with a state-of-the-art telescope and spectrograph indicated that rotation velocity remained high where there was very little luminous matter. Hence, large amounts of non-luminous matter had to be present to hold the luminous stars, gas and dust in their circular orbits. The mass of the dark matter that is required exceeds the mass of luminous matter by a factor of ten.

BIOGRAPHY

Vera Cooper Rubin was born in Philadelphia, Pennsylvania, in 1928. The younger of two daughters, she became fascinated with astronomy at an early age and was encouraged by her parents. Awarded a scholarship to attend Vassar College, she majored there in astronomy. She graduated with a B.A. in 1948 and married Robert Rubin, a graduate student in the Chemistry Department at Cornell University. Giving up a graduate fellowship at Harvard College Observatory, she chose to join him at Cornell where she enrolled in the graduate program of the small Astronomy Department. The Department did not then have a strong research program, but she had the opportunity to study physics with Hans Bethe, Richard Feynman and Philip Morrison.

In terms of her later achievements, the subject matter of her course work and her thesis topic seem prescient. She studied galactic dynamics under Martha Stahr (later Carpenter), at the time a recent Ph.D. from the University of California, Berkeley. This subject, which deals with the motion of stars and gas in a galaxy, would later prove to be one of her major observational interests and would eventually produce conclusive evidence for the existence of dark matter in galaxies. In her Master's thesis, "Evidence for a rotating universe as determined from an analysis of radial velocities of external galaxies," she addressed the question: do the galaxies with known radial velocities exhibit any large-scale motions other than the Hubble expansion? George Gamow had speculated[4] that "all matter in the visible universe is in a state of general rotation around some centre located far beyond the reach of our telescopes." She returned to this topic again in the mid 1970s, and with her colleagues (see Rubin et al., 1976) obtained evidence for an unexpected large motion of our galaxy, the Rubin–Ford effect.

Rubin received her Master's degree in 1951 shortly after the birth of her first child, David. The family then moved to the Washington, D.C., area where her husband had accepted a position at The Johns Hopkins Applied Physics Laboratory (APL). Almost immediately, she enrolled in Georgetown University, which offered the only graduate program in astronomy in the Washington area at that time. She took courses offered by John Hagen (the new field of radio astronomy) and Karl Kiess (spectroscopy) among others, and learned to measure spectra by working with Charlotte Moore Sitterly, the eminent spectroscopist at the National Bureau of Standards. Through pure chance, Rubin, who had never corresponded with George Gamow, was contacted by him. Gamow had learned of Rubin's M.A. thesis from her husband's office mate at APL, Ralph Alpher, the junior author with Gamow of the Big Bang theory. In later discussions, Gamow asked Rubin if there was a scale length that could provide a measure of the turbulent state of the early universe. This question led to her thesis, "Fluctuations in the space distribution of the galaxies," under the direction of Gamow, a professor of physics at George Washington University. A second child, Judy, was born in 1952. Between caring for two young children, graduate studies and writing a thesis, family life during this period was hectic. Fortunately, all of the graduate astronomy classes met in the evening. She received her Ph.D. degree from Georgetown University in 1954 at a graduation ceremony attended by her husband and children.

From 1955 to 1965 Rubin was affiliated with the Georgetown College Observatory, initially as a Research Associate and later as an Assistant Professor. Two more children were born, Karl in 1956 and Allan in 1960. Her research interests, which continued to be in galactic and extragalactic astronomy,[5] dealt with observations and data gathered by others. Her only observing experience was acquired as an undergraduate and with small telescopes at the Georgetown College Observatory. However, the opening of Kitt Peak National Observatory in 1963 marked a watershed in American astronomy and in Rubin's career as an observational astronomer. Kitt Peak Observatory provided access to large telescopes under a peer-review system, and Rubin took full advantage of the opportunity,[6] observing for the first time in the Spring of 1963 at the 36-inch telescope, the then largest on Kitt Peak mountain. She has continued to observe at The National Observatories throughout her career.

Rubin and her husband, with their four children, spent the 1963–4 academic year on sabbaticals at the University of California San Diego in La Jolla, California. Rubin joined the research group of Margaret and Geoffrey Burbidge (see Chapter 25), which was engaged in observing velocity fields in external galaxies. Following the sabbatical in California, Rubin returned to Georgetown University. At that time she was interested in identifying bright O- and B-stars in the anticenter direction in our galaxy so that their velocities could be measured and the rotation curve of our galaxy be extended. In 1963, Allen Sandage, a Carnegie astronomer, invited Rubin to apply for telescope time at Palomar Observatory. On the printed proposal form that was sent to her, the word, "usually," was penciled into the statement, "Due to limited facilities, it is not possible to accept applications from women." She subsequently became the first woman to be granted use of a Palomar Observatory telescope. In 1965, Rubin moved to the Department of Terrestrial Magnetism (DTM) of the Carnegie Institution of Washington. The small staff included the radio astronomers and physicists Drs. Bernard Burke, Ken Turner and Kent Ford. Ford had just completed the development of an RCA image-tube spectrograph system as part of a program conceived by Merle Tuve, the director of DTM. Tuve's goal was to demonstrate the superiority of the image intensifier system over the unaided photographic plate in astronomical research and, with National Science Foundation support, to distribute working systems to observatories interested in acquiring them. Rubin chose to work with Ford in using and further developing the image-tube spectrograph system. Although it had not been Tuve's intention in hiring Rubin, he had also launched a new center for research in extragalactic astronomy.

In their first major observing program with the DTM spectrograph, Rubin and Ford mapped the velocity field of the Andromeda Galaxy, the other large spiral of our local group, completing the study in 1970. During periods when Andromeda was not observable, they explored the capabilities of the DTM spectrograph by observing more distant galaxies to obtain velocity-field maps and spectral-line intensities. Rubin and her colleagues continued the observing programs, which led to the large-scale motion result of 1976 and the optical rotation curve evidence for the existence of significant amounts of dark matter clumped around spiral galaxies in 1980. She has continued her observing career utilizing the CCD electronic detector systems, which surpassed the DTM spectrograph c. 1984. (See Important publications.)

This short biography emphasizes the professional side of Rubin's life. She has also served on editorial boards of journals, governing boards of national observatories and the National Science Board. At the same time, Rubin has regarded her personal family life to be of equal importance, and proudly displays her children's names at the top of her CV. In the late 1960s when she traveled to observatories or meetings, she routinely cooked a large turkey, which was intended to serve the family in her absence. This practice ceased when she began to find most of an uneaten turkey upon her return. Through example, she has shown that it is possible to have a successful career in science and, with her husband, to raise four children: David, Ph.D. Geology; Judy, Ph.D. Cosmic Ray Physics; Karl, Ph.D. Mathematics and Allan, Ph.D. Geology. Rubin is a role model for young women aspiring to careers in science and an outspoken supporter of the rights of women in science. In her many lecture visits at academic institutions, she routinely meets with women and men students to listen, to talk and to give career advice and encouragement. Her principal advice component is: "Don't give up."

Rubin's contributions to astronomy, astrophysics and cosmology have been recognized in many ways. In 1996, she was awarded the Gold Medal of the Royal Astronomical Society, the first woman to be so honored after Caroline Herschel in 1828. In 1995, she gave the Henry Norris Russell Lecture before the American Astronomical Society, and in 1985, gave an Invited Discourse at the 19th General Assembly of the International Astronomical Union meeting in Delhi, India. She is a member of the National Academy of Sciences, the Pontifical Academy of Sciences, and the American Philosophical Society. She has been awarded numerous honorary D.Sc. degrees, from Creighton University in 1978 to Princeton University in 2005. She received the Weizmann Women and Science Award (1996), the Gruber International Cosmology Prize (2002), the Bruce Medal of the Astronomical Society of the Pacific (2003), and the Watson Prize of the National Academy of Sciences (2004).

NOTES

1. Penzias, A. A. and Wilson, R. W., *Astrophys. J.*, **142** (1965) 419.
2. Fixsen, D. J., *et al.*, *Phys. Rev. Lett.*, **50** (1983) 620; and Lubin, P. M., *et al.*, *Astrophys. J. Lett.*, **298** (1989) L1.
3. Hewitt, A., Burbidge, G. and Fang, L. Z., *Observational Cosmology* (Dordrecht: Reidel, 1987) p. 823.
4. Gamow, G., Rotating universe? *Nature*, **158** (1946) 549.
5. Rubin, V. L., *et al.*, Kinematic studies of early-type stars. 1. Photometric survey, space motions, and comparison with radio observations. *Astronom. J.*, **67** (1962) 491.
6. Rubin, V. C., *Bright Galaxies, Dark Matters* (New York: Springer-Verlag/ AIP Press, 1997) pp. 88–9.
7. Summarized in: Burbidge, E. M. and Burbidge, G. R., *Galaxies and the Universe*, eds. A. Sandage, M. Sandage and J. Kristian, (Chicago: University of Chicago Press, 1975) p. 81.

IMPORTANT PUBLICATIONS

Many of these important publications have been annotated by Rubin for the archive: "Contributions of twentieth-century women to physics," CWP (http://cwp.library.ucla.edu).

Rubin, V. C. and Ford, W. K., Rotation of the Andromeda Nebula from a spectroscopic survey of emission regions. *Astrophys. J.*, **159** (1970) 379.

Rubin, V. C., The dynamics of the Andromeda Nebula. *Sci. Am.*, **228** (1973) No.6, 30.

Rubin, V. C., Ford, W. K., Thonnard, N., Roberts, M. S. and Graham, J. A., Motion of the Galaxy and the Local Group determined from the velocity anisotropy of distant Sc I galaxies, I. The data. *Astronom. J.*, **81** (1976) 687.

Rubin, V. C., Thonnard, N., Ford, W. K. and Roberts, M. S., Motion of the Galaxy and the Local Group determined from the velocity anisotropy of distant Sc I galaxies, II. The analysis for the motion 1976. *Astron. J.*, **81** (1976) 719.

Rubin, V. C., Ford, W. K. and Thonnard, N., Rotational properties of 21 Sc galaxies with a large range of luminosities and radii, from NGC 4605 (r = 4 kpc) to UGC 2885 (r = 122 kpc). *Astrophys. J.*, **238** (1980) 471. (It is noted, in Abt, H., *Publ. Ast. Soc. Pac.*, **87** (1985) 1050, that this paper ranks as the most cited paper of all those based on observations with the four large US telescopes. 1980–1.)

Rubin, V. C., The rotation of spiral galaxies. *Science*, **220** (1983) 1339.

Rubin, V. C., Burstein, D., Ford, W. K. and Thonnard, N., Rotation velocities of 16 Sa galaxies and a comparison of Sa, Sb, and Sc rotation properties. *Astrophys. J.*, **289** (1985) 81.

Rubin, V. C., Graham, J. A. and Kenney, J. P. D., Cospatial counterrotating stellar disks in the Virgo E7/S0 galaxy NGC 4550. *Astrophys, J. Lett.*, **394** (1992) L9.

Rubin, V. C., A century of galaxy spectroscopy. *Astrophys. J.*, **451** (1995) 419.

Dark matter in the universe. *The Scientific American Presents* (special quarterly issue: *Magnificent Cosmos*), **9** (1998) No.1, 106.

FURTHER READING

Irion, R., The bright face behind the dark side of galaxies. *Science*, **295** (2002) 960.

Peebles, P. J. E., *Principles of Physical Cosmology* (Princeton: Princeton University Press, 1993). (This book contains background material and a discussion of both large-scale streaming motions and the ubiquity of dark matter on all length scales.)

Rubin, V., *Bright Galaxies, Dark Matters* (New York: Springer-Verlag/AIP Press, 1997). (In addition to much autobiographical material, this book contains a selection of Rubin's less technical articles and talks, some not previously published.)

32 Mildred Spiewak Dresselhaus (1930–)

G. Dresselhaus
Francis Bitter National Magnet Laboratory, Massachusetts Institute of Technology

F. A. Stahl
Department of Physics and Astronomy, California State University, Los Angeles

FIGURE 32.1 Photograph of Mildred Spiewak Dresselhaus.

SOME IMPORTANT CONTRIBUTIONS TO PHYSICS

Mildred Dresselhaus is best known for her contributions to carbon science, including graphite, graphite intercalation compounds (GICs), carbon fibers, ion-implanted graphite, liquid carbon, fullerenes and carbon nanotubes. In addition, she has studied Column V semimetals extensively and started the field of low-dimensional thermoelectric materials. Some of her earlier works on graphite and carbon fibers have recently attracted wider attention because of the current interest in fullerenes and carbon nanotubes for possible applications to nanoscience and nanotechnology.

Out of the Shadows: Contributions of Twentieth-Century Women to Physics,
eds. Nina Byers and Gary Williams. Published by Cambridge University Press.
© Cambridge University Press 2006.

In her earliest work in carbon science, during the 1960s at the MIT Lincoln Laboratory, she showed how magneto-reflection could be used to elucidate the electronic structure of graphite. She thereby determined both the electronic dispersion relations and the Fermi surface of graphite. Her analytical method also served for the semimetals bismuth, antimony and arsenic. In the mid sixties, with the guidance of Professor Ali Javan, Dresselhaus and her student Paul Schroeder built a laser that enabled them to study circularly polarized magneto-reflection spectra. They discovered asymmetries between the graphite valence and conduction bands, which led to the reassignment of the electron and hole locations and led to the presently accepted band structure for graphite. Over the next decade, she and her group improved the accuracy of their model with the use of many different experimental techniques.

Following the discovery of superconductivity in GICs at Bell Labs in 1965, and then the de Haas–van Alphen effect in 1971, Dresselhaus and her group turned to the study of GICs using magneto-reflection, X-rays, the de Haas–van Alphen effect and Raman and infrared spectroscopy. Their results demonstrated that the electronic and lattice structure for the graphite host material was only slightly perturbed by intercalation. Dresselhaus next directed their attention to the intercalate layer, using Raman scattering to show that the properties of a bromine intercalate layer were those of a low-dimensional system, a single layer of a two-dimensional molecular crystal embedded in a weakly interacting graphite host. Further measurements with X-ray scattering, in collaboration with Professor Robert Birgeneau, led to the discovery of a temperature-dependent structural phase transition induced by the large difference in the thermal expansion coefficients of the graphite host and the bromine guest layer. Dresselhaus contributed further to low-dimensional physics with her work on magnetism and superconductivity in GIC intercalate layers.

Early in the 1980s, Dresselhaus and her students, in collaboration with Professor Morinobu Endo, discovered that many effects in graphite and its intercalation structures could be studied more sensitively using carbon fibers. In their electronic properties, carbon fibers approximated single crystal carbon flakes and were very convenient for transport measurements. Vapor-grown carbon fibers were found to be related to nanotubes; those with small diameters actually are multi-wall nanotubes. This work led to further fundamental research by Dresselhaus and Endo, including experiments on activated carbon fibers, fluorine-intercalated carbon fibers and carbon aerogels, in which two-dimensional localization phenomena could be studied under a wide variety of conditions. Results of this research contributed to practical applications, such as are used in lithium-ion battery products.

Further research by the Dresselhaus group in the early 1980s, with the collaboration of Dr. T. Venkatesan, was carried out on ion-implanted graphite to study controlled defects and their annealing. Their pulsed-laser heating of carbon led to new information about liquid carbon, which was shown to be metallic. This work showed further that the debris released from the graphite surface by the laser pulse consisted of large clusters of carbon atoms, later identified as fullerenes by Smalley, Kroto, Curl and others using mass spectroscopy.

The Dresselhaus group began work on fullerenes in 1991, primarily focused on fullerene-based carbon nanotubes, one atomic layer in thickness. Her early work stimulated further theoretical activity, including Dresselhaus's paper jointly with R. Saito and M. Fujita, predicting that single-wall carbon nanotubes could be either semiconducting or metallic, depending on their diameter and chirality. More recent work on nanotubes includes Raman spectroscopy on individual isolated single-wall carbon nanotubes, built on two prior discoveries showing that the light-scattering mechanism was a resonant process between the laser excitation and the nanotube's electronic energy states, depending on the nanotube diameter. Her other finding showed that the metallic and semiconducting nanotubes could be distinguished by the different line shapes in their Raman spectra. The observation of Raman spectra by her group at the single nanotube level was made possible by the singularities in the electronic density of states for a one-dimensional system. The structural characterization of individual nanotubes makes it possible to study many other physical properties of nanotubes, again as a function of their diameter and chirality.

In a career spanning close to a half-century, Mildred Dresselhaus has been the author or co-author of nearly 1000 research papers and review articles, as well as four books, in a field of physics that overlaps other sciences and technology. She has directed the research of about 65 Ph.D. students and about 30 postdoctoral fellows, in addition to collaborating with her husband, Gene Dresselhaus, and other faculty colleagues.

BIOGRAPHY

Mildred Spiewak was born on November 11, 1930, in Brooklyn, New York, to immigrant parents, for whom the Depression was not an easy time. Her childhood was marked by precocity, particularly in music and mathematics. She attended Hunter College High School and then went on to Hunter College, both at that time highly selective institutions for girls. Attracted to the prospect of a teaching career, she began her college program majoring in education. However, her insights in science and mathematics soon drew the attention of faculty members, particularly future Nobelist Rosalyn Sussman Yalow (see Chapter 27), who persuaded her to change her major to physics and mentored her from her sophomore year onward. She received the Bachelor's degree in 1951, and also won a Fulbright fellowship for study abroad. She spent a year at Cambridge University, and returned to Cambridge, Massachusetts, where she earned a Master's degree at Radcliffe College in 1953.

For her doctorate she went to the University of Chicago, where her research entailed microwave studies on superconductors in a magnetic field. In particular, she discovered an anomalous relationship between the superconductor's surface impedance and the applied magnetic field strength, anomalous because it did not fit the 1957 theory of Bardeen, Cooper and Schrieffer (BCS), just published and highly acclaimed. She submitted her Ph.D. papers for publication in 1958.

That same year she married Gene Dresselhaus, and joined him at Cornell where Gene was a junior faculty member and Mildred an NSF postdoctoral fellow. Both were strongly interested in condensed matter physics. In 1960, determined to collaborate, the Dresselhauses left Cornell, which adhered to anti-nepotism rules, and went to the MIT Lincoln Laboratory, where the professional climate was not restrictive.

From 1960 to 1967, Mildred Dresselhaus worked on magneto-optical effects in semiconductors and semimetals. During the 1960s, the Dresselhaus family was augmented by the births of four children, which necessitated strict time management in her lab activity. In 1967, Mildred transferred to a faculty position as a professor of electrical engineering, starting with a one-year visiting appointment. In 1973, she was appointed to the Abby Rockefeller Mauzé endowed chair, which she held until 1985. From 1977 to 1983, she served as director of MIT's Center for Materials Science and Engineering. Her position was broadened in 1983 to include a professorship in physics as well as engineering. In 1985, she was named Institute Professor, the highest faculty rank at MIT.

Many other offices, honors and other forms of recognition have been awarded to Mildred Dresselhaus. In 1984, she became only the second woman elected president of the American Physical Society (APS) in nearly a century. She subsequently served as president of the American Association for the Advancement of Science (AAAS) in 1997-8. She was elected to the National Academy of Engineering (1974), the American Academy of Arts and Sciences (1974), the National Academy of Science (1985) and the American Philosophical Society (1995), and served as the Treasurer of the National Academy of Sciences (1992-6), the first woman officer elected to serve the Academy. In 1990, she received the National Medal of Science, and in

1999, she received the Nicholson medal from the APS in recognition of her work as a mentor, teacher and researcher, and for her national and international outreach in science.

In 2000, Mildred Dresselhaus was nominated by President Clinton to serve as Director of the Office of Science for the US Department of Energy. In 2001, she received the Compton medal from the American Institute of Physics (AIP), and in 2003 she was named chairman of the AIP board. In 2004, she received the Founders Medal of the Institute of Electrical and Electronics Engineers (IEEE), and in 2005, the Heinz Award for Technology, the Economy and Employment. Distributed over her career are 21 honorary doctorates, as well as six international visiting professorships and exchange programs. She has consistently worked on behalf of women in physics and related fields, locally at MIT, nationally in scientific organizations and internationally.

Throughout her life, music has been a major interest for Mildred Dresselhaus, who became an accomplished violinist in the course of her adolescence and early adulthood, and continues playing actively. She has characteristically shared her musical as well as her scientific interests with her family, students and friends.

IMPORTANT PUBLICATIONS

Dresselhaus, M. S. and Mavroides, J. G., The Fermi surface of graphite. *IBM J. Rese. Dev.*, **8** (1964) 262.

Dresselhaus, M. S. and Dresselhaus, G., Intercalation compounds of graphite, *Advances in Phys.*, **30** (1981) 139–326; Reprinted, *Adv. Phys.* **51** (2002) 1–186.

Light scattering in graphite intercalation compounds. In *Light Scattering in Solids III*, **51** eds. M. Cardona and G. Güntherodt, *Topics in Applied Physics* (Berlin: Springer-Verlag, 1982) pp. 3–57.

Dresselhaus, M. S., Dresselhaus, G., Sugihara, K., Spain, I. L. and Goldberg, H. A., *Graphite Fibers and Filaments* (Berlin: Springer-Verlag, 1988). Vol. 5 of Springer Series in Materials Science.

Dresselhaus, M. S. and Kalish, R., *Ion Implantation in Diamond, Graphite and Related Materials* (Berlin: Springer-Verlag, 1992). Vol. 22 of Springer Series in Materials Science.

Dresselhaus, M. S., Dresselhaus, G. and Eklund, P. C., *Science of Fullerenes and Carbon Nanotubes* (New York, NY: Academic Press, 1996).

Rao, A. M., Richter, E., Bandow, S., *et al.*, Diameter-selective Raman scattering from vibrational modes in carbon nanotubes. *Science*, **275** (1997) 187–91.

Saito, R., Dresselhaus, M. S. and Dresselhaus, G., *Physical Properties of Carbon Nanotubes* (London, UK: Imperial College Press, 1998).

Jorio, A., Saito, R., Hafner, J. H., *et al.*, Structural (*n, m*) determination of isolated single-wall carbon nanotubes by resonant Raman scattering. *Phys. Rev. Lett.*, **86** (2001) 1118–21.

Dresselhaus, M. S., Lin, Y.-M., Cronin, S. B, *et al.*, Low dimensional thermoelectricity. In *Semiconductors and Semimetals: Recent Trends in Thermoelectric Materials Research III*, ed. T. M. Tritt (San Diego, CA: Academic Press, 2001) pp. 1–121.

FURHTER READING

Ambrose, S. A., Dunkle, K. L., Lazarus, B. B., *et al.*, *Journeys of Women in Science and Engineering: No Universal Constants* (Philadelphia: Temple University Press, 1997) pp. 105–8.

Appell, D., Profile: Mildred S. Dresselhaus, aspirations in science and civics. *Scientific American*, March issue (2002) pp. 38–9.

Hargittai, M. and Hargittai, I., *Candid Science IV: Conversations with Famous Physicists* (London, UK: Imperial College Press, 2003).

Hartman, M. S., *Talking Leadership; Conversations with Powerful Women* (New Brunswick, NJ: Rutgers University Press, 1999) pp. 65–78.

Svitil, K. A., The most important women in science. *Discover*, **23** (2002) 52–7.

33 Myriam P. Sarachik (1933–)

Jonathan R. Friedman

Department of Physics, Amherst College

FIGURE 33.1 Photograph of Myriam P. Sarachik.

CONTRIBUTIONS

Professor Myriam P. Sarachik has devoted her career to low-temperature physics, making many significant contributions to solid-state physics. Her experimental work has advanced the understanding of superconductivity and the effects of magnetic impurities in metallic alloys. She has been a key player in the field of metal–insulator transitions in semiconductors as well as in two-dimensional systems. In addition, she and collaborators are credited with the discovery of

Out of the Shadows: Contributions of Twentieth-Century Women to Physics,
eds. Nina Byers and Gary Williams. Published by Cambridge University Press.
© Cambridge University Press 2006.

a new phenomenon: resonant magnetization tunneling in a single-molecule magnet.

Some of the most interesting properties of materials are revealed at low temperatures. This is because at higher temperatures microscopic randomness tends to mask or even suppress quantum effects. Thus, at low temperatures, many normal metals become superconductors, the resistance of certain types of materials increases in quantum steps and gases of noninteracting atoms condense into a single quantum state. Depending on the effect being studied, "low temperature" can mean anything from a few degrees (kelvin) above absolute zero down to microkelvin (or even nanokelvin for laser-cooled atomic gases!).

Early work

Sarachik's doctoral work in the late 1950s provided an important experimental test of the then-new Bardeen–Cooper–Schreiffer theory of superconductivity,[1] the first microscopic theory of the phenomenon. In this theory, electrons, interacting with each other through phonon vibrations in the material, combine to form pairs. A fixed amount of energy is required to break a pair, a quantity known as the superconducting energy gap. Because of the presence of this gap, the theory predicted that just about every measurable electromagnetic quantity in the superconductor should be an exponential function of the ratio of the energy gap to the temperature. Sarachik's doctoral work provided the first systematic measurements of the magnetic field penetration depth in superconducting lead and how the quantity depends on temperature. The penetration depth is the distance that a magnetic field penetrates into a superconductor before it is screened out by currents flowing in the superconductor. From the data, they were able to deduce a value of the energy gap that was in reasonable agreement with the directly measured value, thus providing one of the first experimental confirmations of the new theory.

As a postdoctoral researcher at AT&T's Bell Laboratories, Sarachik established experimentally that the presence of local

364 OUT OF THE SHADOWS

magnetic moments in a metallic alloy causes the electrical resistance of the alloy to increase at low temperatures.[2] When a magnetic impurity, such as iron, is put into a nonmagnetic metal, it sometimes retains its magnetic character, forming what is known as a local magnetic moment. For most metallic materials the electrical resistance decreases as the temperature is lowered because there is less scattering between the moving electrons and the phonon vibrations of the atoms in the material; at low temperatures, the resistance eventually reaches a constant value. Anomalously, some metallic materials do not reach this plateau, but instead the resistance has a minimum, increasing again as the temperature is lowered. Professor Sarachik, by carefully studying alloys of different chemical composition, established a one-to-one correspondence between the presence of local magnetic moments in the alloys and the anomalous minimum. Contemporaneously, theorist J. Kondo was studying the interaction between local magnetic moments and conduction electrons in the metal. He predicted that the moment would magnetically polarize nearby conduction electrons in the material. This effect, most prominent at low temperatures, would increase the magnetic scattering that occurs in the material and thereby cause the resistance to increase at low temperatures, instead of reaching a plateau. Kondo cited Sarachik's work – even before it was published – as experimental support for his theory, now widely known as the Kondo effect.

Metal–insulator transition

As a Professor at the City College of New York, Sarachik has studied in depth the metal–insulator transition in semiconducting materials, and, more recently, in two-dimensional systems. Most materials can be classified in one of two categories: metals and insulators. Metals conduct electricity well and their conductance (the reciprocal of resistance) tends to increase as the temperature is lowered. Insulators allow very little current to flow and their electrical conductance tends toward zero at low temperatures.

To study the transition between one type of behavior and the other, much attention has been paid to semiconductors. As the name "semiconductor" implies, it lies in some middle ground between metal and insulator. This is because semiconductors, although technically insulators in their pure state, can be turned metallic through the addition of a small amount of impurities. As impurities are added to a semiconductor (making it "doped"), electrons are either added to or removed from the material, making the material more conducting. (In the latter case, it is the "holes" left behind that conduct.) When studying semiconductors with impurity concentrations very near the metal–insulator transition, it is necessary to go to very low temperatures to determine whether a particular sample is an insulator (conductance decreases to zero) or a metal (conductance remains nonzero). And, the closer one gets to the transition, the lower in temperature one needs to go in order to make that determination. The metal–insulator transition is considered to be well defined only in the limit of absolute zero temperature.

The underlying physics of the metal–insulator transition has been (and continues to be) of great interest to physicists. The presence of impurities in a semiconductor can have several effects. One is due to the fact that the impurities introduce somewhat mobile charge carriers (electrons or holes) into the material. These charge carriers interact with each other electromagnetically; e.g., they repel each other. Another effect has to do with the fact that the impurities are randomly interspersed throughout the material, which gives rise to more scattering of the flowing current. Other effects are related to the electron's magnetic properties. Which of these mechanisms is primarily responsible for driving the metal–insulator transition has been one of the key questions in solid-state physics in the latter half of the twentieth century.

While the debate still rages on many fronts, Professor Sarachik's work has shed a significant amount of light on the question.[3] Carefully studying the behavior of materials very near the transition, Sarachik's research group was able to identify which mechanism drives the transition in various materials. By applying a magnetic field or pressure to a sample they could change how it behaves near the transition, elucidating the dominant mechanism for the material.

For a more detailed discussion of some of the Sarachik group's studies of the metal–insulator transition in doped semiconductors, see Box 33.1.

Metal–insulator transition in two dimensions
While metals seem ubiquitous in the everyday world, it has long been an accepted theoretical result that no metal can exist in less than a three-dimensional world. It was then surprising when in the early 1990s Sergey Kravchenko and collaborators at the University of Oklahoma found evidence of a metallic state in a two-dimensional system. Two-dimensional systems are achieved using a sandwich of materials in which the electrons are forced to flow only very near the interface between two pieces of the sandwich. In 1995, Kravchenko joined the Sarachik research group, which began trying to understand the nature of this unique system.

In the last few years, the Sarachik group has done numerous studies of these two-dimensional "metals," most involving the use of magnetic fields.[4] One major result of these studies is that the metallic state seems to be destroyed when a magnetic field is applied parallel to the plane in which the electrons are confined. This implies that the conducting phase has something to do with the spin of the electrons.

Electrons can react to a magnetic field in one of two ways. The field can cause the trajectory of the electrons to bend so that they tend to flow in circles that are perpendicular to the direction of the field; this is known as orbital coupling. The field can also influence the spin of the electrons, causing those electrons with their spin pointing in the same direction as the field to have higher energy, while those whose spin points opposite to the field have lower energy. When the field is applied parallel to the two-dimensional plane of the electrons, orbital motion is prohibited since the electrons cannot flow in circles that leave the plane. Thus, the experimental results imply that the root of the unexpected metallic phase involves the electrons' spin. Current experiments are attempting to discern the precise physical mechanism by which the spins form the conducting state.

Resonant tunneling of magnetization

Tunneling is a fundamentally quantum-mechanical phenomenon. In most cases of tunneling, a particle (such as an electron) escapes from a potential well by passing through a region where it is energetically forbidden from being. Throughout most of the history of modern physics, it was thought that tunneling could only occur for small atomic or subatomic particles. Thus, a boulder stuck in the shallow valley between two mountain peaks does not spontaneously "tunnel" through one of the mountains and end up rolling down the opposite face to the lowlands, even though it would be energetically favorable to get to lower ground. On the other hand, if this same situation is reduced to the atomic scale, quantum mechanics allows a "boulder" the size of an electron or atom actually to tunnel through a barrier.

In the 1980s it was suggested theoretically that under suitable circumstances some macroscopic objects should be able to tunnel. One such object is the magnetic moment of a small magnetic particle. In this case, the tunneling is not of the position of the particle from one place to another. Rather, the magnetic moment tunnels from pointing in one direction to pointing in another. That is, the magnet may start out with its north pole pointing up and its south pointing down. Without the physical magnet itself moving, the magnetic moment may tunnel so that now the north pole points down and the south up.

Sarachik and collaborators (including this author) discovered the first unambiguous evidence for tunneling in a molecule-size magnet.[5] The molecule they studied is a relatively small magnet: it has a magnetic moment about 20 times larger than that of a single electron. Because of the symmetry of the molecular structure, the magnetic moment finds it energetically favorable to point in one of two directions, called "up" and "down." The molecule has a series of quantized energy states, some associated with the up direction and others with down. An applied magnetic field changes the energy of these states and at certain values of field, an up state and a down state will become "resonant" – they will have the same energy.

At these particular values of field, resonant tunneling occurs. That is, tunneling between up and down occurs only when the magnetic field makes two energy states resonant. The experimental mark of this phenomenon is the fact that the measured probability for the magnetic moment to switch from up to down is much larger at the resonance fields than at other fields. Essentially, this means that tunneling can be turned on and off by changing the magnetic field, providing convincing evidence that the switching of the direction of the magnetic moment is caused by tunneling and not by some other mechanism.

Box 33.1 A technical discussion of the metal–insulator transition

Symmetry plays an essential role in all of physics and the metal–insulator transition is no exception. The symmetry of a physical system near the transition determines the "universality class" for the transition. For example, there is a magnetic-field universality class (defined by the breaking of time-reversal symmetry), a spin-orbit universality class and a magnetic-impurity universality class.

For each universality class, there are theoretical predictions for the critical behavior of the system, the quantitative manner in which the transition is approached. In general, as the transition is approached from the metallic side the zero-temperature conductivity obeys:

$$\sigma(T \to 0) \propto (t - t_c)^{\mu}$$

where μ is called the critical exponent, t is some control parameter such as impurity concentration or uniaxial stress and t_c is the value of t at which the transition occurs. In general, the universality class for the transition determines the value of μ and so it is this quantity that must be determined with great experimental precision to learn the deep physics of the transition.

The critical exponent μ is determined by first plotting the conductivity of a sample as a function of $T^{1/2}$ and then extrapolating the data to zero temperature to obtain $\sigma(T \to 0)$ for each sample. Finally, this quantity is plotted as a function of impurity concentration and fit to the above expression to obtain μ.

One major result of the Sarachik group's research was the demonstration that a magnetic field could change the universality class of a system. In boron-doped silicon, μ was found to increase from ~0.65 at zero magnetic field to ~1.0 at 7.5 Tesla, as expected for the magnetic-field universality class.

When the transition in the same system is tuned using uniaxial stress rather than dopant concentration, the Sarachik group found $\mu \sim 1.6$, indicating that the stress has changed the physics.

Scaling is an important tool for understanding the physics of the metal–insulator transition.
For example, according to the theory of electron–electron interactions, the magnetoconductance – the difference between the conductivity in a field H and in zero field – should be of the form

$$\Delta\sigma(H, T) \equiv \sigma(H, T) - \sigma(0, T) = T^{1/2} F(H/T)$$

where $F(H/T)$ is a complicated function that depends only on the ratio H/T. This means that if $\Delta\sigma(H,T)/T^{1/2}$ is plotted against H/T, all of the data for all values of H and T should fall on a single curve. This is indeed what the Sarachik group found for the magnetoconductance of boron-doped silicon. Since other mechanisms, such as disorder, violate this form of scaling, this result implies that electron–electron interactions strongly dominate the physics of the metal–insulator transition in that system.

BIOGRAPHY

Many women who attempted to have a career in science in the United States during the 1950s and 1960s faced extraordinary obstacles. So, it is fair to ask: what characteristics give someone the determination and courage to overcome these adversities? In the case of Professor Sarachik, perhaps one major influence was a childhood as a World War II refugee. "Being different was something I had to get used to and make my peace with," says Sarachik.

Myriam Sarachik (née Morgenstein) was born August 8, 1933, in Antwerp, Belgium. When she was six, her orthodox Jewish family fled the country to escape from the advancing Nazi army. When they reached Calais, France, the Germans were already taking that city and the family returned to Antwerp some weeks later. Nearly a year later, they attempted to escape again, but were caught trying to cross into unoccupied France. The family was incarcerated for several months in camps in German-occupied France before escaping. Eventually making their way to Cuba, they spent five-and-a-half years there as refugees and emigrated to New York City in 1947.

After graduating from the prestigious Bronx High School of Science, Sarachik enrolled in Barnard College. She took a physics class at neighboring Columbia University and found it an unexpected challenge, especially her first mid-term. "It was the first and only thing I ever failed," she recalls. Instead of being deterred from the subject, however, she became resolved to succeed in physics and finished the class with a B.

Sarachik went on to attend graduate school at Columbia University, doing her dissertation on the superconducting penetration depth of lead with Professor Richard Garwin; she received her Ph.D. in 1960. From 1962 to 1964 she held a postdoctoral research position at AT&T's Bell Labs in New Jersey and in 1964 became an Assistant Professor at the City College of New York (part of the City University system), where she has remained ever since.

As a young scientist, Sarachik faced significant opposition to her desire to pursue a career in physics. "It was really difficult towards the beginning," she says, "especially when I had a baby. Nobody would give me the time of day." Several people actively encouraged her to leave the field. Notably, however, some of these same people, when faced with her steadfast refusal to quit, agreed to help her in various ways, like writing letters of recommendation or assisting her in finding jobs.

While there were several people who did aid Sarachik in her early career, she does not consider any of them role models. Ironically, she says her father, a businessman, came closest to filling the part. "I came from an orthodox family. As a woman, I was not supposed to be doing any of this," she says. Nevertheless, she found some implicit encouragement from him. "I was doing what he would have loved to do himself."

The tough times getting established ended for Sarachik after a few years at City College. "The potential barrier was enormous," she says, "but once I got over it, it was smooth sailing." Sarachik's early years at City College were devoted to studies of magnetic impurities in materials at what low-temperature physicists now call "high temperatures" – temperatures above 1 K.

Following a family tragedy, Sarachik left research for several years. After she returned, she started work in the field in which she was to earn much of her reputation: the metal–insulator transition in semiconducting materials. At the prompting of a funding agent from the National Science Foundation, Sarachik applied for and received a grant for a dilution refrigerator, a device that gave her the ability to study samples at temperatures as low as 0.02 K. Armed with this, a large-field magnet, some good samples and eager students, Sarachik turned out a series of seminal papers elucidating the nature of the metal-insulator transition and the roles played by impurity concentration, magnetic field, temperature and pressure in the physics that underlies the transition.

Primarily due to these advances, Sarachik received a number of prestigious honors during her "banner year" of 1994–5. In 1994, she was elected as a member of the National Academy of Sciences, becoming one of four professors in the entire City University system to receive membership. Then, in 1995, she was promoted to Distinguished Professor, the highest professorial rank at the college.

Later that year she received the New York City Mayor's Award for Excellence in Science. Mayor Rudolph Giuliani personally made the award in a high-profile ceremony at the mayor's mansion. In addition to these prestigious honors, Sarachik is also a Fellow of the American Physical Society (APS), the American Association for the Advancement of Science, the American Academy of Arts and Sciences and the New York Academy of Sciences. In 2001, she became APS Vice President and, in accordance with the Society's bylaws, she served as President in 2003. In 2005, she shared the APS's Oliver E. Buckley Condensed Matter Physics Prize and received the L'OREAL–UNESCO Award for Women in Science.

Outside of science, Sarachik has also worked to promote human rights. She serves on the Board of Directors of the Committee of Concerned Scientists, the Human Rights of Scientists Committee of the New York Academy of Sciences and has previously been a member and chair of the APS Committee on the International Freedom of Scientists.

If there is a secret to Professor Sarachik's success as a physicist, it is a combination of dedication and what she calls "a good nose" for where the interesting questions lie. Without employing any extraordinary experimental apparatuses or techniques, she has consistently been able to make decisive contributions in several areas of research. "I've learned to pick the problems where intuition gets me a long way," she says.

Added to this is a continual eager interest in her work. Asked to name her most important contribution to physics, Sarachik replies, "I always think that what I'm working on now is the best one."

NOTES

1. Sarachik, M. P., Garwin, R. L. and Erlbach, E., Observation of the energy gap by low-temperature penetration depth measurements in lead. *Phys. Rev. Lett.*, **4** (1960) 52.
2. Sarachik, M. P., Corenzwit, E. and Longinotti, L. D., Resistivity of Mo-Nb and Mo-Re alloys containing 1% Fe. *Phys. Rev.*, **135A** (1964) 1041.

3. Dai, P., Zhang, Y. and Sarachik, M. P., The effect of a magnetic field on the critical conductivity exponent near the metal–insulator transition. *Phys. Rev. Lett.*, **67** (1991) 136. Sarachik, M. P., Transport studies in doped semiconductors near the metal–insulator transition. In *The Metal-Nonmetal Transition Revisited: A Tribute to Sir Nevill Mott, FRS* (London, UK: Francis and Taylor Ltd., 1995). Bogdanovich, S., Dai, P., Sarachik, M. P. and Dobrosavljevic, V., Universal scaling of the magnetoconductance of metallic Si: B. *Phys. Rev. Lett.*, **74** (1995) 2543.

4. Simonian, D., Kravchenko, S. V., Sarachik, M. P. and Pudalov, V. M., Magnetic field suppression of the conducting phase in two dimensions. *Phys. Rev. Lett.*, **79** (1997) 2304. Abrahams, E., Kravchenko, S. V. and Sarachik, M. P., Metallic behavior and related phenomena in two dimensions. *Rev. Mod. Phys.*, **73** (2001) 251. Vitkalov, S. A., Zheng, H., Mertes, K. M., Sarachik, M. P. and Klapwijk, T. M., Small-angle Shubnikov–de Haas oscillations in silicon MOSFETs. *Phys. Rev. Lett.*, **85** (2000) 2164.

5. Friedman, J. R., Sarachik, M. P., Tejada, J. and Ziolo, R., Macroscopic measurement of magnetization tunneling in high-spin molecules. *Phys. Rev. Lett.*, **76** (1996) 3830. Mertes, K. M., Zhong, Y., Sarachik, M. P., *et al.*, Abrupt crossover between thermally-activated relaxation and quantum tunneling in a molecular magnet. *Europhys. Lett.*, **55** (2001) 874.

FURTHER READING

Hysteresis steps demonstrate quantum tunneling of molecular spins. *Physics Today*, **50**:1 (1997) 17–19.

Metal–insulator transition unexpectedly appears in a two-dimensional electron system. *Physics Today*, **50**:7 (1997) 19–21.

Wasserman, E., *The Door in the Dream* (Washington, DC: Joseph Henry Press, 2000).

34 Juliet Lee-Franzini (1933–)

Paolo Franzini
Istituto di Fisica, Università degli Studi di Roma, La Sapienza, Italy

FIGURE 34.1 Photograph of Juliet Lee-Franzini.

CONTRIBUTIONS

Juliet Lee began her career in physics studying K-mesons (or kaons) with the help of nuclear emulsions, at the time when the first hints for parity violation were being found in kaon decays. This technique was pioneered by Marietta Blau (see Chapter 10) some thirty years before and had become a powerful tool in the study of elementary particle physics. That somehow, by studying the little black silver

Out of the Shadows: Contributions of Twentieth-Century Women to Physics,
eds. Nina Byers and Gary Williams. Published by Cambridge University Press.
© Cambridge University Press 2006.

grains marking the path of very short-lived particles, one could shed light on particles whose nature was so much a mystery that they were called "strange particles," gave her a first taste of the wonders of experimental particle physics.

The 1955 Columbia stack included about 5000 positive kaon decays, which were used for the first precision determinations of lifetimes and energy spectra of various decay modes. There she found the first example of the radiative decay $K^+ \to \pi^+ \pi^0 \gamma^0$. Since the gamma ray (photon) was not observable in nuclear emulsion, she had to compute the spectrum of the positively charged pion to confirm this decay was the most likely one. Nuclear emulsions had contributed much to particle physics in the forties and early fifties but they were rapidly becoming obsolete and replaced by bubble chambers and electronics.

In 1957, parity violation had just been established by Chien-Shiung Wu (Chapter 24). This caused a great renewal of interest in the study of weak interactions, and in Columbia that came to mean studying muons. Muons were very mysterious particles whose existence among elementary particles seemed superfluous. Commenting on this, I. I. Rabi allegedly one day asked: "who ordered that?" At this time, nuclear emulsions and kaon decays lost some of their appeal in favor of muons. Now, there is a problem with muons: they like to decay in a complicated way i.e., into an electron, a neutrino and an antineutrino. At first it would appear more economical for the muon just to spit out a photon and change into an electron. Well it is not really that easy when you look carefully at the problem. Juliet calculated that the muon should decay the apparently easier way at least once every one hundred thousand times. Nobody had seen a single muon decay that way, but the results were marginal. She chose the problem as her thesis topic and got to work together with Marcel Bardon and David Berley, in Nevis, where the cyclotron made muons in abundance.

They constructed scintillation counters, assembled the electronics and finally obtained the first significant limit – that neutrinoless muon decays occur less than once in ten million times. This shows, conclusively, that something prevents the annihilation of the neutrino with an anti-neutrino, which is contrary to the idea of matter and antimatter.[1] She wrote up this experiment in 1959 as her Ph.D. thesis while completing a companion experiment searching for the decay of a muon into three electrons in the Columbia University bubble chamber, with fellow student Nicholas Samios. Again they obtained the significant null result that they could find no neutrino-less decay. The only way out of this dilemma was to postulate the existence of different kinds of neutrinos. This evidence against virtual neutrino–antineutrino annihilation led Leon Lederman, Melvin Schwartz and Jack Steinberger to construct the first large high-energy physics experiment. This was the Brookhaven neutrino spark chamber experiment, which indeed confirmed that there were different types of neutrinos. One is the electron type neutrino or ν_e, and another the muonic neutrino ν_μ. Today we say that there are neutrinos of different flavors. Including the tau-neutrino ν_τ, we believe there are three flavors of neutrinos.

She continued her studies of the muon at Nevis Laboratory for five years, measuring first its helicity in collaboration with Paolo Franzini, whom she had just met, and M. Bardon. They used a fanciful wheel of scintillators and lead scatterers, which Lederman, while allocating the funds, called the most expensive apparatus at Nevis. Then, in 1962, with Allan Sachs, John Peoples and Peter Norton, they measured the Michel parameter ρ that is an important property of muon decays. They used sonic spark chambers placed inside a 180° spectrometer, another technical first.[2] The accuracy of that measurement, of about three parts per million, has yet to be superseded. The extraction of the results involved ultra-delicate understanding of the electron spectrum around its end point. This required that she spend more time doing radiative correction calculations in close consultation with theoretical experts, for the design of the experiment and later for extraction of the result, than in the actual data taking. This series of experiments was fundamental towards establishing the V-A theory of the weak interactions upon which our present understanding of the natural laws of physics, the so-called Standard Model, is built. At the State University of New York at Stony Brook, in 1963, Juliet Lee, later Lee-Franzini, started a bubble chamber group and played a leading role in transforming Stony Brook into an important university for particle physics. Her laboratory studied photographs exposed to particle beams produced at the Brookhaven National Laboratory Alternating Gradient Synchrotron, and the Princeton Pennsylvania Accelerator (PPA). The experimental emphasis was back to kaons and to hyperon studies. Some of the early support for the theory of quark mixing, which in those days was the Cabibbo form of universality of the weak interaction, came from her laboratory.

In the seventies, Juliet and Paolo with young collaborators performed a series of experiments at Fermilab to study proton–proton inelastic diffractive scattering. Inelastic diffractive scattering becomes possible at very high energy because an outgoing excited proton differs in wavelength from the target proton by a small fraction of the target proton radius. Coherence then becomes possible just as in light scattering. This phenomenon is also known as diffraction dissociation. The extremely good resolution of the Stony Brook solid-state proton spectrometer allowed resolving the detailed structure of the mass spectrum. This spectrum was composed of a very large number of nucleon resonances of high and low masses almost overlapping each other.[3] Juliet designed a next-generation experiment at FNAL to resolve fully and confirm her identification of a whole series of nuclear resonances, but then the march of high-energy physics intervened. Both experimetal findings and theoretical insights had convinced most physicists that six, rather then four, quarks ought to exist. Indirect proof for the next quark, the b-quark, was obtained in 1977 in the form of a very heavy object, an upsilon decaying into two muons. This object, the Υ, is now understood to be a quark–anti-quark bound state with the quark the so-called b-quark. The complete understanding of the properties of the new quark required refined studies of production and decay of new states of matter.

Box 34.1   Discovery of the upsilon

In 1977, the upsilon, Υ, a bound state of a beauty quark b and its anti-quark $\bar{b}$, was discovered at Fermilab. To be strictly correct, three narrow states of mass about ten times that of a hydrogen atom had been observed, partially resolved. That beauty quarks and anti-quarks were involved was a very likely explanation, on the basis of expectations and past experience. However, beauty is hidden inside the Υ because the beauty of the b-quark cancels that of the anti-b-quark, thus their existence was not quite proved.

Box 34.2   Discovery of 4S

Early in 1980, a fourth quasi-bound S-wave Υ (4S) was discovered. The observation of high-energy electrons in its decay gave conclusive evidence of the existence of b and anti-b-quarks bound with ordinary quarks to form B-mesons. CUSB also discovered higher S-wave states Υ (5S, 6S, 7S) and the six P-wave bound states χ_bs. The in-depth studies of their gluonic, photonic and leptonic decays and transitions increased tremendously our understanding of the strong, electromagnetic and weak interactions of the beauty quark. In order to ferret out the physics of this new "hydrogen atom" of particle physics Juliet re-acquainted herself with many aspects of almost classical quantum mechanics, and participated in vigorous discussions with Cornell theorists. Use of coupled-channel methods allowed her to determine the masses and widths of the higher γ states and infer the excited B-meson (B^*) and the strange B-mesons (B_S, B_S^*) production thresholds long before the latter were observed by CUSB.

From 1979 to 1991, Juliet's research was centered mainly at the Cornell Electron Storage Rings (CESR) at Cornell University. Since 1983 she has been co-spokesperson of the Columbia–Stony Brook (CUSB) experiment. CUSB made dozens of highly significant contributions in the area of upsilon spectroscopy.[4] She investigated the excited states of the upsilon (b–anti-b bound states) and whether isolated b-or anti-b-quarks could be observed. Some of her ideas are contained in a theoretical paper she co-authored with A. I. Sanda, S. Ono and N. A. Torqvist. Detailed fine structure and hyperfine structure calculations gave her confidence in the level splittings she detected, even when they were at the limit of being experimentally resolvable, and only verified some years later. She studied three-gluon decays of upsilons, the two-gluon transitions between them, as well as the two-gluon decays of χ_bs. She found confined gluons as credible as their free cousins, the photons.[5]

With the existence of the charge 1/3 b-quark firmly established by upsilon spectroscopy and B-meson decay studies, it was thought necessary that its charge 2/3 partner, the t-quark, also exists. Unfortunately, its mass was not predictable, masses remaining one of the mysteries of physics. The Fermilab Tevatron, constructed under the leadership of Rich Orr and Helen Edwards (see Chapter 35), was a nice place to look for it since lower-energy machines gave null results. At Stony Brook, starting in 1983, she worked with vigor in the design of a detector for this purpose. This venture began as a 50-person "small" experiment named LAPDOG. By 1992 it had became the large DØ collaboration, 500 physicists strong. Accustomed to working with a small (CUSB had up to 30 members, 12 on the average) group of protean workers, Juliet found the highly structured and compartmentalized style of large collaborations not congenial. She left DØ after it had completed its primary mission of producing the top quark in 1995.[6]

Aside from finding the elusive Higgs particle, another intriguing puzzle exists in particle physics, the understanding of CP violation. CP violation has so far been observed only in the kaon system. Juliet was recruited in 1991, with Paolo, by the Frascati Laboratory to help in the realization of a new, state-of-the-art giant detector (later called KLOE), which would operate at the world's first kaon factory DAΦNE.

KLOE is meant to study CP and other discrete symmetries, not only by traditional methods but also through kaon interferometry. Her vigorous labor on behalf of KLOE physics was certainly more than partly responsible for the success of the small KLOE collaboration, some 100 physicists, in being the first to observe collisions in 1999 at this particle factory.[7]

BIOGRAPHY

Juliet Lee was born in Paris, France, on May 18, 1933, in 19 Rue des Ecole, just behind the Sorbonne. Her parents were pursuing their Ph.D. degrees at the University of Paris: the father in international law and the mother in French literature. Her early infancy was spent on a French farm, where milk was liberally laced with red wine.

At the start of the Sino-Japanese war in 1937, her father received a diplomatic posting at Hanoi, Vietnam, where the family stayed until 1940 when they made their way to a suburb of Chonqing, the wartime capital of China. Life during that period was harsh and almost destitute. For seven years she attended school very sporadically and was occasionally tutored in the "four books and five jings" of the Chinese classics by literati in exile. Not being hampered by knowledge of the spoken Chinese tongue, she found it a good exercise in figuring out an alien logic.

In mid 1947, the Chinese government sent her father as first secretary to the United Nations in New York. The then outstanding New York City public school program found her an enthusiastic and adept pupil. Within five-and-a-half years she completed the liberal arts curriculum, danced the ballet, tutored inner city youngsters in algebra, worked first as a multilingual guide at the United Nations and then worked as a cook and waitress in the restaurant her father acquired when the Chiang Kai-shek government he represented was overthrown. The restaurant happened to be in the Columbia University neighborhood and she took to dropping in and listening in on graduate physics classes. She was drawn by how the workings of the hurly-burly real world can be so neatly described with mathematics.

With the bankruptcy of her family's financial resources, and with the non-availability of physics fellowships for people like herself, she began attending physics graduate school part-time while supporting herself and family by dancing and by scanning nuclear emulsions at the Columbia Nuclear Emulsions Laboratory, which was conveniently located on the 13th floor of the Pupin physics laboratory. Its location was felicitous for obtaining a high quality, albeit temporary, scanning crew of out-of-work writers, actors and musicians. She found the process of preparing photographic emulsions for exposure to cosmic rays and later to accelerator beams, developing and mounting them on glass plates, interesting tinkering. She learned very early to perform complicated calculations. She corresponded or discussed with theoreticians and was at ease in the theoretical community as well as in the laboratory. In 1957, when she finally got a research assistantship, she also got caught up in the "parity violation" fever. Her thesis work took her to Columbia's Nevis Cyclotron Laboratories in Westchester, where the cyclotron was hidden amongst the greenery of a pastoral estate. After completion of her thesis research, she continued as a postdoctoral fellow at the Nevis Laboratories.

She took a year off to indulge in her curiosity of the heavens, with a theoretical fellowship from the National Academy of Science, but returned to particle physics in 1963 when she joined the faculty of SUNY at Stony Brook as its first experimental particle physicist. She ceased to be known as Juliet Lee, in mid 1964, when she married Paolo Franzini. He has been her collaborator since then. Paula Franzini was born in 1965 and raised in physics laboratories around the country, receiving a large part of her education from Juliet. She also became a physicist with over three dozen theory publications before becoming an artist (Ph.D. Stanford 1987. First exposition, Galerie de L'Atelier Circulaire 1999).

During her stay at Stony Brook, Lee-Franzini was always active in experimental physics, in teaching and in the professional life in general. All through 1963–91, she commuted weekly during semester time between teaching duties at Stony Brook and the research laboratories. Her teaching activities embraced a wide spectrum: from novel presentations of physics to the general public, interesting pre-medical students in physics by using bio-physiological examples, co-founding the department's Undergraduate Electronics Laboratory and Advanced Graduate Laboratory, to teaching elementary particle physics at both the undergraduate and graduate level, finally producing over a dozen Ph.D. students, most of whom are now leaders in the field. Her departmental and university activities were numerous. She importantly contributed to the establishment of a first-rate university within 25 years. In her research, she has been a Principal Investigator since 1964, successively of the NY Research Foundation, the US Atomic Energy Commission and the US National Science Foundation.

Since 1991, she has been devoting her major efforts to the physics of DAΦNE, at Frascati. She seems to have come a full circle of living, working and imbibing with people whose language she barely speaks but with whom she communicates mainly by role modeling and enthusiastic empathy.

Honor
Fellow of the American Physical Society.

Jobs and positions
1960–2 Research Associate, Nevis Laboratories, Columbia University
1962–3 Fellow, National Academy of Science–National Research Council
1963–4 Assistant Professor of Physics, State University of New York at Stony Brook
1965–74 Associate Professor of Physics, State University of New York at Stony Brook
1974–93 Professor of Physics, State University of New York at Stony Brook
1980–1 Visiting Professor of Physics, Cornell University
1994– Adjunct Professor of Physics, State University of New York at Stony Brook

1991–6 VIP Physicist, Laboratori Nazionali di Frascati dell'INFN
1996– Director of Research, Laboratori Nazionali di Frascati dell'INFN

Education
B.A. 1953 Hunter College
M.A. 1957 Columbia University
Ph.D. 1960 Columbia University

Professional service
Board of Directors, The Research Foundation of the State of New York, 1986–91
Executive Committee of the Division of Particles and Fields, American Physical Society, 1987–9
Nominating Committee, American Physical Society, 1989–91
Nominating Committee of the Division of Particles and Fields, American Physical Society, 1990–1

NOTES
1. Lee-Franzini, J., Bardon, M. and Berley, D., Upper limit for the decay mode $\mu \rightarrow e + \gamma$. *Phys. Rev. Lett.*, **2** (1959) 357.
2. Lee-Franzini, J., Bardon, M., Norton, P., Peoples, J. and Sachs, A., Measurement of the momentum spectrum of positrons from muon decay. *Phys. Rev. Lett.*, **14** (1965) 449.
3. Lee-Franzini, J., P–P inelastic scattering at small momentum transfer. In *High Energy Collisions*, ed. C. Quigg (New York: American Institute of Physics, 1973); Lee-Franzini, J., Childress, S., Franzini, P., *et al.*, Proton-proton inelastic scattering at 0.5 TeV. *Phys. Lett.*, 65I3 (1976) 177.
4. Lee-Franzini, J., Quark Spectroscopy with a new flavor. *Survey in High Energy Physics*, **2** (1981) 3.
5. Lee-Franzini, J., *et al.*, Observation of P-wave bound states. *Phys. Rev. Lett.*, **49** (1982) 1612 (with the CUSB Collaboration); Lee-Franzini, J. and Franzini, P., Physics at CESR. *Physics Report 81C*, **3** (1982); Lee-Franzini, J. and Franzini, P., Upsilon resonances. *Ann. Rev. Nuc. Part. Sci.*, **33** (1983) 1; Lee-Franzini, J., Heintz, U., Lovelock, D. M. J., *et al.*, Hyperfine splitting of B-mesons and B-production at the $\gamma(4S)$. *Phys. Rev. Lett.*, **65** (1990) 2632.

6. Lee-Franzini, J., *et al.*, Observation of the top quark. *Phys. Rev. Lett.*, **74** (1995) 2632 (with the Dø Collaboration).

7. Lee-Franzini, J., The status of DAΦNE and KLOE. In *The Second DAΦNE Physics Handbook*, eds. L. Maiani, G. Pancheri and N. Paver (Frascati: S.I.S. INFN, 1995) pp. 761–806; Lee-Franzini, J. and Franzini, P., CP violation in K-system. *Survey in High Energy Physics*, **13** (1998) 1.

35 Helen Thom Edwards (1936–)

John Peoples, Jr.
Fermi National Accelerator Laboratory, Batavia, IL

FIGURE 35.1 Photograph of
Helen Thom Edwards.

IMPORTANT CONTRIBUTIONS

Helen Edwards made a major impact on high-energy physics through
her contributions to the construction and operation of the Fermilab
Main Ring and the Tevatron. (See Box 35.1.) The Main Ring, a 400 GeV
proton synchrotron, was the engine that provided the high-energy
proton beams that enabled many important discoveries in particle
physics: e.g., the discovery of the b-quark, precise measurements of

Out of the Shadows: Contributions of Twentieth-Century Women to Physics,
eds. Nina Byers and Gary Williams. Published by Cambridge University Press.
© Cambridge University Press 2006.

Box 35.1 Synchrotrons

Big proton and electron synchrotrons enabled the particle physics experiments that provided critical insights for the formulation and subsequent verification of the basic theory of quarks, leptons and bosons, which is called the Standard Model. Without these synchrotrons this theory might not have come into existence, and certainly would remain untested. The Tevatron is a synchrotron that can accelerate protons and antiprotons to 980 GeV. After the acceleration cycle is complete, the protons are made to collide either with a fixed target external to the synchrotron or with a counter-rotating beam of 980 GeV antiprotons within the Tevatron. The latter mode presently provides the highest collision energies in the world and the former enables the production of the highest energy collisions with a fixed target to create a very intense source of meson beams that can be used to form intense, high-energy beams of neutrinos, photons and mesons. Protons are accelerated in a synchrotron by radio frequency cavities that increase the proton energy a small amount each time a single turn around the nearly circular guide path of 6.28 km circumference is completed. With a very large number of turns, the protons acquire very high energies. To guide these protons, the strength of the magnetic guide field is increased in synchronism with the increasing proton energy. The peak energy is limited by the strength of the guide field and the circumference of the synchrotron. The Main Ring was the first big proton synchrotron on the Fermilab site. Later when the Tevatron was operational it became the injector for the Tevatron. The Tevatron is able to accelerate the protons to much higher energies. It is the first proton synchrotron that successfully exploits superconductivity to make a much stronger guide field than can be achieved using magnets made with iron pole pieces and copper coils.

Box 35.2   Slow resonant extraction

Slow resonant extraction of beams from synchrotrons was developed in the late sixties. Physicists realized then that the cross-sections for reactions that would provide the clearest understanding of the internal structure of protons, neutrons and mesons were very small, and that their observation required intense beams of particles. The only detectors that could handle such intense beams were electronic counter detectors, which were in their infancy. Since electronic detectors have a finite time resolution, it was crucial to make the time structure of the pulses of the extracted electron beams as uniform as possible. If these occur in intense bunches, two or more collisions may occur within the time resolution of the detector and interfere with the desired observations.

Resonant extraction exploits the fact that the amplitude of the transverse oscillations that an individual particle makes with respect to the central orbit of the synchrotron grows exponentially, i.e., becomes resonant, when the number of oscillations per turn (the tune) is a multiple of a simple fraction like a third or a half. Since the increase of the transverse amplitude per turn may approach a centimeter or more before it hits the beam tube walls, a carefully placed electrostatic septum can deflect the particle onto an orbit that takes it cleanly out of the machine. This extraction process is controlled by exciting a set of magnets, such as a sextupole or an octupole, to make the tune a function of amplitude. The tune of the large amplitude particles is increased until they approach the resonance; then they leave the synchrotron after several more turns. By slowly increasing the strength of the sextupole or quadrupole magnets, all of the particles can be smoothly extracted.

Prior to the realization of the possibility of slow resonant extraction, internal targets were used to produce photons and secondary electrons. The radiation damage created by internal targets severely limited the intensity of the beams that could be delivered to an experiment. Resonant extraction made it possible to bring the beam out of the machine with very little loss and slowly deliver it to the experiment where it mattered.

the neutrino cross-sections and the determination of properties of the hyperons. She was responsible for completing and commissioning the Main Ring resonant extraction system. (See Box 35.2.) In 1972, resonant extraction of a particle beam from a synchrotron was in its infancy. The application of this technique to the Main Ring was very difficult, given the magnetic rigidity of its proton beam. Her ideas improved the quality and stability of the Main Ring beam, and she led the development of the specialized components that made it possible to efficiently extract and split the extracted beam into proton beams, which were then transported to experimental areas. Her improvements were essential for the generation of the nine high quality external proton beams. She spearheaded the integration of these components into the Main Ring and the External Beam Switchyard, and led the Switchyard group between 1972 and 1977. Her pervasive hands-on involvement in all aspects of this work is characteristic of her style, and this style would carry over to her leadership of the construction and operation of the Tevatron.

In 1977, Helen Edwards and several others at Fermilab formed the Underground Parameters Committee, the team that would produce the design of the Tevatron accelerator systems and build the Tevatron. For the next six years, she provided the technical leadership that enabled the Tevatron to reach 800 GeV and become the first operational superconducting proton synchrotron. Initially, the Tevatron was operated in the fixed-target mode and the techniques that Helen Edwards developed for resonant extraction and primary beam splitting for the Main Ring were improved and used in the Tevatron with great success. The Tevatron enabled very substantial advancements in the precision of experiments that had been done with the 400 GeV Main Ring. These experiments made significant contributions to the measurement of the properties of charmed and strange particles, the internal structure of nucleons and the strength of the electroweak currents, and ultimately, in 2000, one experiment detected the tau neutrino, the last of the four members of the third generation of quarks and leptons to be observed.

FIGURE 35.2 A photograph of the Tevatron at Fermilab.

The Tevatron, however, had its greatest impact on high-energy physics when it was used as a proton–antiproton collider. Run in this mode, the last and heaviest quark, the top quark, was discovered at Fermilab in 1995. Until the CERN Large Hadron Collider begins operation toward the end of the first decade of the twenty-first century, the Tevatron will remain the highest-energy collider in the world. Helen Edwards led the technical effort that transformed the Tevatron into this very-high-energy collider.

None of these accomplishments would have been possible without her ability to lead a large team of engineers, physicists and technicians. One person working in isolation could not have created a great machine like the Tevatron. It required the sustained work of more than 500 people for a decade to achieve its greatness. The Tevatron is the telescope that allows physicists to peer into the deepest reaches of inner space, just as the great telescopes of the mid twentieth century allowed astronomers to observe the outer reaches of our universe.

BIOGRAPHY

Helen Thom, later Helen Edwards, was born in Detroit, Michigan, on May 27, 1936. She attended Kingswood School at Cranbrook in Bloomfield Hills, Michigan, and the Madeira School for Girls in Washington, D.C. The Madeira School had a reputation for teaching rigorous science and mathematics so it was not surprising it awakened her interest in these disciplines. She remembers that her biology and mathematics teachers helped her take her first steps toward a career in science. These first steps were taken with great difficulty because she was afflicted with dyslexia and this condition made reading difficult. She mastered a subject only through intense concentration, and her ability to concentrate on the essentials characterized her later approach to scientific and technical problems.

Helen entered Cornell University in the fall of 1953 with the intent of majoring in physics. She was one of about a dozen physics majors, and the only woman.

Throughout her scientific career she would be the only woman working on the big accelerators at Fermilab. Her classes typically contained about 20 students since the engineering physics majors also attended the classes taught by the Physics Department. The Cornell Physics Department was a Mecca for young people wanting to pursue a career in physics in the early fifties. She completed the requirements for her Bachelor's degree in physics in 1957, and because she enjoyed physics and Ithaca she enrolled in the Cornell graduate program in physics.

Helen's first research project was a measurement of the development of electromagnetic showers under the supervision of Ken Griesen. The results were the basis of the thesis for her Master's degree – a detour that only women graduate students were required to make on the way to their Ph.D.! While this extended the duration of her graduate studies, the project provided an excellent introduction to accelerators. Cornell was a good place to be because graduate students not only had the opportunity to build their experimental apparatus, they had the opportunity to operate the synchrotron that created the beams for their experiments. The energy and intensity of the synchrotron were increased continuously, and ultimately the energy reached 2.2 GeV. In order to reach higher energy and intensity, existing components were pushed to their limits and new, better components were added. The approach was to try something new, adapt it if it worked, or discard it if it didn't work. This Edisonian approach created an environment in which students, postdocs and faculty worked closely together to produce beams for their experiments.

After she completed her thesis project and received her M.A. in physics, Helen joined Boyce McDaniel's group and began the research for her Ph.D. thesis. The work was at the frontier of the field that was evolving out of nuclear physics and becoming high-energy physics. Her doctoral thesis research examined the details of the angular distribution of the reaction products in the photoproduction reaction $\gamma + P \rightarrow \Lambda^\circ + K^+$. The Λ° and the K^+ were detected with scintillation counters, and the K^+ velocity was measured by time of flight using a technique that compared the arrival time of the K$^+$ in the detector with the phase of the accelerator RF system. This opportunity to gain hands-on experience with the techniques of the big accelerators of the day extended her understanding of the principles of operation of these machines. By the time she had completed her Ph.D. she was well trained in the techniques of accelerator physics as well as particle physics. After completing her Ph.D. in 1966 she elected to stay at Cornell and work on the new electron synchrotron that Robert Wilson and Boyce McDaniel were planning to build.

In the early sixties, Wilson, McDaniel and Don Edwards, together with other members of the Newman Laboratory of Nuclear Studies (LNS) staff at Cornell, searched for an ambitious machine concept as a replacement for the aging synchrotron. They chose a 10 GeV electron synchrotron. This was a daring choice because it would become the highest-energy electron synchrotron in the world. Big machines were being built or had been built only at national laboratories such as Brookhaven, CERN, DESY and SLAC.

In 1963, Helen married Don Edwards who was the principal beam designer for the 10 GeV synchrotron and who had played a key role in converting the weak-focusing 1 GeV synchrotron into a strong-focusing machine. This started a partnership that would have a strong impact on Cornell, Fermilab and DESY.

The National Science Foundation agreed to provide the funding for the synchrotron. However, in order to build it for the promised price, almost the entire LNS staff had to build everything at Cornell. Perhaps because of this challenge, the project attracted a number of young and rising accelerator physicists including Helen Edwards. She joined the project immediately after completing her Ph.D. in 1966, and, like all LNS staff members, was deeply involved in building and testing the myriad state-of-the-art accelerator components. She was put in charge of the commissioning of the 10 GeV machine, and when that was complete she led the commissioning of the slow resonant extraction system. (See Box 35.2.) In addition to improving the efficiency of the detectors, resonant extraction of the intense internal beam held the promise of limiting the build-up of unwanted radioactivity in the synchrotron magnets. The radiation damage created by internal targets severely limited the intensity that could be delivered to an experiment, and resonant extraction opened the possibility of bringing the beam out of the machine with very little loss. In order to make resonant extraction work effectively, the builders had to gain a very deep understanding of the dynamics of the circulating beam. Helen Edwards gained that understanding in the course of building and commissioning the slow extraction system of the 10 GeV synchrotron.

By 1967, it was clear that Cornell would build this large modern accelerator ahead of schedule within a very modest budget of $11.3 million. Given the imminence of success, it was not surprising that the Universities Research Association offered Robert Wilson the directorship of the National Accelerator Laboratory (NAL). This was where the action in big proton accelerators would be in the USA for the next 30 years, and the project attracted many of the best accelerator builders in the world. Don Edwards left for NAL in 1969, and as soon as the 10 GeV slow extraction system was commissioned Helen joined Don and their other Cornell colleagues who had already left for NAL.

Helen Edwards joined the Booster Section as the Associate Head. She led the commissioning effort and the Booster was quickly completed and put into service. By early 1971, the Booster was working well enough to deliver 8 GeV protons to the Main Ring whenever they were needed. The action moved to the Main Ring and Helen Edwards joined the Main Ring commissioning effort. This was a demanding, strenuous effort because the Main Ring was only partially installed and the beam line enclosures to the experimental areas were still under construction. The pace of construction and commissioning of the project was incredibly fast; only three-and-a-half years elapsed from ground breaking in December 1968 in a farmer's corn field, to the first circulating beam in the 6.28 km Main Ring in March 1972.

Since the next step in the progression of steps toward the delivery of the beam to experiments was the extraction of the proton beam from the Main Ring, Helen Edwards was a natural choice to lead that effort in view of her experience with slow resonant extraction at Cornell. The implementation of slow extraction in the Main Ring was much more challenging than it had been at Cornell, since some of the key components like electrostatic septa were not proven and had to be developed at Fermilab during commissioning. The extraction efficiency had to be at least 98% in order to keep the beam-induced radioactivity to a manageable level. This was much more demanding than the 80% to 90% efficiency that had been achieved at the 10 GeV Cornell Synchrotron. The time structure of the slow extracted beam had to be uniform over the 1.5 second spill length. By 1974, the septa were working and all three experimental areas were receiving protons simultaneously. Ultimately, the extracted proton beam would be split into nine external beams by cascading many of these electrostatic septa, and this would give the experimental program immense flexibility.

The peak energy of the Main Ring, the switchyard and the experimental areas was raised from 200 GeV to 300 GeV, and then from 300 GeV to 400 GeV, and by early 1975 the standard operating energy was 400 GeV. Helen Edwards and her group met the challenges and routinely supplied the experimental areas with intense beams of 400 GeV protons. While the primary use of some of the proton beam was to produce secondary and tertiary beams of mesons, neutrinos, muons and photons, some experiments used a fraction of the primary proton beam directly. Leon Lederman discovered the family of upsilon mesons with such an experiment in 1977. That discovery provided the first and conclusive experimental evidence for the b-quark, the first member of the third generation of quarks to be discovered.

When Leon Lederman became the Director of Fermilab in the fall of 1978, the Department of Energy had finally warmed to the Fermilab concept for the Tevatron. The Underground Parameters Committee had made significant progress toward a detailed plan to make the Tevatron a complete machine. Everything was the biggest and newest of its kind. In 1981, the Accelerator Division and the Energy

Saver Division were merged under the leadership of Rich Orr and Helen Edwards, and this reorganization accelerated the development and installation of the large cryogenic systems and the specialized accelerator systems in and around the Main Ring enclosure. Doubler Magnet production had started in mid 1979, and by July 1983 a beam of 512 GeV protons was circulating in the Doubler. After a short period of operation at 400 GeV to complete the 400 GeV experimental program, 800 GeV fixed target operations began in January 1984 and continued, with frequent interruptions for the construction of collider detector halls and other facilities needed for the collider, until 2000.

As the Tevatron fixed-target operation became routine, Helen Edwards turned her full attention to transforming the Energy Doubler ring into the Tevatron Collider. While others were building the Antiproton Source, the Accelerator Division began the difficult task of building and installing the systems that would make the Tevatron a collider while simultaneously providing 800 GeV protons to the experimental program. It didn't take very long to produce the first collisions.

In the fall of 1985, the Antiproton Source began to deliver a few antiprotons to the Main Ring. A few weeks later, Helen Edwards's team had accelerated them to 150 GeV in the Main Ring, and then transferred them to the Tevatron, which already contained a load of protons. The counter-rotating beams of protons and antiprotons were brought into collision at 800 GeV, just before the deadline for the start of the demolition of the Main Ring enclosure at straight section D-Zero. The construction was needed to make way for the colliding beam hall for the D-Zero experiment.

During the 15-month hiatus in accelerator operations created by the construction of the D-Zero collision hall, Helen Edwards led the effort to build and integrate the additional equipment such as low beta insertions that would not only make the Tevatron the highest-energy collider in the world but a collider that could produce large numbers of detectable W and Z bosons. Precision measurements of the basic properties of the W and Z bosons were among the high-priority goals of the high-energy physics community. The highest-priority goal was the discovery of the top quark. The achievement of these goals required luminosity and more luminosity. Since the initial luminosity in 1987 was about a factor of ten less than the design luminosity of 10^{30} cm^{-2} sec^{-1}, she initiated a program of incremental improvements to increase the luminosity. The program was successful and over the next several years the luminosity steadily increased, and by 1989 the design luminosity had been reached. Through her efforts, a clear plan was laid out that Fermilab would follow over the next decade to increase the luminosity still further. The discovery of the top quark in 1995 by the two collider-detector teams working at Fermilab was only possible because the Tevatron luminosity had exceeded the original luminosity goal by a factor of 20. The path to the successful Tevatron, the first superconducting synchrotron and superconducting collider, was not a simple one. There were many technological detours and dead ends. Nevertheless, Fermilab reached the end of the path successfully and first, through the leadership of Helen Edwards.

By 1987, the Superconducting Super Collider (SSC) had become the great challenge for accelerator builders. Helen Edwards, like many of the leading accelerator physicists in the US, had begun contributing to the SSC design on a part-time basis as early as 1983. Until 1987 she served as Deputy Head of the Fermilab Accelerator Division, which was led by Rich Orr. After that she served as the division head, until she joined the SSC as their Technical Director in January 1989. Congress had appropriated the funds for construction of the SSC in the fall of 1988, but the approval had been very contentious. The SSC began to lose support in Congress almost immediately. When the projected cost began to grow, its opponents in Congress gained the upper hand, and eventually terminated further funding. Since Helen Edwards had not planned to serve as Technical Director indefinitely, she returned to Fermilab in mid 1991 in order to pursue her interests in accelerator technology.

After returning to Fermilab, she turned to the application of superconducting radio-frequency (RF) techniques to very-high-energy electron accelerators. Bjorn Wiik, who later became the Chairman of the DESY Directorate, invited Helen and Don Edwards to join the TESLA Test Facility Project (TTF) in 1991. DESY, which was the German national synchrotron laboratory, had begun research and development (R&D) on superconducting RF in 1990, with the ultimate goal of building TESLA, a 500 GeV accelerator in the center of mass electron–positron linear collider. The initial goal of the TTF project was the construction of a 150 MeV, 1.3 GHz superconducting RF linac that could serve as a test bed for TESLA R&D. She began to split her time between Fermilab and DESY in order to work on the TTF project and she served as the TTF project manager from 1992–4. After 1994, while she continued to split her time between DESY and Fermilab, she started an accelerator research program at Fermilab with the goal of producing a low-emittance electron beam by illuminating a photocathode with a laser. This technique, which was invented by others, is at the forefront of accelerator technology and will be one of the ingredients of a cost-effective, X-ray free electron laser. She successfully built the low-emittance photoinjector that was used in the TTF and she worked with the DESY team throughout the commissioning of the TTF as a free electron laser (FEL). The German government recently authorized DESY to begin the construction of an X-ray light source, which will be a scaled-up version of TTF. Helen Edwards continues to be very active in accelerator technology and accelerator system design technology and the construction of accelerators for high-energy physics.

Over her career, Helen has been the recipient of many honors for her contributions to the development of accelerators for high-energy physics. She is a member of the National Academy of Engineering, a member of the American Academy of Arts and Science and a Fellow of the American Physical Society. She was a member of the US Department of Energy High Energy Physics Advisory Panel. She was awarded the US Particle Accelerator School Prize for Achievements in Accelerator Physics and Technology in 1985, the E. O. Lawrence Prize in 1986 and a MacArthur Fellowship in 1988. The latter award is given to "extraordinarily talented individuals to discover and create, free of financial constraints." She, together with Richard Lundy, Rich Orr and Alvin Tollestrup, was awarded the National Technology Medal in 1989 by President George H. W. Bush for the design, construction and successful operation of the Tevatron. More recently, the American Physical Society awarded her the 2003 Robert Wilson Prize for her pivotal achievement and critical contributions as the leader in the design, construction, commissioning and operation of the Tevatron and for her continued contributions to the development of high-gradient superconducting linear accelerators as well as bright and intense electron sources.

IMPORTANT PUBLICATIONS

Blair, R., Edwards, H., *et al.*, Measurement of the rate of increase of neutrino cross-sections with energy, Fermilab-Pub-83-027-Exp, Fermilab-Pub-83-027-E, 1983. *Phys. Rev. Lett.*, **51** (1983) 343–6.

Edwards, H., The energy saver test and commissioning history. Proceedings, 12th International Conference on High Energy Accelerators, Fermilab (1983) pp. 1–9.

The Tevatron energy doubler: a superconducting accelerator. *Ann. Rev. Nucl. Part. Sci.*, **35** (1985) 605–60.

Edwards, D., Edwards, H., *et al.*, The flat beam experiment at the FNAL photoinjector, LINAC2000-MOB13, Aug 2000 3pp. Published in eConf C00821:MOB13, 2000.

Edwards, H., The superconducting electron–positron collider with an integrated X-ray laser laboratory. DESY 2001-011, March 2001.

Ayvazian, V., Edwards, H., *et al.*, Generation of GW radiation pulses from a VUV free electron laser operating in the femtosecond regime, DESY-01-226, Dec 2001, 5pp. *Phys. Rev. Lett.*, **88** (2002) 104802.

36 Mary Katharine Gaillard (1939–)

Andrzej J. Buras

Technische Universtät München, Garching Germany

FIGURE 36.1 Photograph of
Mary Katharine Gaillard.

SOME IMPORTANT CONTRIBUTIONS

The Standard Model of elementary particles (quarks and leptons) and
their interactions (strong, weak and electromagnetic) was constructed
in the 1960s and in the first years of the 1970s. By the end of 1973,
the basic structure of the Standard Model existed and what followed
were comparisons of its predictions with experimental data. During
the first five years of these efforts, Mary K. Gaillard made outstanding
contributions. For these achievements she received the prestigious
J. J. Sakurai prize in 1993. Her first paper of this period was the 1974
pioneering calculation with B. W. Lee, of strong interaction effects

Out of the Shadows: Contributions of Twentieth-Century Women to Physics,
eds. Nina Byers and Gary Williams. Published by Cambridge University Press.
© Cambridge University Press 2006.

at short distances in decays of K-mesons into two π-mesons.[1] These mesons are bound states of light quarks and antiquarks. Owing to the property of "asymptotic freedom" in strong interactions, the coupling characterizing the strength of these interactions becomes small at short distances and the effects of strong interactions can be calculated by perturbative methods. This work is the prototype of calculations still going on today of strong interaction effects in weak decays of mesons.

A second paper,[2] written also with Ben Lee in 1974, investigated the rare decays of K-mesons and in particular discussed the transition of a neutral K-meson (K^0) to its antiparticle $(\overline{K}^0)$. This is known as $K^0 - \overline{K}^0$ mixing, a very tiny effect observed experimentally in 1960. Gaillard and Lee found that the size of this transition could be explained by quantum effects involving the charm quark provided its mass was roughly twice the proton mass or roughly four times the K-meson mass. The charm quark was discovered eight months later, and its mass turned out to be only 30% lower than the Gaillard–Lee prediction. The rather poor calculation technology available at the time made this prediction a great achievement. Simultaneously, this work demonstrated that through the study of quantum effects in low-energy processes it is possible to make predictions about the properties of heavy objects, which cannot be produced directly in the collisions at low energies. The Gaillard–Lee prediction of the charm quark mass had a considerable impact on future calculations, which aimed at the predictions of masses of heavy particles such as the top quark, the Higgs boson and supersymmetric particles prior to their discovery. This paper also reported studies of a number of rare decays of K-mesons to π-mesons and leptons in the Standard Model, which had an important impact on future calculations.

A few years later, she wrote two papers on b-quarks and t-quarks in collaboration with Ellis, Nanopoulos and Rudaz.[3] These papers were among the first to analyze quantitatively the phenomena of the violation of CP symmetry in K-meson decays and of $B^0 - B^0$ mixing, in analogy to K^0-K^0 mixing. Both phenomena play a crucial role in the present day phenomenology of mesons.

Another important contribution of Gaillard in this period was the 1975 analysis of the properties of charmed mesons (the bound states involving charm quarks) with the aim of discovering these particles. This review article,[4] written in collaboration with Lee and J. Rosner, is considered a classic paper in charm physics. In the same spirit Gaillard, in collaboration with J. Ellis and D. V. Nanopoulos, made a detailed analysis of the production, decay and observability of the Higgs boson, the sole particle in the Standard Model that, at the time of this writing, has not yet been discovered.

In 1976, Ellis, Gaillard and Nanopoulos pointed out that the experimental signature of gluons,[5] the mediators of strong interactions, would be simultaneous production of three groups of particles (called "jets") in high-energy electron–positron collisions. Two of these jets originate from the production of a quark and an antiquark. The third comes from the emission of a gluon. This three-jet structure was discovered in 1979 at DESY in Hamburg, and its observation gave a great impetus to the phenomenological applications of Quantum Chromodynamics (QCD). Since the probability for the appearance of the three-jet events depends on the strength of interactions characterized by the QCD coupling, these events are well suited for the determination of this coupling. Indeed, the most accurate determinations of QCD coupling have been made through studies of jet events.

It was realized in the 1970s that at very high energies, corresponding to 10^{15} proton masses, the strong, weak and electromagnetic interactions could be unified in a theory ("grand unified theory") characterized by a reduced number of parameters relative to the Standard Model. In particular, with increasing energies, the differences between the strengths of strong, weak and electromagnetic interactions diminish owing to quantum effects so that at very high energies all the couplings characterizing these interactions become equal. In quantum field theory also, masses of quarks and leptons depend effectively on energy through quantum effects and become smaller as the energy increases. The decrease depends on the strength of the interactions a given particle experiences. Since the quarks interact strongly, weakly and electromagnetically and the leptons do not feel strong interactions, the decrease of masses with energy is substantially faster for quarks than leptons. In this context, M. Chanowitz, Ellis and Gaillard pointed out,[6] in 1977, that in grand unified theories (GUTs) the mass of the (at that time) conjectured b-quark and the mass of the τ-lepton, which *was* discovered in 1976, should be equal at very high energies. Knowing the mass of the τ-lepton and including quantum effects, they were able to predict the b-quark mass at low energies. This prediction turned out to be in agreement with the mass of the quark discovered a few months later at Fermilab (Chicago). Following this, a refined and substantially extended analysis of a GUT based on the SU(5)-symmetry (Georgi–Glashow model) by Buras, Ellis, Gaillard and Nanopoulos showed that the successful prediction of the b-quark mass was valid provided there were not more than six quarks.[7] The high-energy experiments performed at CERN at the end of the 1980s confirmed this expectation. Galliard's paper on this can be considered the first extensive analysis of the phenomenological implications of grand unified theories, in particular the decay of the proton, and had a considerable impact on future analyses of GUTs, which tended to dominate particle physics in the 1980s. In fact, the paper is the most cited (more than 1 000 citations) among Gaillard's papers.

An unresolved question in the Standard Model of particle physics is the origin of the symmetry breaking by which particles acquire masses. The simplest implementation of this phenomenon predicts the existence of at least one new scalar particle ("Higgs particle") less massive than about 1000 times the proton mass. Chanowitz and Gaillard showed on very general grounds that,[8] in the absence of such new particles, new interactions would be observable by proton collisions with sufficiently high energy beams. Although there is now indirect evidence in favor of a light Higgs particle, this work influenced many subsequent studies of physics signatures at very-high-energy colliders. In particular, it played a role in the delineation of requirements for accelerator facilities and particle detectors that will be used to explore the breaking of supersymmetry, a symmetry among bosons and fermions. The latter symmetry is expected to be present in case elementary Higgs particles exist. The phenomenological signatures of the Higgs particles were discussed by Gaillard, Ellis and Nanopoulos in another well-known paper.[9] If supersymmetry is correct, it would imply that Einstein's theory of general relativity is embedded in a supersymmetric generalization of gravity known as supergravity. Gaillard's work in the 1990s focused on supersymmetry and supergravity. At present, the only known way to reconcile Einstein's theory with quantum physics is through a supersymmetric theory in higher dimensional space-time, in which the fundamental objects are not particles, but extended objects like strings. Gaillard has been studying supergravity theories that are low-energy limits of the "superstring theory" and working out predictions of this theory for phenomena that may be detected both in accelerator experiments and in cosmological observations. To this end, she has contributed to several technical developments in supergravity theory and applied them to the study of specific models derived from superstring theory. In particular, she has developed a procedure for evaluating quantum corrections to supergravity theories. Using these techniques, Binetruy, Gaillard and Wu constructed a model for supersymmetry breaking,[10] which is consistent with observations.

BIOGRAPHY

Mary K. Gaillard grew up as Mary K. Ralph in Painesville, Ohio, in a family of teachers and preachers, where a college education was taken for granted. Having always enjoyed math, she discovered physics as a senior in high school, and chose it as her college major.

Gaillard's college choice was largely dictated by family finances; she had a full scholarship from Hollins College in Virginia, which had about one physics major every two years. However, there was an excellent physics professor, Dr. Dorothy Montgomery, at Hollins at the time. Montgomery had been at Yale, but was let go when her husband died, and moved to Virginia to raise her children. Montgomery arranged for Gaillard to work in a cosmic-ray laboratory in Paris during a "Year Abroad" program, and encouraged her to apply for a summer student program at Brookhaven National Laboratory. These experiences convinced Gaillard that she wanted to become a high-energy physicist. At Brookhaven she worked with a group of high-energy experimentalists from Columbia University, and then chose Columbia for graduate school. She preferred – and was more talented at – solving problems on paper than making apparatus work, but in the graduate school culture it was considered pretentious and risky to consider doing theory, so she assumed she would become an experimentalist. Gaillard became a theorist somewhat by default. She married a Frenchman, Jean-Marc Gaillard, and went to France to finish graduate work. She was told that she had little chance to be accepted by the theory group at Orsay, south of Paris, that offered the graduate courses, and that she should get a job in a lab as most of her colleagues did. However, she was refused by every lab in the Paris area; most of them took only students from the top two "Grandes Ecoles" – which at the time admitted only men. Finally, her husband's boss decided to hire her, but that was thwarted when she happily told the medical examiner that she was pregnant – she couldn't be hired because of the lab's radiation hazards. After completing her exams, however, she was admitted into the French Centre National de Recherche Scientifique (CNRS) as a member of the Orsay theory group. By that time, she was living near Geneva, Switzerland, since her husband had got a job at CERN, the European Center for Nuclear Research. She became a visitor to the CERN theory group and was given a desk in a crowded basement office. Her office space improved over the 18 years she spent at CERN, but her status as an unpaid visitor never changed. However, she gradually moved up the ranks of the French CNRS ladder, attaining the highest rank of Director of Research in 1980. In 1979 she started a theory group at a high-energy physics lab in Annecy, France, near Geneva.

Her early research experience was rather unusual because she was initially a student in a research institution with no students – at least not in theory. At the same time she had three small children – all born before she got her doctoral degree. As a consequence of her unusual situation, most of her early scientific contacts were with her husband's experimental colleagues, and her research was directed at issues closely related to the analysis and interpretation of experimental results.

A major turning point in Gaillard's career was a year spent with her family at Fermilab in 1973-4, where she began working with Ben Lee, the newly appointed head of the theory group there, who was tragically killed in an automobile accident a few years later. When she returned to CERN, she began a six-year collaboration with John Ellis, Dimitri Nanopoulos and other CERN theory postdocs. By the end of the 1970s, her work had become widely cited and Gaillard was invited to lecture around the world. She was mentoring many postdocs at CERN, and became increasingly dissatisfied with her situation as an unpaid visitor there. Together with changes in her personal life, this discomfort led her to accept the offer of a professorship at the Berkeley campus of the University of California, becoming the first woman faculty member of the Physics Department. She and her present husband, theoretical physicist Bruno Zumino, have been on the Berkeley physics faculty since 1981.

Gaillard's work has been recognized by the Prix Thibaud of the Academy of Arts and Sciences of Lyons, the E. O. Lawrence Memorial Award of the US Department of Energy and the J. J. Sakurai Prize of the American Physical Society (APS). She is a member of the National Academy of Sciences and a fellow of the American Academy of Arts and Sciences. She has served on numerous physics advisory committees and editorial boards both in the United States and in Europe. She is deeply committed to the promotion of basic science, science education and the inclusion of underrepresented groups. Among her committee service in the APS were efforts on behalf of women in physics; shortly before leaving Europe she wrote a report on women physicists at CERN. In 1996, she was appointed by President Clinton to a six-year term on the National Science Board. In this capacity she recently chaired a Task Force on math and science education that has culminated in Board recommendations for achieving a national consensus on educational standards in these fields.

NOTES
1. Gaillard, M. K. and Lee, W. B., Delta I = 1/2 rule for nonleptonic decays in asymptotically free field theories. *Phys. Rev. Lett.*, **33** (1974) 168.
2. Gaillard, M. K. and Lee, B. W., Rare decay modes of the K mesons in gauge theories. *Phys. Rev.*, **D10** (1974) 897.
3. Gaillard, M. K., Ellis, J. R. and Nanopoulos, D. V., Left-handed currents and CP violation. *Nucl. Phys.* **B109** (1976), 213; Gaillard, M. K., Ellis, J. R., Nanopoulos, D. V. and Rudaz, S., The phenomenology of the next left-handed quarks. *Nucl. Phys.*, **B131** (1977) 285.
4. Gaillard, M. K., Lee, B. W. and Rosner, J. L., Search for the charm. *Rev. Mod. Phys.*, **47** (1975) 277.
5. Gaillard, M. K., Ellis, J. R. and Ross, G. G., Search for gluons in e^+e^- annihilation. *Nucl. Phys.*, **B111** (1976) 253.
6. Gaillard, M. K., Chanowitz, M. S. and Ellis, J. R., The price of natural flavour conservation in neutral weak interactions. *Nucl. Phys.*, **B128** (1977) 506.
7. Gaillard, M. K., Buras, A. J., Ellis, J. R. and Nanopoulos, D. V., Aspects of the grand unification of strong, weak, and electromagnetic interactions. *Nucl. Phys.* B, **135** (1978) 66.
8. Gaillard, M. K. and Chanowitz, M., The TeV physics of strongly interacting Ws and Zs. *Nucl. Phys.*, **B261** (1985) 379.
9. Gaillard, M. K., Ellis, J. R. and Nanopoulos, D. V., A phenomenological profile of the Higgs boson. *Nucl. Phys.*, **B106** (1976) 292.
10. Gaillard, M. K., Binetruy, P. and Wu, Y.-Y., Modular invariant formulation of multi-gaugino and matter condensation. *Nucl. Phys.*, **B493** (1977) 27.

37 Renata Kallosh (1943–)

Andrei Linde
Physics Department, Stanford University

Michael Gutperle
Physics Department, University of California, Los Angeles

FIGURE 37.1 Photograph of
Renata Kallosh.

Renata Kallosh is a Professor of Physics at Stanford University, USA.
She is a leading expert in theoretical and mathematical physics today.
Her research is on quantum theory of fundamental interactions,
including quantum gravity. She is an author of 180 papers on particle physics, supersymmetry, supergravity, string theory, theory of
black holes and cosmology. This list contains the most complicated

Out of the Shadows: Contributions of Twentieth-Century Women to Physics,
eds. Nina Byers and Gary Williams. Published by Cambridge University Press.
© Cambridge University Press 2006.

parts of theoretical physics, many of which are still "under construction."

The scientific career of Renata Kallosh began in the mid sixties. From the very beginning she worked at the frontier of theoretical physics. At that time many people believed that one should compose the set of the most general axioms of quantum theory, and the rest will automatically follow. Renata spent several years studying this axiomatic approach and it became the subject of her Ph.D. dissertation. However, soon thereafter she realized that true progress in the field could be reached only within the context of gauge theories of elementary particles.

According to the Standard Model of electroweak interactions, introduced by S. L. Glashow, S. Weinberg and A. Salam at the end of the sixties,[1] there is no fundamental difference between weak and electromagnetic interactions. Weak interactions are mediated by vector particles similar to photons, but these particles are extremely heavy due to the spontaneous breakdown of gauge invariance. This symmetry breaking is responsible for the differences we see between weak and electromagnetic interactions. At present, realistic theories of elementary particles are based on the principle of spontaneous symmetry breaking, but the way towards the universal acceptance of this was slow and Renata Kallosh was one of the few experts whose work led the way.

The landscape for theoretical physics changed in 1971 when the young physicist Gerard 't Hooft from Holland published a paper[2] that indicated that gauge theories of the type proposed by Glashow, Weinberg and Salam are renormalizable. This meant that there was a consistent way to work with these theories removing the infinite expressions that plague other versions of the theory of elementary particles. Problems of consistency of theories with spontaneous symmetry breaking were resolved by Renata Kallosh and Igor Tyutin,[3] and by 't Hooft and Martinus Veltman.[4] The paper by Kallosh and Tyutin was written earlier than that of 't Hooft and Veltman but was published later in a Russian journal, and remains relatively unknown. The discovery of renormalizability of gauge theories with spontaneous symmetry breaking was a crucial step in development of the unified theories of fundamental interactions. In 1999 't Hooft and Veltman received a Nobel Prize for this discovery.

The next step towards unification of theories of fundamental interactions was the discovery of supergravity.[5] This theory is based on a special symmetry between bosons and fermions, called supersymmetry. (See Box 37.1.) Supergravity gave physicists a way of realizing

---

Box 37.1   Supersymmetry

Progress in the understanding of the fundamental constituents of matter and their interactions has been intimately tied to uncovering the underlying symmetries in the laws of physics. Elementary particle theory is based on three principles: quantum mechanics, special relativity and gauge symmetry. It is a consequence of relativity and quantum mechanics that there are two distinct types of particles, fermions and bosons, with half integer and integer spin respectively. Fermions are the basic constituents of matter (like electrons or quarks) whereas the particles mediating forces between matter are bosons (like photons or gluons).

Supersymmetry is a symmetry that relates bosons and fermions. It has been shown that supersymmetry is the unique extension of the symmetries of particle physics consistent with the principles of quantum mechanics and relativity. Supersymmetry predicts the existence of new particles ("superpartners") of the known particles in the Standard Model. These particles have not been observed yet and there are only indirect hints that supersymmetry is a symmetry of nature, since supersymmetry is broken at low energies. At present, several particle physics experiments are searching for evidence for supersymmetry.

If supersymmetry is a symmetry in nature, it also has to be applied to the theory of gravitation. The resulting theory is called supergravity and naturally provides a unification of the space-time symmetries and gauge symmetries.

a dream of Einstein. It unifies symmetries of the theory of elementary particles with the symmetries of space-time. But at the quantum level a supergravity theory is difficult to formulate. In simpler cases, quantization of gauge theories involves the introduction of auxiliary fields called Faddeev–Popov ghosts. They interact with other fields in the Feynman diagrams used to make calculations in quantum field theory, but they do not interact with each other. Kallosh found that quantization of supergravity requires the introduction of Faddeev–Popov ghosts of a new type, which interact with each other. In addition, she introduced another family of ghosts, which later were called Kallosh–Nielsen ghosts.[6]

Before the introduction of supersymmetry, the theory of quantum gravity had a major problem. Quantum corrections calculated contained infinite expressions as ambiguous as the infinities that stood in the way of the development of the theory of weak interactions prior to the unified electroweak theory with spontaneous symmetry breaking. In this respect, the quantum theory of supergravity seemed better than the theory of quantum gravity because in the first and second order of perturbation theory the infinities disappeared. In the third order, the theory was largely unexplored because of the enormous complexity of the calculations involved. It was found that infinities do appear in the simplest versions of the theory, but there was a hope that in the most advanced theory, $N = 8$ supergravity, all infinities will not appear at least until the seventh order of perturbation theory. However, investigations of Kallosh demonstrated that infinities may appear in $N = 8$ supergravity, and in all other versions of supergravity, starting from the third order of perturbation theory. This result was important in turning the attention of experts away from the theory of supergravity to the theory of superstrings.[7]

The theory of superstrings (see Box 37.2) was introduced in 1985 by Michael Green and John Schwarz, and since that time has become a major tool for investigation of the theory of fundamental interactions including gravity.[8] The basic idea of the string theory is that the coordinates of space-time are not just labels, but are dynamical variables on a two-dimensional worldsheet. They have their own dynamics. The usual treatment of space-time is a low-energy approximation to string theory. Supersymmetry plays a prominent role in string theory, and it is desirable to respect all of its symmetries in quantum calculations. This was extremely difficult. In 1988, Kallosh found some ways to overcome this difficulty and to quantize Green–Schwarz superstring theory in a manifestly supersymmetric way.[9]

Box 37.2 String theory

String theory is a promising candidate for a unified quantum theory of all forces, including the interactions of the Standard Model and gravity. String theory evades the problem of infinities, which plagues other attempts to quantize gravity, by replacing point-like objects and interactions with one-dimensional strings. Different particles manifest themselves as excitations of a string, and interactions are replaced by the splitting and joining of strings.

Historically, string theory emerged out of the dual model theory of nuclear interactions in the early seventies. The work of Green and Schwarz in the eighties established supersymmetric string theory as a consistent, perturbatively finite quantization of gravity. It was found that there are five superstring theories (called Type I, Type IIA, Type IIB, $SO(32)$ heterotic and $E_8 \times E_8$ heterotic), which all have seemingly different properties. (See Fig. 37.2.)

A major advance in the understanding of string theory occurred in 1995 with Witten's discovery that the five different superstring theories are limits of an underlying theory, which is called "M-theory." Furthermore, these theories are related by duality symmetries. This so-called "duality-revolution" has spurred an enormous interest in string theory in the past decade.

A duality symmetry relates two different but equivalent descriptions of the same physical system. For example, in supersymmetric gauge theory, one can either treat gauge bosons (gluons) or magnetic monopoles as the fundamental fields. Which description is more

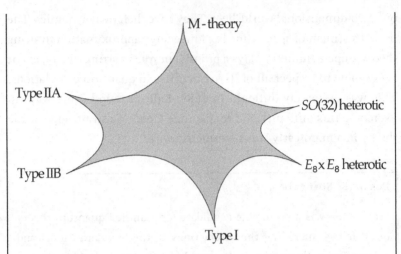

FIGURE 37.2 A schematic representation of the moduli space of
M-theory. The different superstring theories are limits of parameters
(moduli) of M-theory.

appropriate depends on the value of a parameter of the gauge theory,
namely the coupling constant. There is a map, a duality transformation, which exchanges electric and magnetic fields and inverts the
coupling constant. In the same way the various string theories are good
descriptions for some values of the parameter space of string theory and
they are related by duality transformations.

String theory has achieved many beautiful results. Among them
are the statistical interpretation of the entropy of black holes, deep
mathematical results in the theory of algebraic surfaces and many
new insights in the theory of supersymmetric gauge theories. These
advances are to a large extent theoretical and mathematical. Whether
string theory indeed describes nature is an open question.

In 1990, Kallosh began investigations in a new field: black
holes in supergravity and string theory. Black holes were relatively
unexplored at that time, but have become a major field of study in
theoretical physics. This new approach to fundamental interactions,
which Kallosh adopted, is known now as M-theory, or nonperturbative

string theory. Different regimes of this theory correspond to different versions of string theory and supergravity. There are some nontrivial relations between versions of string theory and supergravity that cannot be understood by the usual methods but sometimes become manifest when one studies black holes. In this sense, the theory of black holes provides a unique way of studying the underlying structure of the theory of fundamental interactions. Renata's pioneering work played an important role in this investigation.

Black holes can exist only if their electric and magnetic charges are sufficiently small. The maximally charged black holes are called extremal black holes. Kallosh and her colleagues have found that extremal black holes in string theory and supergravity are supersymmetric.[10] Such black holes have amazing properties. A most interesting quantity describing the properties of a black hole is its entropy. Kallosh has shown that supersymmetric extremal black holes possess the exceptional property that all higher order quantum gravity corrections to the entropy of extreme black holes vanish. This was a signal that supersymmetric black holes may play an important role in quantum gravity.

One of the most important results of her study of black holes was the general supersymmetric black hole entropy formula.[11] This formula has some remarkable properties. First of all, the entropy depends only on quantized values of charges and does not depend on various continuous parameters. A second important property of the entropy is duality. This property comes from the duality properties of the equations of motion of supergravity theory. Such duality invariant black holes cannot be described in the framework of the perturbative string theory. One needs string theories and their dual partners to find the states of the fundamental M-theory described by extremal black holes. Therefore, supersymmetric black holes help us to understand the nonperturbative structure of M-theory. In the studies of black holes, Kallosh with colleagues has found a new phenomenon, enhancement of unbroken supersymmetry near the black-hole horizon.[12] Simultaneously, it has been realized that near the horizon of extreme black holes, and in the more general cases, supersymmetric branes, the geometry of space can be represented as a product of anti-de Sitter space and a sphere. This observation about the anti-de Sitter space near the extreme black-hole horizon as well as near the horizon of some particular stringy branes, called D3 branes, was more recently reformulated as a particular duality relation between supergravity theory in the bulk and some conformal field theories at the boundary of the anti-de Sitter space. This is known as the Maldacena conjecture,[13] which is widely believed to be a significant step towards understanding of quantum gravity.

Over the last few years Renata Kallosh worked on problems of M/string theory, supergravity and cosmology. Recent cosmological observations support the theory of cosmic inflation. This is a challenge for M-theory since it is difficult to derive inflationary cosmology from this theory. One of the possibilities to implement inflation in M-theory, studied by Kallosh and her collaborators, was via the motion of D-branes towards each other. The distance between branes plays the role of the scalar field responsible for inflation.[14]

One of the most interesting recent results obtained by Kallosh is related to the theory of dark energy. Cosmological observations indicate that several billion years ago the universe entered a stage of accelerated expansion. The simplest way to explain the observational data is to assume that the vacuum has a small positive energy density (cosmological constant). This would imply that the universe is going to continue its accelerating expansion forever. However, it is extremely difficult to find a realization of this scenario in the context of M-theory. This problem was solved only recently by Kallosh and her collaborators. They have found that M-theory provides two different possibilities. The first one is that the vacuum energy is time dependent, gradually switching from positive to negative values. These theories lead to a prediction of a global collapse of the universe within the next 10–20 billion years. The second class of theories, which is called KKLT construction,[15] is one of the most important developments in string theory during the last several years. It leads to a prediction of a practically unlimited accelerated expansion of the universe.

In order to find which of these possibilities is realized in nature, one should compare the cosmological observational data with the the-oretical predictions. One can show that some of the models predicting the collapse of the universe within the next five billion years can be ruled out by the existing observational data. The investigation of the stability of the universe on a greater time scale will require a much more detailed observational study of the present stage of acceleration of the universe.

These results may have important implications. One may ask: "Why should anybody care about the M-theory? Why should anybody care about precise measurements of cosmological parameters? Why do we need to spend billions of dollars for the development of science?".

Now one can add something new to the existing arguments. The theories studied by Kallosh are at the forefront of modern physics. These theories may look very complicated and abstract, but without their development and without cosmological observations we will be unable to know the future of the universe and the future of mankind.

BIOGRAPHY

Renata Kallosh was born on January 16, 1943, in Chernowitz. This is a small town in southwestern Ukraine, situated on the Prut River in the Carpathian foothills. Chernowitz first belonged to the Polish-Lithuanian kingdom, then to Turkey, to Austria-Hungary and to Romania. Finally, in 1940, the town was acquired by the USSR. Renata's parents worked in the Chernowitz university; they were linguists. Her father died soon after World War II, when Renata was only three years old. Her mother took care of the family, which also included Renata's older brother Shandor. A few years later Shandor moved to Lvov, then to Moscow, and became a composer. Renata's childhood in postwar USSR was not very easy, but her natural tal-ents helped her not only to survive but even to enjoy life. She loved sports, and held a second place in Ukraine in gymnastics. But she loved physics and mathematics even more.

In 1959, she went to Moscow and became a student of the Moscow State University. She graduated in 1965 and became a graduate student at the Theoretical Department of the Lebedev Physical Institute, Moscow. This is one of the best scientific organizations of Russia. Among other people working there was Vitaly Ginzburg, one of the authors of the theory of superconductivity, and the famous physicist and dissident, Andrei Sakharov.

In 1968, she got her Ph.D. for work on axiomatic field theory and current algebra. At that time Igor Tamm was the head of the Theoretical Department of the Lebedev Institute. When he attended the first talk given by Renata at a theoretical seminar, he immediately recognized her talent and offered her a position at the Lebedev Institute. She became the first woman scientist working in the Theoretical Department.

Tamm was a Nobel Prize winner for his work on Tamm–Cherenkov radiation. He had a half-serious theory that women from East Europe with a name beginning with the letter K have a good chance of becoming great scientists. There were two examples confirming this rule: Sofia Kovalevskaia, a famous Russian mathematician, and Irène Curie (in Russian transcription her name begins with K). He sincerely believed that Renata Kallosh was going to provide the third example and become a distinguished physicist. This was one of the many theoretical predictions of Igor Tamm that came true. In 1981, Renata became a Professor of Physics at the Lebedev Physical Institute, one of the leading experts in gauge theories, quantum gravity and supergravity.

In 1989–1990 Renata Kallosh spent two years at CERN, Switzerland, working there together with her husband, Andrei Linde, one of the authors of inflationary cosmology. One evening she received an email message from the USA offering her and Andrei Linde two Professor positions at Stanford University. "What is Stanford? What shall we do?" she asked her colleague, a physicist from Paris. He gave her an answer consisting of two mutually contradicting parts: "Do you seriously consider moving from Europe to this cultural desert? But many people would die to get an offer like that." A decision was taken not without the influence of Renata's sons, Dimitri and Alex. Dimitri, a computer wiz, told his parents that he emailed many questions on computer science to different parts of the world, and the best answers always came from Stanford.

In Autumn, 1990, Renata Kallosh and Andrei Linde accepted positions of Professors of Physics at Stanford, and they continue working there now.

NOTES

1. Glashow, S. L., Partial symmetries of weak interactions. *Nucl. Phys.*, **22** (1961) 579; Weinberg, S., A model of leptons. *Phys. Rev. Lett.*, **19** (1967) 1264; Salam, A., Weak and electromagnetic interactions. In *Svartholm: Elementary Particle Theory, Proceedings of the Nobel Symposium Held 1968 at Lerum, Sweden, Stockholm, 1968,* 367.
2. 't Hooft, G., Renormalizable Lagrangians for massive Yang–Mills fields. *Nucl. Phys.*, **B35** (1971) 167.
3. Kallosh, R. E. and Tyutin, I. V., The equivalence theorem and gauge invariance in renormalizable theories. *Sov. J. Nucl. Phys.*, **17** (1973) 98.
4. 't Hooft, G. and Veltman, M., Regularization and renormalization of gauge fields. *Nucl. Phys.*, **B44** (1972) 189.
5. Ferrara, S., Freedman, D. Z. and van Nieuwenhuizen, P., Progress toward a theory of supergravity. *Phys. Rev.*, **D13** (1976) 3214; Deser, S. and Zumino, B., Consistent supergravity. *Phys. Lett.*, **65B** (1976) 369.
6. Kallosh, R. E., Modified Feynman rules in supergravity. *Nucl. Phys.*, **B141** (1978) 141.
7. Kallosh, R. E., Counterterms in extended supergravities. *Phys. Lett.*, **B99** (1981) 122.
8. Green, M. B., Schwarz, J. H. and Witten, E., *Superstring Theory* (Cambridge, UK: Cambridge University Press, 1987).
9. Kallosh, R. E., Quantization of Green-Schwarz superstring. *Phys. Lett.*, **195B** (1987) 369.
10. Kallosh, R., Linde, A., Ortin, T., Peet, A. and Van Proeyen, A., Supersymmetry as a cosmic censor. *Phys. Rev.*, **D46** (1992) 5278.

11. Ferrara, S., Kallosh, R. and Strominger, A., N = 2 extremal black holes. *Phys. Rev.*, **D52** (1995) 5412; Ferrara, S. and Kallosh, R., Supersymmetry and attractors. *Phys. Rev.*, **D54** (1996) 1525; Ferrara, S. and Kallosh, R., Universality of supersymmetric attractors. *Phys. Rev.*, **D54** (1996) 1525; Ferrara, S., Gibbons, G. W. and Kallosh, R., Black holes and critical points in moduli space. *Nucl. Phys.*, **B500** (1997) 75.

12. Chamseddine, A., Ferrara, S., Gibbons, G. W. and Kallosh, R., Enhancement of supersymmetry near 5d black hole horizon. *Phys. Rev.*, **D55** (1997) 3647.

13. Maldacena, J., The large N limit of superconformal field theories and supergravity. *Adv. Theor. Math. Phys.*, **2** (1998) 231.

14. Herdeiro, C., Hirano, S. and Kallosh, R., String theory and hybrid inflation/acceleration. *High Energy Physics*, **0112** (2001) 027. Dasgupta, K., Herdeiro, C., Hirano, S. and Kallosh, R., D3/D7 inflationary model and M theory. *Phys. Rev.*, **D 65** (2002) 126002.

15. Kachru, S., Kallosh, R., Linde, A. and Trivedi, S. P., de Sitter vacua in string theory. *Phys. Rev.*, **D68** (2003) 046005.

38 Susan Jocelyn Bell Burnell (1943–)

Ferdinand V. Coroniti and Gary A. Williams
Department of Physics and Astronomy, University of California, Los Angeles

FIGURE 38.1 Photograph of Jocelyn Bell Burnell. (Credit: The Open University, courtesy AIP Emilio Segrè Visual Archives.)

SOME IMPORTANT CONTRIBUTIONS

In 1967, Jocelyn Bell Burnell discovered a rapidly pulsating radio source – quickly named pulsars – thus providing the first evidence for a 30-year-old suggestion that one of the end states of stellar evolution could be neutron stars. The impact was immediate and profound, as pulsars joined the recently discovered quasars and cosmic microwave background radiation in the emergence of modern astrophysics. Today, pulsars are still a field of active research, and, despite over 30 years of effort, a complete understanding of their physics remains elusive.

As part of the University of Cambridge's pioneering program in radio astronomy in the early 1960s, Professor Anthony Hewish designed a novel radio telescope consisting of 2048 full-wave dipoles arrayed in a 470 m E–W and 45 m N–S configuration and operated at

Out of the Shadows: Contributions of Twentieth-Century Women to Physics, eds. Nina Byers and Gary Williams. Published by Cambridge University Press.
© Cambridge University Press 2006.

81.5 MHz with a 1 MHz bandwidth. The telescope's purpose was to search for rapid time variability or scintillations in the signals from extragalactic radio sources; as radio waves travel close to the sun, electrons in the solar corona and solar wind scatter the waves, thereby revealing or allowing limits to be placed on the then unknown spatial extent of these sources. After two years of construction the telescope went into operation in July 1967, and Jocelyn Bell Burnell, then a graduate student, was in charge of monitoring the first survey and analyzing the measurements.

In early August, Bell Burnell noticed an unusual signal consisting of short bursts whose temporal structure was faster than the time resolution of the signal processors. After improving the temporal resolution, the signals were found to repeat approximately every 1.3 seconds. The measurements were acquired at midnight, and therefore could not be the hoped for scintillations of extragalactic radio sources. The repetition time was totally unexpected, since no astronomical object had ever been observed to vary on such short time scales. By mid-August, after eliminating at Hewish's suggestion the possibility that the signal was some source of interference or telescope malfunction, Bell Burnell established that the source was located at a fixed right ascension and declination, and therefore was extraterrestrial in origin. In November, a systematic study of these strange signals was started. After correcting for the Doppler shift due to the Earth's motion, the repetition period was refined to be $1.337\,329\,5 \pm 0.000\,002\,0$ seconds, and the burst duration was found to be 0.016 seconds. This timing accuracy was comparable to the best atomic clocks of that period. Concerned that the strange signals might still be an anomaly, Hewish asked Bell Burnell to search for other examples. By December, she had discovered a similarly varying source at a different location in the sky, and two additional sources were found shortly thereafter.

The announcement of the discovery of pulsars was made in the February 24, 1968 issue of *Nature*,[1] and the report of three additional pulsars followed in the April 13, 1968 issue of *Nature*.[2] In both scientific and media impact, the discovery of pulsars ranks with Eddington's announcement of the deflection of starlight by the sun, thereby confirming Einstein's prediction from General Relativity, and the more recent discovery by the Cosmic Background Explorer (COBE) of the temperature fluctuations in the microwave background radiation; however, unlike these latter two, which were anticipated, no one expected to discover rapidly pulsating radio sources. The media freely speculated about the possible detection of extraterrestrial intelligence. The astronomical community commenced a flurry of research that resulted in the discovery of many more pulsars including the 33 ms pulsar in the Crab Nebula, the remnant of the 1054 AD supernova, and the 89 ms pulsar in the Vela supernova remnant.

In their February *Nature* paper, the Cambridge group argued that the short pulsar period implied a small physical dimension for the source, and suggested that only white dwarfs and the then putative neutron stars were likely source candidates; the authors slightly favored the neutron star interpretation, but only considered the pulsational modes of oscillation. The possible existence of neutron stars dates to the 1930s when Baade and Zwicky speculated that supernovae might leave a neutron star remnant.[3] Gamow and Landau considered the formation of a neutron core in massive stars that had ceased nuclear burning,[4] and Oppenheimer and Volkoff analyzed the gravitational stability of degenerate neutron stars,[5] estimating the upper limit to their possible mass. Within a few months of the discovery paper, Gold and Paccini convincingly argued that pulsars could only be rapidly rotating, highly magnetized ($\sim 10^{12}$ Gauss) neutron stars, and, along with Gunn and Ostriker,[6] predicted (subsequently confirmed) the slowing down of their rotation due to energy losses from magnetic dipole radiation. Over the next few years, Goldreich and Julian,[7] and Sturrock demonstrated that the strong vacuum electric field ($\sim 10^{13}$ V/cm) near the highly conducting neutron star's surface would cause field emission,[8] and that the synchrotron curvature photons that are radiated by the accelerated electrons would break down the highly magnetized QED vacuum to produce pairs.[9] Thus, within a brief three years, the discovery of pulsars confirmed the existence of neutron stars as one of the three end states of stellar evolution, and established a new paradigm of rotationally powered high-energy astrophysical objects.

BIOGRAPHY

Susan Jocelyn Bell (Burnell) was born in Belfast, Northern Ireland, on July 15, 1943. Her Quaker family lived in a remote country house, where the four children in the family were raised by nannies. Her father G. Philip Bell was an architect, so there were plenty of model houses and model cars for the children, and his library included a number of books on science. He also took the children on frequent visits to the Armagh Observatory about 50 km from Belfast, where he had a job designing a new addition to the observatory. They got to know the staff members there, which inspired a considerable interest in astronomy. The children attended a fairly primitive country school in a neighboring village, where the educational standards were unfortunately not very high. Bell Burnell was also a slow starter, and as a consequence at age 11 she failed the nationwide exam that would have put her onto a college-prep track in the school system. Her parents debated the expense of private school, but since the Quakers valued education and promoted women's rights, in 1956 they enrolled her in the Mount School, a Quaker women's boarding school in York, England. Bell Burnell thrived there, and although the school was not strong in the physical sciences, she liked her physics class. After graduating in 1961, she enrolled in the University of Glasgow in Scotland as a physics major, and was the only woman in her classes. She did very well, and earned a Bachelor of Physics degree with honors in 1965.

Bell Burnell was accepted into graduate school in the astronomy department at Cambridge in 1965, and began helping to build the new radiotelescope designed by Hewish. Stringing 120 miles of cable over 1000 nine-foot poles in a 4.5 acre array was not an easy task in the cold and rain of the British winter. When the telescope was put into operation it quickly began generating huge amounts of data, over 100 feet of chart paper every day. Digging out the tiny pulsar signals from the interference of other sources was a major feat, requiring keen observation, since such rapidly varying signals were just not expected. After making more precise measurements of the signals, Hewish and his team began speculating as to what the ultimate source might be. At one point they wondered whether the signal could be a beacon from intelligent life, a fanciful "little green men" theory. This theory was dashed when in December, 1967, Bell Burnell found the second pulsar in a different part of the sky, since it would be utterly unlikely that the little green men could set up two such beacons. The discovery of the second pulsar, and then quickly two more, provided complete certainty that the signals Bell Burnell had seen were a real effect, and the team submitted their famous paper to *Nature* in January, 1968.

Bell Burnell had discovered the second pulsar on the eve of a trip to Ireland to announce her engagement to Martin Burnell, and the two were married at the end of 1968. Bell Burnell finished up her Ph.D. thesis, which was primarily concerned with the original topic Hewish had given her, a study of scintillating radio sources (pulsars were only mentioned in an appendix).[10] She moved with her husband to southern England, where he had a job as a local government official, and she was able to obtain a teaching position at the nearby University of Southampton. Her research there was in the field of gamma-ray astronomy, a field quite different from the radio astronomy of her thesis work.

In 1973, Bell Burnell and Hewish were awarded the Michelson medal from the Franklin Institute in Pennsylvania for their discovery of pulsars. In 1974, however, the Nobel Prize was awarded only to Hewish and to his mentor, Sir Martin Ryle, who had originally pioneered the development of radio astronomy at Cambridge. The fact that Bell Burnell was left out of the award has sparked many long debates about how the Nobel Committee makes its decisions on crediting discoveries.[11] It is certainly clear that Hewish and Ryle were highly deserving of the honor, since without their contributions the innovative telescope that Bell Burnell monitored would never have existed. On the other hand, the fact that she was able to dig out a tiny, completely unexpected signal was a major discovery that others could easily have missed.

In 1973, a son was born to the Burnells. Bell Burnell took time off to care for the child, and then in 1974 took a part-time job at the Mullard Space Science Laboratory in Surrey. Initially she worked as a programmer, but later became an associate research fellow and took on the responsibility of analyzing the data coming in from an English X-ray satellite. X-ray astronomy was just becoming an interesting field at that time, with such interesting topics as X-ray bursts and variable X-ray sources,[12] and the satellite data proved to be quite significant. During this period Bell Burnell also taught part-time in the Open University, which is a distance-learning institution for adults who have not been able to attend regular universities.

In 1982, Bell Burnell moved to Scotland, again following her husband. She took a part-time position as senior research fellow at the Royal Observatory in Edinburgh, and also continued part-time in the Open University. In 1986, she became the project manager for the James Clerk Maxwell Telescope, a submillimeter telescope in Hawaii. The administrative load and constant travel, however, soon weighed her down, and in 1990 she switched to managing the EDISON infrared satellite for a year.

The Burnells had divorced in 1989, and in 1991 Bell Burnell moved to the city of Milton Keynes to become a full-time professor of physics at the Open University. In this position, she was only the third woman at the rank of professor in the physical sciences in England. She continues to carry out research in infrared and submillimeter astronomy.[13] Currently, Bell Burnell is serving as a dean at the University of Bath.

In an interview Bell Burnell commented on the difficulties a woman faces in combining a scientific career with family responsibilities:[14]

A lot of my working life has been driven by family circumstances. I worked part-time for 18 years and was married to a peripatetic husband who moved around an awful lot, so I sought whatever job I could get in astronomy or physics wherever he was . . . Although we are now much more conscious about equal opportunities I think there are still a number of inbuilt structural disadvantages for women. I am very conscious that having worked part-time, having had a rather disrupted career, my research record is a good deal patchier than any man's of a comparable age . . . The life experience of a woman is rather different from that of the male . . .

In addition to the Franklin Medal, Bell Burnell was also awarded the Oppenheimer Memorial Prize in 1978 from the Center for Theoretical Studies in Florida, the Tinsley Prize in 1987 from the American Astronomical Society and the Herschel Medal in 1989 from the Royal Astronomical Society. She has also received honorary doctorates from Heriot-Watt University, University of Warwick, University of Newcastle and Cambridge University, and in 1999 was named a Commander of the British Empire (CBE) "for her services to astronomy."

NOTES

1. Hewish, A., Bell, S. J., Pilkington, J. D. H., Scott, P. F. and Collins, R. A., Observation of a rapidly pulsating radio source. *Nature*, **217** (1968) 709.
2. Pilkington, J. D. H., Hewish, A., Bell, S. J. and Cole, T. W., Observations of some further pulsed radio sources, *Nature*, **218** (1968) 126.
3. Baade, W. and Zwicky F., *Phys. Rev.*, **45** (1934) 138.
4. Gamow, G., 1936, *Atomic and Nuclear Transformations*, Oxford (second edition), p. 234. Landau, L., *Nature*, **141** (1938) 333.
5. Oppenheimer, J. R. and Volkoff, G. M., *Phys. Rev.*, **55** (1939) 374.
6. Gold, T., *Nature*, **218**, (1968) 731; *Nature*, **221** (1969), 27. Paccini, F., *Nature*, **219**, (1968) 145.
7. Goldreich, P. and Julian, W. H., *Astrophys. J.*, **157** (1969) 869; *Astrophys. J.*, **160**, (1970) 971.

8. Sturrock, P. A., *Astrophys. J.*, **164** (1971) 529.

9. Erber, T., *Rev. Mod. Phys.*, **38** (1966) 626.

10. Hewish, A. and Burnell, S. J., Fine structure in radio sources at metre wavelengths. I. The observations *Mon. Not. R. Astron. Soci.*, (MNRAS), **150** (1970) 141.

11. Wade, N., The discovery of pulsars: a graduate student's story. *Science*, **189**, (1975) 358. Reed, G., The discovery of pulsars: was credit given where it was due? *Astronomy*, **11** (1983) 24.

12. Mason, K. O., Bell Burnell, S. J. and White, N. E., Observations of X-ray bursts from the vicinity of 3U 1727–33. *Nature*, **262** (1976) 474.

13. Fender, R. P., Bell Burnell, S. J., Waltman, E. B., Pooley, G. G., Ghigo, F. D. and Foster, R. S., Cygnus X-3 in outburst: quenched radio emission, radiation losses and variable local opacity, *Mon. Not. R. Astron. Soc.*, **288** (1997) 849.

14. Lambourne, R., Interview with Jocelyn Bell Burnell, *Education and the Profession*, **31** (1996) 183.

IMPORTANT PUBLICATIONS

Mitchell, R. J., Dickens, R. J., Burnell, S. J. B. and Culhane, J. L., The X-ray spectra of clusters of galaxies and their relationship to other cluster properties, Mon. Not. R. Astron. Soc., **189** (1979) 329.

Courvoisier, T. J.-L., Bell Burnell, S. J. and Blecha, A., Optical and infrared study of the three quasars OX 169, NRAO 140 and 3C 446, *Astron. Astrophys.*, **169** (1986) 43.

FURTHER READING

Gunn, and Osriker J., *Nature*, **221**, (1969) 454.

McGrayne, S. B., *Nobel Prize Women in Science: Their Lives, Struggles, and Momentous Discoveries* (New York: Birch Lane Press, 1993) p. 359.

39 Gail Hanson (1947–)

David G. Cassel
Physics Department, Cornell University

FIGURE 39.1 Photograph of
Gail G. Hanson.

IMPORTANT CONTRIBUTIONS TO PHYSICS

When she was a young postdoctoral research associate at the Stanford
Linear Accelerator Center (SLAC), Gail Hanson discovered the phenomenon called "jets" in elementary particle interactions. This was
one of the important discoveries leading to the establishment of particles called quarks as fundamental constituents of matter, and to the
extremely successful theory called the Standard Model that provides

Out of the Shadows: Contributions of Twentieth-Century Women to Physics,
eds. Nina Byers and Gary Williams. Published by Cambridge University Press.
© Cambridge University Press 2006.

the framework for understanding experimental results in elementary particle physics. Identification and study of jets are major tools used for obtaining most physics results from experiments at the highest energies available today.

Hanson was a member of the SLAC–LBL (Stanford Linear Accelerator and Lawrence Berkeley Laboratory) collaboration, which built and utilized a detector called the Mark I to study annihilations of electrons (e^-) and their antiparticles called positrons (e^+) at the SPEAR e^-e^+ storage ring at SLAC. The detectors used in elementary particle physics experiments are very large, very complex and always push the state of the art in technology. As a result, these experiments are possible only through the collaborative effort of many physicists.

Detectors and collaborations have evolved substantially in size and complexity since Hanson discovered jets. The size of a typical modern detector now measures tens of meters in each dimension and involves millions of sensitive elements. The collaborations that build and utilize these detectors involve hundreds of physicists at all stages in their careers, but the basic things that need to be done and the modes of operation have remained much the same since the SLAC–LBL collaboration. Even though hundreds of individuals are involved in these collaborations, it is still true that work is accomplished and discoveries are made by individuals like Hanson who are working in these collaborations.

A short introduction to elementary particle physics at the time of Hanson's discovery of jets will help in understanding her work and appreciating its importance in the development of our understanding of quarks and the structure of matter.

One of the goals of modern physics is to understand the nature of the fundamental elementary particles, i.e., the particles that combine to form all of the wonderful varieties of matter found in our world. By the mid twentieth century, it was well known that visible matter is composed of atoms, which in turn are composed of a massive atomic nucleus surrounded by a cloud of light electrons. It was also understood that the atomic nucleus is composed of protons and neutrons. By

1960, about a dozen subatomic particles including the proton and neutron were identified and thought to be the fundamental elementary particles. Most of these particles were divided into three main groups called leptons, mesons and baryons; the names are derived from the Greek words meaning light, medium and heavy. The electron (e) is a lepton, as are a heavier charged particle called the muon (μ) and lighter neutral particles called neutrinos. The mesons identified by that time were the π and K mesons, and the baryons included the proton (p) and neutron (n) as well as several more massive particles with similar properties. The mesons and baryons are collectively known as hadrons.

As protons and neutrons were examined more closely, for example by studying the scattering of high-energy electrons by protons or deuterium nuclei, it became clear that they have some sort of internal structure. In other words, protons and neutrons are composed of some smaller and more fundamental particles. Furthermore, in the 1960s, a host of new mesons and baryons were discovered – by now scores of each of these types of particles are known. With the discovery of each new hadron, it became more and more difficult to claim that they were all elementary. It was much more plausible to assume that hadrons were composed of more fundamental particles, and that the rich variety of properties observed in these new hadrons resulted from different ways of combining these more fundamental particles.

Two theorists, M. Gell-Mann and G. Zweig, proposed that all of the known mesons and baryons could be understood as combinations of particles called quarks (q). The baryons are composed of three quarks bound together, and mesons are composed of a quark bound with an antiquark – an antiparticle of a quark ($\bar{q}$). This picture was extremely successful; indeed, only three kinds of quarks (called up, down and strange) were required to describe the many different properties that were observed in all of the mesons and baryons known at that time. Furthermore, the theory received important confirmation in an experiment at SLAC in which very-high-energy electrons were scattered by protons. This experiment revealed that the internal structure of protons and neutrons was due to very small "point-like" particles. The observations could be understood readily if the proton and neutron were composed of the quarks of Gell-Mann and Zweig.

The theory that mesons and baryons were composed of quarks faced at least one serious difficulty: a vast amount of experimental effort was expended on trying to find free quarks – quarks not bound with other quarks or antiquarks – and none were found. The simplest theory of quarks and antiquarks required that they be charged in order to account for the existence of charged mesons and baryons, and that their charges be either positive or negative fractions (1/3 or 2/3) of the electron charge. None of the experimental efforts resulted in a confirmed discovery of any particle with this very clear "signature." Full acceptance of quarks as the constituents of mesons and baryons was hampered by the failure of experiments to reveal the existence of free quarks.

In parallel with this experimental development, attempts to understand the forces between quarks or between quarks and antiquarks led to the notion that free quarks could not exist; they could only appear in the bound combinations – mesons and baryons – already observed. The reason for this unusual phenomenon – called "confinement"– was that the potential energy of a quark and antiquark would *increase* linearly with their separation. (The same would be true if an attempt were to be made to separate the three quarks in a baryon.) This property of the strong force between quarks contrasts dramatically with other known fundamental forces, such as electric and gravitational forces, where the potential energy between particles that attract each other *decreases* with the separation. The increase in potential energy between a quark and an antiquark would lead to a new phenomenon not observed elsewhere in physics: if an attempt were made to separate a quark and antiquark, the potential energy would become so large that a new quark and antiquark would be produced. This occurs because the total energy of the combination of the new quark with the original antiquark and the new antiquark with the original quark would be less than the original potential energy of the original pair at the large separation. Taking the next step, an attempt to separate a quark and antiquark very quickly would result in the production of several mesons along the line joining the original quark and antiquark. This process is called "fragmentation" of the original quark and antiquark pair. Going one step further, if quark and antiquark were produced traveling away from each other with a large amount of energy, this picture suggests that the result would be two jets of mesons traveling away from each other along directions close to those of the original quark and antiquark.

This picture, that jets of particles would result from an attempt to separate a quark and an antiquark produced at high energy, suggested that the observation of jets would be the closest thing to observation of free quarks and antiquarks. Theoretical physicists, particularly R. P. Feynman, advocated searching for jets in the annihilations of electrons and positrons as a means of verifying that quarks exist with the properties required to fulfill their roles in the structure of mesons and baryons. Theoretical calculations showed that several properties of the jets could then be used to verify that quarks are the source of jets: the rate of production of jets in $e^- e^+$ annihilations, the distribution of the polar angle of the jet axis (the angle of the jets with respect to the directions of the incoming e^- and e^+ beams) and the azimuthal angle (around the $e^- e^+$ beam line) of the jets with respect to a particular fixed direction perpendicular to the beam line.

By 1975, the SPEAR storage ring at SLAC had become a major player in the international elementary particle physics program. Evidence was accumulating that the J/ψ particle discovered at SPEAR and at Brookhaven National Laboratory in 1974 was a meson composed of a fourth quark and its antiquark. In the same year, a new lepton, called the tau (τ) lepton, was discovered. At SPEAR, electrons and positrons were accelerated to (equal) high energies of a few GeV (1 GeV is 10^9 electron volts), stored in a system of magnets (the storage ring) and made to collide head-on in a detector. The total energy of the colliding electrons and positrons could be as high as 7.4 GeV, which turned out to be just high enough to discover jets.

Hanson was one of about 35 physicists studying the events produced in the state-of-the-art detector (later known as the Mark I detector) at SPEAR. She was interested in searching for evidence of jets produced in collisions at the highest energy available at the time. A few years later, when energies a factor of two or more higher became available at the PETRA collider at the DESY laboratory in Hamburg and at the PEP collider at SLAC, jets became easy to identify – they could be observed simply by looking at pictures of the tracks that particles made in detectors. However, at the energies available at SPEAR, evidence for jets was much less obvious and finding them required a subtle and creative analysis of the data.

Since jets were not readily visible to the eye, the first step was to find a method of determining the jet axis from the tracks of the particles in the detector. Hanson chose a method suggested by the theoretical physicists J. D. Bjorken and S. J. Brodsky. One advantage of this method was its analogy to the familiar method of determining the best axes for describing rotations of a solid body. This choice was inspired because other physicists had an intuitive feeling for what it meant from their experience studying solid body rotations, a subject included in the education of every undergraduate physics major. Quantities derived from the choice of the jet axis were then used to define a variable called "sphericity" (S). The value of S would be near 1 for events with an isotropic or spherical distribution of tracks, while the value of S would be near 0 if all tracks were aligned very close to the axis determined by the algorithm. Hanson showed that the average sphericity of events decreased substantially as the total energy of the colliding electrons and positrons varied over the accessible range, from 3.0 to 7.4 GeV. This in itself was evidence that events were becoming more jet-like as energy increased, a behavior expected from the picture of how jets were formed.

More convincing evidence was provided by the distributions of S for events at each of the collision energies. These distributions at four energies were compared to the distributions expected from two models. One model, "isotropic phase space," assumed no correlations among the particles in the event, other than those imposed by the conservation laws and special relativity. The other (jet) model modified this simple picture by assuming that the angles of the particles with respect to the direction of the original quark–antiquark system tended to become smaller – in a well-prescribed manner – as the total energy of the particles increased. At the lowest energy, both models were indistinguishable from each other and from the data. Hanson showed that – as the total energy increased – the data were substantially different from the predictions of the isotropic model, but agreed very well with the predictions of the jet model.

Although these observations established that the distributions of particles in $e^- e^+$ collision events became more jet-like as the collision energy increased, one other major piece of the puzzle was required to establish that the jets arose from the fragmentation of the quarks of Gell-Mann and Zweig. An essential ingredient of the theory was that quarks must have intrinsic angular momentum or spin of $\frac{1}{2}$ in units of the fundamental constant $\hbar$. Today, every graduate student studying elementary particle physics learns that production of a pair in $e^- e^+$ annihilation leads to a polar angle distribution of $(1 + \cos^2\theta)$ for the direction of the quarks that are produced. Since jets are supposed to follow the directions of the pair, observation of this polar angle distribution would have been conclusive evidence that the jets originate from the production of spin $\frac{1}{2}$ objects, rather than from some isotropic process, or – for that matter – production of spin 0 objects.

Unfortunately, the limited polar angle acceptance of the Mark I detector precluded using this polar angle distribution to determine whether or not the observed jets originated from spin $\frac{1}{2}$ particles. However, the detector had excellent acceptance in azimuthal angle ϕ and – if the spins of the e^- and e^+ were aligned perpendicular to the plane of the storage ring before annihilation – the azimuthal distribution would contain a term proportional to $\cos(2\phi)$. This alignment of spins is called "polarization." Fortunately, at the highest SPEAR energies, the e^- and e^+ were polarized perpendicular to the storage ring plane due to the emission of synchrotron radiation that they produced because they were traveling in circular orbits. Existence of this polarization was established in a paper published by the same collaboration a few months earlier. In this paper, the expected $\cos(2\phi)$ term was observed in the azimuthal distribution of muons (which were known to have spin $\frac{1}{2}$) produced in $e^-\ e^+$ annihilations. Hanson showed that the azimuthal distribution of the jets had an unmistakable dependence on $\cos(2\phi)$ at the highest energies. Indeed, at the highest energies the measured size of this term was consistent with its maximum possible value.

The paper describing Hanson's discovery of jets in $e^-\ e^+$ annihilation is a remarkable example of how all of the essential properties of an important phenomenon were established in its first observation.

Jets were confirmed a few years later as higher energy $e^-\ e^+$ colliders came on line. As mentioned earlier, jets become narrower with higher energy so they are even visible, without sophisticated analysis, in pictures of tracks in a detector. Furthermore, the $(1 + \cos^2\theta)$ distribution expected from production of $q\ \bar{q}$ pairs in $e^-\ e^+$ colliders became well-established using detectors with larger polar angle acceptance. Jets were somewhat harder to find in high energy proton–antiproton ($p\ \bar{p}$) collisions but eventually they were also found in this environment. Most of the physics results reported from the highest-energy colliders – both $e^-\ e^+$ and $p\ \bar{p}$ – are based on the study of jets. Nearly all of the measurements anticipated at the Large Hadron Collider (the very high energy p–p collider under construction at the CERN laboratory) will be based on studies of the jets in the events that will be produced and detected.

Important discoveries in physics usually depend on the right person with the right ideas, interests, talent and experience being at the right place at the right time. In the case of quark-fragmentation jets produced in collisions of elementary particles, Hanson, a young woman near the start of her distinguished career in elementary particle physics, was that right person.

Box 39.1   How experimental elementary particle physicists collaborate

Experimental elementary particle physics collaborations are organized in a manner that is quite unique in human activity. Although there are many variations, these collaborations generally function in the following manner.

A collaboration is composed of many physicists who belong to groups of modest size based in universities or national laboratories. These groups are led by one or more senior professors or laboratory staff members. They include physicists at many stages in their careers, ranging from beginning graduate students, to postdoctoral research associates, to very senior professors and members of the group. These groups crystallize to form a collaboration based on a set of common physics goals and a common idea of how to construct a detector to accomplish these goals. Each individual group is independently funded and can choose where it wishes to expend its effort. Hence, these collaborations are managed and function by common consent and agreement rather than by direction from a governing body that controls resources, the mode of governance that is typical in industry.

Each group provides detector components that are integrated with components from the other groups to form the complete detector. The expertise and labor of physicists in the group, along with those of engineers and technicians, go into the design and construction of the equipment and its installation in the detector. After the detector is completed, the members of the collaboration go through a process of commissioning and calibrating the detector in order to ensure that the individual components are working properly and to prepare the whole system for taking data.

In the taking of data, large quantities of raw data in the form of digitized electronic signals from the sensitive elements of the detector are collected and stored on computer discs or magnetic tapes. Then, the next job of the collaboration is to process these signals with the help of high-speed computers to derive kinematic quantities such as the momenta and energies of the particles produced in the particle interactions that occurred in the detector.

The final stage, the goal of the whole effort, is to study correlations among these particles in order to measure some important physical quantity – or better yet – to discover a new and important phenomenon like the jets that Hanson discovered. This is accomplished by a few members of the collaboration who work together to study these correlations to measure some interesting phenomenon. Other members of the collaboration then carefully study the work of these individuals and provide constructive criticism to ensure that the results are accurate. Often, other colleagues check the results by measuring the same phenomenon using different techniques. The results are then approved by the whole collaboration for publication, with all collaborators listed as authors.

BIOGRAPHY

Gail G. Hanson was born Gail Leslie Gulledge in Dayton, Ohio, February 22, 1947. She has a younger sister, Phyllis, who lives in Meridian, California. She married Andrew Jorgen Hanson in 1968, and they had two children, Russell, born in 1981, and Sonya, born in 1987. She was divorced in 1998. Russell is a graduate student at Boston University and Sonya is an undergraduate student at the University of Southern California.

Hanson studied physics at the Massachusetts Institute of Technology where she received her B.S. degree in 1968 and her Ph.D. degree in 1973. She was a pioneer in studying $e^- e^+$ annihilation, which was the subject of her thesis based on work done at the Cambridge Electron Accelerator.

After finishing her Ph.D., Hanson became a postdoctoral research associate at the Stanford Linear Accelerator laboratory. She joined the SLAC–LBL collaboration working with the detector later named Mark I at the SPEAR $e^- e^+$ storage ring. This collider facility and the detector were state of the art at the time. The SPEAR/Mark I program ranks among the most productive efforts in elementary particle physics. Hanson participated in the contribution of the SLAC–LBL collaboration to the discovery of the J/ψ particle and in the discovery of the τ lepton. In parallel with this effort, she searched for evidence of jets of particles in the events observed in the Mark I detector.

In 1975 – only two years after receiving her Ph.D. – Hanson demonstrated that jets were produced in e^-e^+ annihilations and showed that they arose from spin $\frac{1}{2}$ particles produced in the annihilation. This discovery confirmed theoretical predictions that quark–antiquark pairs produced in high energy interactions would result in jets of particles traveling along the directions of the original quark and antiquark. It was also an important confirmation of the existence of quarks, which cannot be observed more directly.

The significance of Hanson's discovery of jets was recognized by the American Physical Society when it awarded the 1996 W. K. H. Panofsky Prize to her for this discovery. The citation for Hanson's prize reads:

> Gail Hanson and Roy Schwitters are honored for their separate contributions, which together provided the first clear evidence that hadronic final states in e^- e^+ annihilation, which are largely comp-
> osed of spin 0 and spin 1 particles, originate from the fragmentation of spin $\frac{1}{2}$ quarks.

> Gail Hanson observed hadron jets and determined the jet axis by developing and applying the sphericity analysis to the hadrons in e^-e^+ events. She showed that events become more jet-like with increasing energy, contrary to what one expects from a simple phase space production mechanism. Using the beam polarization, she showed that the observed azimuthal distribution of the jet axis was that expected from the production of spin $\frac{1}{2}$ quarks that fragment into hadrons.

Hanson shared the prize with Schwitters, who had provided evidence that electrons and positrons were polarized when they collided in

SPEAR at high energies, and that spin $\frac{1}{2}$ particles were the parents of mesons that were produced in annihilations. This polarization was used by Hanson in demonstrating that her jets arose from spin $\frac{1}{2}$ parents.

Hanson remained in staff positions at SLAC until she moved to Indiana University to become Professor of Physics in 1987. In 1997, she was honored by Indiana University with the title of Distinguished Professor of Physics. In 2002, she became a Distinguished Professor of Physics at the University of California, Riverside.

At SLAC, she participated in further research with the Mark II collaboration at the PEP e^- e^+ collider. At Indiana, she joined the OPAL experiment at the high energy e^- e^+ collider LEP at CERN. In OPAL, she held the important post of Physics Coordinator and participated in the studies leading to the OPAL evidence for the existence of the B_s and Λ_b particles.

Looking toward the future, Hanson is a member of the CMS collaboration, one of the two major elementary particle physics experiments being prepared for the Large Hadron Collider. She is also involved in studies for a μ^- μ^+ collider in the more distant future.

Hanson is a Fellow of both the American Physical Society and the American Association for the Advancement of Science. She has also provided advice to the elementary particle physics community via service on numerous distinguished panels and committees.

IMPORTANT PUBLICATIONS

Litke, A., *et al.*, Hadron production by electron-positron annihilation at 4 GeV center-of-mass energy. *Phys. Rev. Lett.*, **30** (1973) 1189.

Augustin, J.-E., *et al.*, Discovery of a narrow resonance in e^+ e^- annihilation. *Phys. Rev. Lett.*, **33** (1974) 1406.

Hanson, G., *et al.*, Evidence for jet structure in hadron production by e^+ e^- annihilation, *Phys. Rev. Lett.*, **35** (1975) 1609.

Perl, M. L., *et al.*, Evidence for anomalous lepton production in e^+ e^- annihilation. *Phys. Rev. Lett.*, **35** (1975) 1489.

Acton, P. D., *et al.*, Evidence for b-flavoured baryon production in Z^0 decays at LEP, OPAL Collaboration. *Phys. Lett. B*, **281** (1992) 394.

Evidence for the existence of the strange b-flavoured meson B_s^0 in Z^0 decays, OPAL Collaboration. *Phys. Lett. B,* **295** (1992) 357.

FURTHER READING

Gottfried, K. and Weisskopf, V. F., *Concepts of Particle Physics, Volume 1* (Oxford: Clarendon Press, 1984).

Riordan, M., *The Hunting of the Quark* (New York: Simon and Schuster/Touchstone, 1987).

Lederman, L., *The God Particle* (Boston: Houghton Mifflin, 1993).

40 Sau Lan Wu

David B. Cline
*Department of Physics and Astronomy, University of California
Los Angeles*

FIGURE 40.1 Photograph of
Sau Lan Wu.

IMPORTANT CONTRIBUTIONS TO PHYSICS

Sau Lan Wu has spent most of her career on the study of electron–
positron collisions with powerful detectors. She has initiated some of
the key observations in the field of elementary particle physics; she
has studied some of the most important particles in elementary particle physics, including the J/ψ, gluon, Z and the first possible evidence
for the Higgs boson. In fact, she is associated with the discovery of the
gluon and the J/ψ particle. She has also made important contributions to the first observation of the large CP nonconservation in the
b system.

Out of the Shadows: Contributions of Twentieth-Century Women to Physics,
eds. Nina Byers and Gary Williams. Published by Cambridge University Press.
© Cambridge University Press 2006.

To understand Dr. Wu's contribution, we need first to take a look at the current Standard Model of elementary particles. This includes families of quarks and leptons. Quarks are the building blocks of protons and neutrons, and leptons are particles like electrons and neutrinos.

The J/ψ particle was the first example of a particle that has charmed quarks, which belongs to the second quark family. The discovery of this particle earned two Nobel Prizes: to B. Richter of Stanford University and to S. C. C. Ting of Massachusetts Institute of Technology (MIT). Sau Lan Wu was a key member of Ting's team. After this discovery, she gave numerous seminars and colloquia on this new J/ψ particle in the United States and Europe.

In 1977, Sau Lan Wu joined the faculty of the University of Wisconsin. At the same time, she formed her own group, which together with experimental physicists from Germany, Great Britain and Israel, became the TASSO Collaboration to work at PETRA. PETRA (Positron–Electron Tandem Ring Accelerator) was a 2.3-kilometer electron and positron storage ring at the DESY Laboratory in Hamburg, Germany, while TASSO (Two-Arm Spectrometer SOlenoid) was one of the detectors using this accelerator.

Besides taking an active part in the construction of the Cerenkov counters for the TASSO detector, she decided that the most interesting possible physics from PETRA was to find the gluon and therefore concentrated on developing methods for its discovery.

Let us continue with the description of the Standard Model. Besides the quarks and the leptons, the Standard Model also contains the particles that are responsible for the interactions between them. Perhaps the most far-reaching conceptual advance during the nineteenth century is that charges interact with each other through the intermediary of the electromagnetic field, or, in modern language, by exchanging photons. But not all interactions are electromagnetic. In particular, since there is no way for protons and neutrons to be bound together by exchanging photons (there being no attractive force), the existence of nuclei requires some other interaction.

What is responsible for this new interaction is called the gluon: just as charged particles interact through the exchange of photons, the quarks inside protons and neutrons interact through the exchange of gluons. Technically, these exchanged particles responsible for the interactions are called gauge particles. At the time the TASSO detector became operational, the only gauge particle observed experimentally was the photon.

The simplest way to produce a gluon is by the gluon bremsstrahlung process, where a colliding electron and positron give rise to a quark, antiquark and gluon:

$$e^+e^- \rightarrow q\bar{q}g$$

analogous to the photon bremsstrahlung process involving creation of oppositely charged muons and a photon, $e^+e^- \rightarrow \mu^+\mu^-\gamma$. Just as the $e^+e^- \rightarrow q\bar{q}$ gives a two-jet event, the gluon bremsstrahlung process leads to a three-jet event. The question was: how could these three-jet events be found?

First, Wu gave an estimate for the total energy of e^+ and e^- needed to observe the three jets. Her result was 22 GeV, easily reachable by PETRA. Next, she made the important observation that, by the conservation of energy and momentum, the three jets in $e^+e^- \rightarrow q\bar{q}g$ must be coplanar. As a consequence, the search for the three-jet events can be carried out in the two-dimensional "event plane" – the plane formed by the momenta of q, $\bar{q}$ and g. Using this observation, Wu and her postdoc, Georg Zobernig, published in May, 1979, "A method of three-jet analysis in e^+e^- annihilation." Their method not only finds the three-jet events, but also identifies the three jets; thus, the characteristics of the jets can be studied in detail.

A month later, using their methods, Wu and Zobernig found the first three-jet event from PETRA at a center of energy of 27 GeV. Indeed it is the first such event in the world. This event, referred to as gold-plated because the three jets are very clearly separate, was shown by Björn Wiik in his talk "First Results from PETRA" at the Neutrino 79 International Conference of 1979 at Bergen, Norway; it

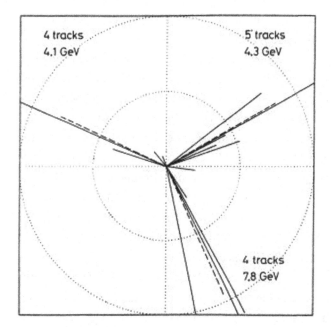

FIGURE 40.2 The first three-jet event. Solid lines represent individual particles; dashed lines indicate the central axis of each of the three jets.

is also shown in Fig. 40.2 here. Shortly thereafter, Wu and Zobernig found a number of additional three-jet events.

The significance of the three-jet events is as follows. Since e^+ and e^- are fermions, and two fermions cannot turn into three fermions, the three emerging jets cannot all be from quarks or antiquarks. Therefore, the presence of three-jet events in e^+e^- annihilation implies the discovery of a new particle, a new boson.

Two months later, the other collaborations working at PETRA confirmed this discovery of TASSO. A year later, in 1980, the spin of the gluon was determined by Wu and her TASSO collaborators to be indeed 1, as a gauge particle should. Thus, the gluon is the second gauge particle experimentally observed directly. For this discovery, Wu was awarded, together with Söding, Wiik and Wolf also of the TASSO Collaboration, the 1995 High Energy and Particle Physics Prize of the European Physical Society "for the first evidence for three-jet events in e^+e^- collisions at PETRA," which gives the first direct observation of the gluon.

In 1983, the gauge particles for weak interactions, the W and the Z, were discovered at CERN, Geneva, Switzerland; the author was on the team that made that discovery. This completed the experimental observation of all the gauge particles in the Standard Model.

The next step in Dr. Wu's career was an equally bold one. Shortly after the discovery of the gluon, Wu and her group, together with institutes and universities mainly from Europe, formed the ALEPH Collaboration to design a detector for the planned Large Electron Positron Accelerator (LEP) at CERN. When ALEPH was formed, Dr. Wu's group was the only US group in the collaboration. At that time, the most exciting experimental problems were: (1) to determine the number of families of quarks and leptons and (2) to search for the Higgs boson. At LEP, this number of families can be determined by studying the line shape of the Z, since this line shape is sensitive to the number of species of light neutrinos.

Since the number of families is a most important quantity, the ALEPH Collaboration decided to have two parallel efforts: one using the calorimeter and the other using charged tracks. Since Wu and her group participated in the design and construction of the tracking chamber – the novel time projection chamber (TPC) – and were largely responsible for the determination of the momenta of the charged particles from the TPC information, it was natural for them to use their expertise to analyze the hadronic events from the TPC tracks to determine the line shape of the Z resonance.

The result from both analyses was that the number of families is three, and it became the most important result from the study of Z at LEP at the time that LEP turned on.

The group of Sau Lan Wu was entirely responsible for the discovery at LEP of two new hadrons: Λ_b and B_s. Since three quarks (strange, up and down) make up the Λ, and similarly (charm, up and down quarks) makes Λ_c^+, it is expected that the combination (bottom, up and down quarks) will make the b-baryon Λ_b. B mesons are mesons with a bottom antiquark $\bar{b}$: thus the antiquark–quark pair ($\bar{b}$, down) forms B^0, ($\bar{b}$, up) forms B^+ and ($\bar{b}$, strange) forms B_s. While B^0 and B^+ were previously found by the CLEO Collaboration at Cornell University, B_s was not seen before LEP. Similarly, no bottom-quark baryon had yet been seen at all. Since LEP was an excellent source of bottom quarks, it was a good place to look for Λ_b and B_s. After their discovery in 1991 and 1992 respectively, their masses and lifetimes were measured in subsequent years.

Let us return to the Standard Model of elementary particles once more. Besides the particles already mentioned earlier, there is one more particle in this Standard Model: the Higgs boson. In the Standard Model, the left-handed and the right-handed components of quarks and leptons have different coupling to the gauge particles. This leads to the following dilemma: since the electron, for example, has a mass of about half an MeV, a Lorentz transformation can change a right-handed electron to a left-handed electron, in apparent contradiction to their different coupling to the gauge particles. This dilemma is resolved by the introduction of the Higgs boson. This Higgs boson is a ubiquitous field that gives masses to all the quarks, leptons, W and Z. Therefore, this Higgs boson occupies a unique position in the Standard Model.

Wu and her group were already deeply involved in the search for the Higgs boson at the beginning of LEP. However, before the latter part of the year 2000, this Higgs boson H was not observed experimentally. In fact, it was the only particle in the Standard Model not observed then. By that time at LEP, the way to look for H was through the process:

$$e^+e^- \rightarrow ZH.$$

Since both the Z and H are unstable, this process leads to various final states. Sau Lan Wu and her group studied all the important final states, including the case where both the Z and the H decay into quark jets, leading to a four-jet final state. This four-jet final state has the major advantage of a large branching ratio of over 50%, but also has a relatively large background and hence requires powerful methods for the analysis. A signal for the presence of the H is an excess of such four-jet events.

Starting in 1995, LEP increased the center-of-mass energy of the electron and positron from 91 GeV to eventually 210 GeV. After the center-of-mass energy at LEP reached 205 GeV in 2000, excess candidates began to show up.

These excess candidates found by Wu and her group gave the first possible experimental observation of the Higgs boson. From these and similar events, the most likely mass for the Higgs boson is 115 GeV/c^2. We look forward to the future confirmation of this Higgs at a mass close to this value, either at the Fermilab Tevatron collider or the Large Hadron Collider (LHC) being built at CERN.

Sau Lan Wu and her group also took part in the BaBar experiment at SLAC, especially the first observation of a large CP violation in B^0 decay. This observation involves the joint effort of the entire BaBar Collaboration, and the main contribution of Wu and her group is on flavor tagging, i.e., the efficient separation of B^0 and $\overline{B}^0$. This is crucial for the measurement of CP nonconservation.

We can put Dr. Wu's work into the following perspective. Few scientists working in experimental elementary particle physics have played such a key role in the discovery of the gluon, the study of the Z particle and the first possible experimental observation of the Higgs boson. These are among the most fundamental of all elementary particles. In particular, the Higgs particle carries the quantum number of the vacuum; this type of particle will also play a crucial role in the early universe back to 10^{-37} seconds from the Big Bang.

BIOGRAPHY

Sau Lan Wu was born in Hong Kong. The award of a full scholarship by Vassar College made it possible for her to pursue her education in the United States. In 1963, she graduated from Vassar with a B.A. degree honored with Phi Beta Kappa and summa cum laude. She then attended the Harvard Graduate School and graduated in 1970 with a Ph.D. degree in physics.

Life was not very easy for Sau Lan Wu as a woman struggling in a male-dominated field. Let me illustrate a few examples of her difficult experience. As early as 1963, just before she graduated from Vassar College and began applying to various graduate schools, her applications were declined because of her gender. For instance, when she applied to Princeton University she received a reply stating that Princeton would not accept women unless they were the wives of faculty, and that she should study physics at another university. When she applied to Caltech, she was told that they only accepted exceptional women because they did not have any dormitories for women. Harvard's Graduate School did accept women and she decided to attend Harvard for her graduate studies in physics.

In her first year at Harvard, she was the only woman in the entering class. She was not allowed to enter the men's dormitory where the male students were doing homework together. After one year, in the spring of 1964, she received her Master's degree at Harvard. She attended the commencement ceremony held in the Harvard yard; at the end of the ceremony, a free lunch was organized for all the alumni. She was removed by a guard and told that for the past 100 years, women were never permitted to attend the lunch. With tears in her eyes, she left the Harvard yard and her male friends who were enjoying the lunch. In 1970, just before she obtained her Ph.D., she applied to ten different institutions, which were most active in high-energy physics; however, she did not even get a single interview. She sadly thought she would have to work without pay in order to continue her research interests in high-energy physics. Finally, with her persistence she was able to secure a postdoc position at MIT.

After holding a research position at MIT., Dr. Wu was appointed assistant professor of physics at the University of Wisconsin-Madison in 1977. She was promoted to a full professor in 1983, and then Enrico Fermi Distinguished Professor of Physics in 1990.

Shortly after going to the University of Wisconsin, Dr. Wu formed a group to carry out experiments at DESY in Germany. Later, she moved her group to CERN, Geneva, Switzerland, in 1986.

The list of awards given to Dr. Wu is impressive and includes:

Outstanding Junior Investigator Award by the US Department of Energy (1980)

Romnes Faculty Award by the University of Wisconsin-Madison (1981), an award given to an outstanding faculty member soon after tenure

Enrico Fermi Distinguished Professor of Physics by the University of Wisconsin-Madison (1990–)

Hilldale Professorship by the University of Wisconsin-Madison (1991–8)

Fellow of the American Physical Society (1992)

The 1995 High Energy and Particle Physics Prize of the European Physical Society for "the first evidence for three-jet events in e^+e^- collisions at PETRA," which gives the first direct observation of the gluon

The 1995 Executive Committee Special Prize of the European Physical Society

She was elected to be a fellow of the American Academy of Arts and Sciences (1996)

She received a Vilas Professorship by the University of Wisconsin-Madison (1998–). Vilas Professorships are the highest ranking professorships at the University of Wisconsin.

One of Dr. Wu's most important contributions has been the training of Ph.D. students, and her record is extremely impressive – up to the year 2002, 34 Ph.D. students since she became a professor at the University of Wisconsin. Eighteen of her former graduate students and former postdocs are faculty members in major universities, mainly in the USA. Seven are permanent staff members in major high-energy physics laboratories.

NOTES

The author wishes to acknowledge contributions to this chapter by Dr. Wu.

1. TASSO Notes No. 84 (1979).

IMPORTANT PUBLICATIONS

Sau Lan Wu has over 200 publications, many of which have large numbers of authors. Listed here are some of the important ones, which she published with a few authors or to which she made very important contributions.

Wu, S. L., Aubert, J. J., Becker, U., *et al.*, Experimental observation of a heavy particle J. *Phys. Rev. Lett.*, **33**, (1974) 1404.

Wu, S. L. and Zobernig, G., A method of three-jet analysis in e^+e^- annihilation. *Z. Physik C, Particles and Fields*, **2**, (1979) 107.

Wu, S. L. and TASSO Collaboration, Evidence for planar events in e^+e^- annihilation at high energies, *Phys. Lett.* B, **86**, (1979) 243.

Wu, S. L., e^+e^- physics at PETRA – the first five years. *Phys. Rep.*, **107**, (1984) 59.

Wu, S. L. and ALEPH Collaboration, Determination of the number of light neutrino species, *Phys. Lett.* B, **231**, (1989) 519.

Evidence for b baryons in Z decays. *Phys. Lett.* B, **278**, (1992) 209.

Observation of the semileptonic decays of B_s and Λ_b hadrons at LEP *Phys. Lett.* B, **294**, (1992) 145.

Wu, S. L., Söding, P., Wiik, B. and Wolf, G., The first evidence for three-jet events in e^+e^- collisions at PETRA – first direct observation of the gluon. *Proceedings of the International Europhysics Conference on High Energy Physics*, July 27–August 2, 1995, Brussels, Belgium. (This talk was presented by Wu on the occasion of the award ceremony for the authors to accept the 1995 EPS High Energy and Particle Physics Prize.)

Wu, S. L. and ALEPH Collaboration, Observation of an excess in the search for the Standard Model Higgs boson at ALEPH. *Phys. Lett.* B, **495**, (2000).1.

Wu, S. L. and BaBar Collaboration, Measurement of CP-violating asymmetries in B^0 decays to CP eigenstates *Phys. Rev. Lett.*, **86**, (2001) 2515.

Wu, S. L. and McNamara, P., The Higgs particle in the standard model: experimental results from LEP *Rep. Prog. Phys.*, **65**, (2002) 465.

Name index

Page numbers in *italics* indicate photographs or figures. Names (page numbers) in **bold** are chapter subjects (pages).

Subject index

Italicized page numbers refer to illustrative material.